2012

THE FIRST FRONTIER

Chautauqua L.

Ft. Presque Isle

Ft. Le Boeuf

Allegheny R.

Ft. Machault

West Branch

Kittanning

Allegheny Front

Logstown

Allegheny R.

Kiskiminetas R.

Ft. Duquesne

Aughwick

Youghiogheny R.

Monongahela R.

Great Valley

Ft. Necessity

Ft. Cumberland

Potomac R.

rie

The

FIRST
FRONTIER

The FORGOTTEN HISTORY
of STRUGGLE, SAVAGERY,
and ENDURANCE
in EARLY AMERICA

Scott Weidensaul

Houghton Mifflin Harcourt
BOSTON   NEW YORK
2012

www.hmhbooks.com

Library of Congress Cataloging-in-Publication Data
Weidensaul, Scott.
The first frontier : the forgotten history of struggle, savagery,
and endurance in early America / Scott Weidensaul.
p.   cm.
Includes bibliographical references and index.
ISBN 978-0-15-101515-3
1. Frontier and pioneer life — East (U.S.) 2. East (U.S.) — History. 3. East (U.S.) — Race
relations — History. 4. North America — History — Colonial period, ca. 1600 – 1775.
5. North America — History, Military. 6. Indians, Treatment of — East (U.S.) — History.
7. Whites — East (U.S.) — Relations with Indians — History. 8. Indians of North
America — East (U.S.) — History. I. Title.
F106.W45 2011   974 — dc23
2011030600

Book design by Brian Moore

Printed in the United States of America
DOC 10 9 8 7 6 5 4 3 2 1

For Amy, patience personified

Contents

Acknowledgments

I am, even more than usual, indebted to my agent, Peter Matson, to whom this subject was of keen interest and whose insight and enthusiasm were mainstays. I was also fortunate to have worked with two peerless editors during the years in which this book was being researched and written — Rebecca Saletan, whose guidance during its formative stages was essential, and Lisa White, who did the heavy lifting as the book came together and moved through editing and production. Lisa's deft editing was a joy.

Any writer owes more than he can express to a good copy editor, and I was especially fortunate to have Barbara Jatkola's critical eye on this manuscript. Copyediting is an unsung but crucial process, and Barb saved me from errors of commission and omission, all with good humor and astonishing attention to detail.

Particular thanks to Pete Seward and Jen Johnson, without whom much would have been impossible, and to Ron Freed for his thoughtful first readings, which helped me enormously. A special tip of the hat to Allister Timms at *Down East* magazine for research assistance on the subject of the "Abenaki navy," and to Professor Earl C. Haag for his expertise in early German / Swiss culture in Pennsylvania.

My wife, Amy, to whom this is dedicated, put up with a lot over the past five years. Thanks, sweetheart.

Author's Note: A Word on Words

When quoting original sources, a writer faces a choice between historical accuracy and readability. In general, I have left the original spelling, capitalization, and punctuation intact, although I have made a few exceptions. I modernized the use of *u* and *v*, which in the sixteenth century were often reversed ("discouered" instead of "discovered"; "vse" instead of "use"), as well as the use of *i* instead of *j* ("iudged" instead of "judged"). Likewise, I modernized the use of *fs* for the double *s* in words such as "addrefs." In rare cases where more extensive editing was necessary to improve understanding, such changes are bracketed in the text or noted in the citation.

Selected Pronunciation Guide

Kəlóskɑpe: KLOOS-geh-bah
Ktəhɑnəto: kTEN-uhn-et-to
Miantonomi: mee-ahn-to-NO-mee
Mol8demak: moh-LAH-de-mahk
Opechancanough: oh-pa-CAN-can-oh
Sattelihu: SAT-a-lee-hyoo
Scarouady: SCAR-roh-ah-dee
Tanaghrisson: tan-ah-GRIS-son
Tsenacommacah: sen-ah-COM-ah-cah

Introduction

On the Hochstetler farm, which in September 1757 sat like an oasis of orchards and fields below the dark forests of the Kittochtinny Hills in eastern Pennsylvania, there was a rhythm to the seasons, and for the young people of the surrounding German community, *ebbelschnitz* time was one of the highlights of autumn.

The apple trees, planted almost twenty years earlier, hung ripe with fruit—a testimony to God's mercy, which, having brought these followers of the Mennonite elder Jakob Ammann out of persecution in Germany and Switzerland, led them to this new land of Pennsylvania, William Penn's "holy experiment" of religious tolerance. Never mind that their neighbors—English, Welsh, Scots-Irish, even their fellow Germans, the Lutherans and Reformeds who worshipped together at the union church down the Tulpehocken Valley—were not always the most welcoming, calling them "Amish" or looking askance at their pacifist ways in these troubled times of war. The Amish community along Northkill Creek—the first of its kind in the New World—was strong and growing. Jacob Hochstetler knew they were blessed every time he looked at his wife, his children, and their prosperous farms.

It was harvest time in the orchards. The best apples would be picked with gloved hands—never letting skin touch skin, which would cause the fruit to spoil—then be carefully packed in boxes of straw to be stored away in the cool, dry root cellar, alongside the potatoes and carrots. In the middle of winter, the fruit would be portioned out, like treasures, crisp and dripping juice as though straight from the tree.

The other, lesser apples would be ground into sauce or boiled in great iron kettles to make creamy apple butter, while the tartest (along with the windfalls) would be home-milled and pressed for cider, stored in casks that went into the root cellar, too.

But enormous piles of apples lay ready this day for *gedaddeschnitz*, dried apples that, when soaked in water, would plump up to make fillings for pies and tarts. In the morning, all the children and teens from the surrounding farms—the Yoders, Hertzlers, Nues, Glicks, Zoogs, and other Amish families—gathered at the Hochstetlers' to help with the chore, as did the Hochstetlers' grown children, John and Barbara, who lived on neighboring farms. Working steadily but happily—with a lot of joking and, among the older ones, whatever discreet flirting they could manage—the kids pared and sliced the apples with sweet-sticky fingers, cutting them into translucent half-moons that Frau Hochstetler and the women laid out on clean sheets to dry in the warm September sun. The smallest children, too young to be trusted with knives, waved switches to chase away the flies. By dusk, the once immense heaps of red, green, and yellow apples had dwindled to nothing but cores. A feast of a meal had been served, and the older folks kept a cautious but sympathetic eye on the happily chatting teens. A "frolic" was part of the bargain at *ebbelschnitz* time, and they had been young once, too.

The conversation that flowed through the darkening evening was almost entirely in German. If Jacob Hochstetler closed his eyes, he could almost imagine he was back in the old country. One tongue he had probably heard only rarely, however, was the swift-tumbling syllables of Lenape, the Algonquian language of the Natives he and the other settlers knew as Delaware Indians. In Lenape, the hills that began just a couple of miles to the north of the Amish farms were *keekachtanemin*, "the endless mountains." The valley itself was *tülpewihacki*, "the land abounding with turtles," and it had been especially beloved by the Lenape. But they'd lost it, just as they'd been forced from the bottomlands along the Delaware River; then from the lower Schuylkill as English Quaker, Irish, and Welsh settlers crowded in; and at last from *tülpewihacki* itself. Now the anger that had been growing among the Lenape for decades had finally led to war.

Like all the inhabitants of the "back parts" of the province, as the frontier was known, the Hochstetlers and their neighbors were nervous. The Delaware, Shawnee, and other tribes of the far Ohio country—refugees from lands in the east, including the Tulpehocken Valley—had renounced their alliances with the British and lifted the hatchet on behalf of the French. Throughout the previous year, the Endless Mountains had not been merely a boundary between settled lands and the wolf-haunted wilderness; they had been a menacing presence, out of which could come an attack at any time. The militia stationed at small frontier blockhouses such as Fort Northkill—a slapdash stockade of ill-fitting logs surrounding a small cabin not far from the Hochstetler farm—hadn't prevented a spate of killings and kidnappings the previous winter and spring. The summer of 1757 had been fairly quiet, though, so perhaps, everyone hoped, the worst was over.

It was long after midnight when the last of the crowd left—a rare reprieve from the to-bed-with-the-sun schedule of a farm family, and the Hochstetlers slept happily but heavily. Toward dawn, one of their dogs began to fuss, and one of the Hochstetler boys, Jacob Jr., sleepily opened the door to investigate.

In the predawn darkness, there was a brilliant orange flash from a musket and a ripping pain in the boy's leg as the round lead ball slammed into it. Somehow he pushed the door closed, dropping the bar, as the family fell from their beds in confusion and fear. Peering outside, they could see eight or ten Indians near the round dome of the bake oven. The two older sons, Joseph and Christian, snatched up their hunting rifles, powder horns, and shot pouches. But their father had not given up his old life in Europe and come halfway around the world to abandon his principles. The Bible said, "Thou shalt not kill," and he forbade his sons—who were skilled hunters and excellent shots—to fire on their attackers.

By now, the house was burning, so Hochstetler herded the family into the cellar. As the fire began to eat through the floorboards, they desperately splashed cider onto the wood to slow the flames. Daylight was coming, and the attackers, worried that they would be caught, began to slip off into the woods. One, a young Indian known by the English name Tom Lions, stopped to pick up a few peaches. He saw the Hochstetlers, choking from smoke, crawl out a small ground-level window, having thought the Indians were gone. Mrs. Hochstetler, "a fleshy woman," was stuck partway out.

Within minutes, it was over. Mrs. Hochstetler was stabbed and scalped, and young Jacob and his sister were killed with tomahawk blows; their father and brothers Joseph and Christian were taken captive. As they were herded away from their burning home, Herr Hochstetler told his sons to fill their pockets with peaches, of all things. Then the raiders uncoiled ropes of braided rawhide or buffalo hair, their ends brightly decorated with tassels and dyed quillwork. Tying these "slave cords" around the necks of the three captives, they marched the men into *keekachtanemin*.

I see *keekachtanemin* every morning when I look out my window. The Kittatinny Ridge, or Blue Mountain, is the first range of the old Kittochtinny Hills, which slant across Pennsylvania from northeast to southwest. There may be no more placid countryside in America than this quiet, Pennsylvania Dutch farmland—a long valley of cornfields and woodlots, bank barns and Holsteins—hemmed in by low-slung Appalachian ridges that fade to blue in the distance. It is the very image of settled, domestic peace.

But the story of Jacob Hochstetler is a reminder of a wilder, darker history here, hiding in plain sight. It is one that harks back to the days when the East was contested ground—fought over by empires and bled for by people who, regardless of their language, color, or birthplace, saw it as their own and worth dying for.

The Indians led the Hochstetlers north into the mountains, avoiding the line of militia forts, such as Northkill and Henry, that had been hurriedly built under the frantic eye of Benjamin Franklin two years earlier, as well as the mountaintop lookout at Fort Dietrich Snyder, on the ridge just a few miles south of where I now sit. They avoided, too, the main footpaths, such as the Tulpehocken Trail, which linked the settlements of Pennsylvania with the council fires of the Six Nations of the Iroquois in New York, and along which the province's Indian diplomat, Conrad Weiser, often traveled. They may have rested their first night in what the locals called the Red Hole, an isolated valley just north of here, accessible through a high notch in the ridge and

from which French, Lenape, and Shawnee parties could make lightning raids.

Those warriors crept beneath tall stands of hemlock, pine, and
chestnut, their passage observed by bull elk moving like pale ghosts
through the shadows, by wolves and mountain lions, lynx and fishers.
It was autumn, when sun-blotting flocks of passenger pigeons roared
through by the billions, shattering tree limbs with their aggregate
weight when they settled down for the night and thickly carpeting the
ground with their droppings like snow.

This valley, where in the autumn the sound of passenger pigeons
has been replaced by the rumbling of combines, is not unique in the
East. Wherever you set foot—on a street in Manhattan as you dodge
traffic; on the soft, freshly turned earth of a Hudson Valley farm; on
the kelpy tide line below a Maine cottage; or in the pine woods and
palmetto thickets of the Carolina Low Country—do not forget that
this was once frontier.

Frontier. The word carries the inevitable scent of the West, of sage-
brush and vast prairie skies, of buffalo beyond number and a peaceful
sheet of smoke hanging low over skin tipis. But before Custer, before
the first Conestoga wagon creaked across the muddy Platte, before
the first trappers pushed up the beaver streams of the Rockies, before
Lewis and Clark and the Corps of Discovery ascended the Missouri,
there was another frontier—one that stretched from the Atlantic coast
inland to the high, rugged ranges of the Appalachians, and from the
Maritimes to Florida. This was the First Frontier.

In the West, the frontier still seems close to the surface. In the East,
the old backcountry is buried beneath roads and strip malls, subdivisions and farms. But even here, if you know where to scratch, you can
uncover the terrain of that lost world where Europeans and Native
Americans were creating a new society, and a new landscape, along
the tidewater and among the forests and mountains—one by turns
peaceful and violent, linked by trade, intermarriage, religion, suspicion,
disease, mutual dependence, and acts of both unimaginable barbarism
and extraordinary tolerance and charity.

That this hybrid society was eventually washed away by the rising tide of European immigration and conquest, to be replaced by an

English colonial system that gave birth, in part of the frontier, to national unity and independence, may today seem inevitable. But for two and a half centuries, beginning with the first regular contacts between Old World and New, the future was anything but preordained.

For two hundred years, Spain was the colonial heavyweight, conquering complex, urban-based chiefdoms in the Southeast and replacing them with a mission system of Indian laborers under the watchful eyes of soldiers and friars. Those missions later collapsed under the predations of Indian slave raiders working with the English colonists, who in turn were all but annihilated by a powerful alliance of tribes that rose up when the English started enslaving them. In Pennsylvania, William Penn's fair-minded Quaker principles forged the Long Peace between colonists and Indians, showing what might have been possible—and yet Penn's sons, who surrendered to avarice and fraud, helped bring on the backwoods war that swept up the Hochstetlers. This and other frontier wars did not break out along predictable cultural or racial fault lines; they were, in the words of one historian, violence "not between strangers [but] between people who had become neighbors, if not kin."

It is a complex story—two and a half centuries of history about which most Americans know virtually nothing. The Seven Years' War, the conflict in which the Hochstetlers were ensnared, may seem impossibly ancient to us, but by the time Jacob's family immigrated to Pennsylvania in the 1730s, Indians and Europeans had been regularly interacting along the eastern seaboard for almost 250 years—the same amount of time that has passed between the Hochstetlers and us.

And in truth, the first tentative engagements occurred well before that—at least a thousand years ago, when the Vikings tried to colonize eastern Canada, and the Basques surreptitiously discovered, as early as the fifteenth century, the great cod and whale fisheries off eastern Canada and New England. It's difficult to say exactly when the tightlipped Basques first arrived; by the time the French and English showed up around 1600, they found Mi'kmaq Indians who were fluent in the Basque trading language and who skillfully sailed Basque-made shallops. One stunned Frenchman saw a Mi'kmaq glide by with an immense red moose painted jauntily on his sail. *The First Frontier: The*

Forgotten History of Struggle, Savagery, and Endurance in Early America looks at how these unimaginably different cultures grew steadily more similar through the centuries and yet remained stubbornly, and in the end tragically, estranged.

Part I, "So Many Nations, People, and Tongues," explores how North America first became inhabited by humans, a story informed by groundbreaking new research and fresh discoveries, which suggest that people occupied the Western Hemisphere thousands of years earlier than anyone once believed. Instead of fur-clad mammoth hunters striding across the Bering Land Bridge, scientists using genetic testing, linguistic analysis, and other techniques are painting a picture of multiple waves of human migration into North America—perhaps by coastal mariners following the food-rich "kelp highway" east around the Pacific Rim, or, more controversially, Ice Age Europeans traveling west along the Pleistocene sea ice, hunting seals like modern Eskimos.

However they arrived, those Paleolithic immigrants eventually forged a kaleidoscope of Native societies, from immensely complex urban cultures that built monumental earthen mounds, to coastal farmers raising maize and squash, to northern hunters stalking moose and caribou. As many as a million people may have lived on the eastern seaboard at the dawn of the sixteenth century. They settled so thickly along the coasts and river valleys that some of the first European explorers wondered whether there was room for anyone else.

At first the Indians welcomed the European visitors, who brought new technologies and goods, sparking trade, intermarriage, a cross-pollination of ideas, and cooperation. Europeans also brought suspicion and discord, rapacity and ruthlessness, as well as one of the worst mass epidemics the world has ever seen. When Europeans as varied as the Swedes, Dutch, Spanish, French, and English established their first beachheads in North America, they encountered a suddenly empty land.

Part II, "Let Us Not Live to Bee Enslaved," examines the colonial explosion of the seventeenth century, especially around Chesapeake Bay, where some of the earliest tensions between Indians and settlers (and between competing colonies) arose, and in New England, where relations began with a long period of peace and mutual cooperation but soon soured, leading to two of the bloodiest clashes ever between Natives and colonists, the Pequot War and King Philip's War—the latter the first regional, pan-Indian uprising against the invaders.

This part also explores the experiences of captives, both white and Native, including an usually quick-witted ten-year-old named John Gyles, who survived blizzards, near starvation, and years of slavery among the Maliseet and French, and Mary Rowlandson, whose seventeenth-century narrative became America's first bestseller. Both her release from captivity and the subsequent publication of her story owed a largely unacknowledged debt to a Harvard-educated Nipmuc, who was himself a victim of war.

Part II ends with an account of the Carolina deerskin and slave trade, which led to the largest Indian revolt in the colonial period. Almost forgotten today, this war changed the face of the South and gave birth to the antebellum plantation system.

Part III, "We That Came out of This Ground," explores the Pennsylvania backcountry, a place where exiles from around the world, and from throughout the battered Indian nations, went in the mid-eighteenth century to start new lives. It follows the intertwined fortunes of a Scots-Irish trader, a German-born frontier diplomat, a French-Iroquois interpreter, and several Native leaders—some war chiefs, some peacemakers—all trying to navigate the increasingly dangerous clash of imperial powers, provincial expansion, and pent-up Indian fury that ignited the Seven Years' War—the first truly world war.

In the end, the story of the frontier is the story of people—not stereotypes, but complex individuals and societies, all trying to make sense of a new kind of world with which none of them had any experience. No one had a monopoly on heroism or unprincipled behavior, which makes the story of the First Frontier at once rich, exhilarating, and heartbreaking.

Language and sources can be minefields for any writer dealing with frontier history. The thorniest issue is what to call the Native people. With hundreds of languages and dialects, they obviously did not have a single term for themselves. Most used phrases that translate to some variant of "real people"—which meant, essentially, "us, not everyone else."

"Indian," "Native American," "Amerindian," "aboriginals," "First Nations" (in Canada), and other terms have been used through the years. None is ideal. For example, none of the Shawnee, Lenape, Miami, or Mingo (Iroquois) who lived in the upper Ohio Valley in the mid-1700s were native to that region; all were refugees, emigrants, or exiles from homelands hundreds of miles away. Though aware of the limitations of these words, I use many of them interchangeably. Despite questions about the political correctness of "Indian," I have not shied away from it, given its universality in the historical record, as well as its continuing acceptance among many contemporary Natives.

Interestingly, "white" is a word the colonists rarely used to describe themselves during the first two centuries of exploration and settlement. They saw their differences with the Indians primarily in religious, rather than racial, terms—as Christian versus pagan—and viewed national, religious, and ethnic divisions among their fellow Europeans as almost more profound and intractable.

"Tribe" is another problematic word, since it implies a rigid ethnic and political division, which was rarely the case in Indian society. Drawing a bright line between, say, the Iroquoian Mohawk and the Algonquian Abenaki to their east is fairly simple, but what about the Pequot and Mohegan of southern New England? Deeply connected by language, intermarriage, tributary status, and disputes over hereditary sachemships, their internal "tribal" politics helped spark the devastating Pequot War of 1637. Historians often sidestep the issue by using the exceedingly broad (and literarily ugly) word "polity" for any social structure, from the clan level to paramount chiefdoms. I use "tribe" sparingly, recognizing that, as William Burton and Richard Lowenthal once said, it is "a convenient, if belabored, category meaning [a] named ethnic unit."

Because Europe was a literate society, while Native America depended on oral traditions, the sources on which we can draw are pitifully lopsided. For every speech in Boston or Philadelphia, for every panicked letter from a worried settler or gut-wrenching recollection by a survivor of captivity, there was a Native echo at the council fires in Onondaga or Logstown, an anguished story in an Abenaki wigwam or the *yihakin* of Tsenacommacah, unrecorded but no less important.

Furthermore, transcribing names from an oral language into a written form creates its own confusion. I spell the name of one eighteenth-century Seneca leader Tanaghrisson, but it appears in historical records as Tanighrisson, Tanacharison, Deanaghrison, Johonerissa, Tanahisson, and Thanayieson, among other renderings, all of which give a sense of how the man actually pronounced his name. Further muddying the waters, many American Indians used multiple names, appropriate for different stages of their lives, and often were given (or asked for) additional names in European languages. For the sake of clarity, I usually refer to an individual by the same name throughout, even though his or her public name may have changed over time.

Place names are similarly confusing. Although choosing to describe a location using its English, French, or Spanish name is, essentially, choosing sides, in many cases we have no record of what that place's original name was. Even when the name is known, using it would not help modern readers identify the location. Even seemingly innocent terms such as "New World" and "Old World" are freighted with the European perspective, but they are often the best terms we have.

The frontier is not gone in the East, but it can be difficult to find. The other day, I drove across the Kittatinny Ridge and turned onto Bloody Spring Road, named for another backwoods attack in 1757. When I turned off onto an unpaved lane, a cloud of dust followed me along the base of the mountain. I got out of the car and listened to the spring birdsong in the woods; this was the site of Fort Northkill, the hapless installation that did nothing to prevent French and Indian attacks on local settlements. The trees are smaller now than I imagine they were in the 1750s, but the chorus of wood thrushes and tanagers was the same that the poorly led, poorly equipped militia would have heard.

From there I drove a few miles south to what had been the Hochstetler farm, to which Jacob—after three years of captivity and privation and a harrowing solo escape through the wilderness of New France and the Ohio country—finally returned. I coasted to a stop near the state historical marker, which proclaims this the site of the first Amish settlement in the United States.

Although an interstate now cuts through part of the fields, and forest has closed in around the homestead itself (where at least one of the original eighteenth-century buildings still stands), most of the land remains a working farm, tilled each spring and harvested each autumn. Squinting a little and looking north, blocking out the rumble of the highway and focusing on the crumpled line of the Kittatinny, I could almost see the valley as it looked on that September day in 1757.

Almost. Alongside the interstate, occupying what had been part of the Hochstetler farm, is a kitschy tourist attraction. Dominating the parking lot, and directly in my line of sight, is an enormous, twelve-foot-tall Amish couple made of fiberglass, happily waving at the highway. A family was posing for a photograph just below the fake farmer's pitchfork. What old Jacob would have made of them, I cannot even begin to imagine.

In other places, though, the frontier seems as close and vivid as if it were still unfolding. One such place is a cluster of islands on the New England coast, whose seaward shores foam with waves breaking white out of the deep blue-black water, the air empty of all but the cries of gulls. Whether one knew it as the dawn-touched edge of the coastal world of *wôbanakik,* at the beginning of the Grubbing Hoe Moon, or as the unexplored "maine land" shore, on June 4 in the year of the Lord 1605, the story that played out there began on a day when everything changed forever.

PART I

"SO MANY NATIONS, PEOPLE, AND TONGUES"

Chapter 1

Mawooshen

SOMEWHERE OVER THE edge of the world, the sun had long since risen out of *sobagwa*, the great sea, but along the margins of *wôbanakik*, nothing was visible in the thick fog, except the ghostly silhouettes of narrow spruces along the shore of the low island and the pale band of gray, barnacle-encrusted rock where the falling tide surged and sucked.

Two long, high-prowed bark *wigwaol* moved along the island's lee, three men in each digging their slender paddles deep to move the sleek canoes forward. As they rounded the island, the sun suffused the fog with a pale golden light, but the paddlers had no time to admire the change, for now the rollers were coming in from the open ocean. The men had to time their strokes to the lifting swells, keeping the canoes angled into the waves so they would not broach. Nevertheless, the lively seas sometimes broke shockingly cold over the bows.

These boats were not the big, seagoing *woleskaolakw*—giant dugouts carved from the trunks of great pines, canoes built to carry fifteen people. Though nearly twenty feet long, these bark canoes were small craft in big waters, and while the men were skilled, they knew disaster lay just a careless moment or unexpected wave away. But a loon called, distant in the fog, and another answered it unseen—long, unearthly wails quavering over the water—and the men smiled at one another. The *mata-we-leh* were the messengers of Kəlóskɑpe, the One Who Made Himself, and perhaps they were asking him to watch over their canoes on this auspicious and frightening day.

For a long time, they moved in a world of mist, their own breath white with the morning chill, seeing nothing beyond the bows of their boats. In the stern of the first canoe sat a *sɑ̀kəmɑ* (leader) named Ktəhɑnəto. Wielding a long steering paddle, he navigated carefully but not blindly, keeping his bearings by the direction of the swells and by listening for the crash of waves on the receding shore behind them.

The fog made the men in the canoes feel as though they remained motionless no matter how hard they paddled, but Ktəhɑnəto never wavered, never appeared to be lost. To one of the men, barely out of his teens, this ability seemed almost supernatural—Ktəhɑnəto's name, after all, meant "doer of great magic"—but it was simply the product of a lifetime spent along the *wôbanakik* coast. Knowing how the currents flowed among these islands on a rising tide, and taking into account the freshening breeze that now twisted the fog into wraiths, Ktəhɑnəto set the course.

Then, with stunning speed, the mist began to split and lift, stripped away by the rising wind and brilliant sun; what was moments before a world cloistered in gray took on the bright edges and sharp shadows of morning.

All around them lay the complexity of rock-hemmed shores, spruce islands, long peninsulas, and deep bays that formed the beautiful edge of *wôbanakik*. Extending before Ktəhɑnəto and his companions, still hazy in the dissipating fog, was a series of steppingstone islands, humped with low hills and black with forest.

They paddled for an hour, passing rocky ledges on which dozens of *askigw* were hauled out, their smooth, mottled pelts shining in the sun. The smaller seals had faces like dogs, but the much larger, grayer ones had long muzzles like those of the big, antlered *moz* of the forest. In autumn, when the *askigw* were fat, they provided meat and oil, and even now, at the beginning of summer, they were good eating. But the men, who had no harpoons and had other things on their minds this morning, didn't stir when a gray *askigw* surfaced near their canoes.

The strengthening sun warmed Ktəhɑnəto's face, glinting off the white bone ornaments in his long black hair. This was the time of day that spoke most directly to how he and his kin saw themselves as a people, how they defined themselves in the world they knew. Because no other nation lived as close to the rising of the sun, all those who lived along this rocky coast called themselves *wapánahki*, "the people of the east." Their beautiful home was *wôbanakik*, "land of the dawn."

The Wapánahki were not a unified people. The tongues spoken by those living far away were similar but subtly different from those of Ktәhanәto and his relatives. Their nearest neighbors to the northeast, along the great bay, were the *panáwahpskek*, "the people who live where the river spreads out," and farther still the *pestamokάdiak*, "the people who live where the pollock are speared." These three peoples were closely bound by ties of blood and marriage, as they were with the *w'olastiqiyik*, "the people of the shining river," whose homes lay north and east of the great mountain *k'ta'dən*.

Ktәhanәto and his kin called themselves *walina'kiak*, "the cove people," because the bay on which they lived was unusually rich in small coves and harbors, even by the bountiful measure of *wôbanakik*. They knew that they inhabited the most beautiful part of the world and felt slightly superior to all other people, especially the *mi'k'makik*, who lived farther north and with whom they sometimes traded, sometimes warred.

Like his companions and his brother Amóret in the bow, Ktәhanәto wore heavy, waterproof moccasins and leggings made from *askigw* skins and a lighter buckskin loincloth, all decorated with open, curving geometric patterns stitched with dyed porcupine quills and bone beads. Copper bangles hung from his belt and wrists. His heavy mantle of tanned *moz* hide, supple and smelling of smoke, which he normally wore tied over one shoulder and belted at the waist, was too heavy and awkward for paddling in summer weather. It, along with its detachable beaver-skin sleeves, was rolled up and stowed carefully in the boat, for use against the nighttime chill. For now, Ktәhanәto and the others were almost naked, enjoying some of the first truly warm days of the year.

It was the Grubbing Hoe Moon, when the women back in the village would begin in earnest to work the moist ground of their small gardens, using *arakehigan* made by hafting the shoulder blades of deer to long, smooth handles. Except for the insects, the little biting *tsé'swak,* this was a relatively easy time of the year for the Wapánahki. A week earlier, on the full moon, the tides had risen and fallen at their monthly extremes, allowing the villagers to wade into the frigid water and gather the first lobsters coming into the shallows, to pry purple-blue mussels and green urchins from the rocks, and to lift the mats of rockweed and find succulent crabs hiding beneath.

The shoreline echoed with the shrieks of children nipped by the crabs, and there was easy laughter among the adults. The laughter was good, because it had been a hungry winter, as winters often were. Ktəhanəto's band, along with several others, had chosen a traditional encampment well inland, hiding among the hills that offered some protection from the worst of the great winter storms that blew in from the northeast, but the *moz* were scarce this year, and while there were plenty of the white hares, *máhtəkʷehsəwak,* to snare, their lean meat filled the belly but didn't stop the hunger.

By the time of the Ice Crusting the Snow Moon, there had been little to eat except old strips of autumn-killed *moz,* hung from the smoke hole, so hard and leathery, it was a morning's work to chew just a few pieces. The people washed them down with a weak tea made from dried sumac and blueberries. "*Woodeénit atók-hagen Kəlóskape* (This is a story of Kəlóskape)," a grandfather would say around the fire at night as the children nestled into their furs. But even the old stories—of how Kəlóskape caught up all the animals in the world in a magic bag of his Grandmother Woodchuck's belly hairs, or how the snowshoe hare lost its tail—couldn't keep the children's minds off the gnawing hunger in their guts or strike the worry from their parents' minds.

Still, Ktəhanəto thought, it had not been as grim as the past several winters, a time of true starvation, when to many it seemed as though the endless cold of the North, where the giant *giwakwa* cannibals lived, was once again stalking the land, never to let summer return. And as if to reinforce that fear, the summers that did finally arrive had been poor, miserly seasons, with frosts that lingered and returned again too soon.

But *wôbanakik* provided. The hunters had eventually found a few small herds of caribou, and one party that trekked to the coast for *askigw* had instead ambushed a great tusked walrus on the beach. It was the first anyone in the band had seen, and its blubbery meat, dragged back to camp on toboggans, had both fed everyone to satiation and provided proof that they were not forgotten.

Wôbanakik provided, as it always did in the fullness of the seasons. In spring, the newly ice-free rivers swarmed with fish: the run of the *sjamej* (Atlantic salmon), fresh and powerful from the sea; the slab-sided shad and sinuous eels that could be corralled in weirs; and the huge *karparseh*, sturgeon, plated monsters the length of two men, that were harpooned at night by the light of birch-bark torches.

The family bands came together in joy and relief at the spring fishing camps, as a silvery tide of millions of shiny, footlong alewives choked even the smallest streams flowing into the sea. Crowding their way up from the ocean to their breeding ponds and lakes, the oily alewives were easy to dip up with woven nets from pools where they gathered to leap small waterfalls. Battling the rocks and water dislodged so many of their sparkling scales that the stream bottoms shimmered like the iridescent linings of mussel shells. The people also gathered oysters at *mardarmeskunteag,* the young shad pool, and added their shells to the mounds that rose like hills where generations long past had also feasted.

The paddlers moved in an easy, long-practiced rhythm, as the islands and hours passed. Ktəhɑnəto and his companions snacked on smoked meat and last summer's dried berries as the falling tide carried them easily down the bay. Now that they were among the outer islands, the air was alive with birds—gulls and white terns screeching discordantly, the little puffins with their bumblebee wings burring through the air. The water heaved with movement as well. There were *askigw* and small, dark porpoises, and once, off toward the open sea, a great, slow-moving *potepe* that rose like an island itself, as black as the rocks and marbled on its head with barnacles. It was shadowed by an immense calf, both slowly cruising north toward *mikmurkeag,* the land of the *mi'k'makik.*

The Moon of the Smelts had passed, and with it the last of the snow, even the dirty, half-rotten drifts hidden in the shade of the deepest spruce forests. The bands had gathered again at their summer villages, patching the wigwam frames with fresh sheets of bark, putting the seeds of squash and beans into the ground as the Sowing Moon grew and waned, ushering in a summer that already seemed more fertile and gentle than those of recent years.

And then came the electrifying news, carried up and down the coast by canoes and runners. Several days earlier, six men gathering birds' eggs had encountered a huge *wigwaol*, bigger than the biggest dugout—an immense canoe with bare trees growing from it, hung with the skins of animals as big as a *potepe* and flapping in the wind, so that from a distance it looked like a *k'chi-wump-toqueh*, a white swan.

Such a *wigwaol* had been seen the autumn before at *panáwahpskek*, and the strange men in it had paid respect to Ktəhanəto's brother, the great *sakəma* Bashabes, to whom dozens of villages looked for guidance. Although the *wenooch*, the strangers, were accompanied by *mi'k'makik*, with whom Ktəhanəto's people sometimes fought, it had been a hospitable meeting, and the first time anyone at *panáwahpskek* had encountered such *wenooch*.

This time, some men from the village at Segoukeag had slept aboard the newest giant *wigwaol*, eating the odd meat and other food they were given. They found some of it very good, especially the *ushcomono*, the round green berries that the *wenooch* ate hot. But the yellow water that the strange men gave them, and that the *wenooch* themselves seemed to enjoy, tasted like piss, the Wapánahki said. One of the visitors came to Segoukeag and was given a dance, but the rest seemed suspicious and stayed aboard the *wigwaol*, gesturing for furs and tobacco to trade. Go to see the Bashabes, the people from Segoukeag told them, but the *wenooch* did not listen.

The sun was climbing high in the morning sky when the canoes carrying Ktəhanəto and his companions rounded the last of the small islands, facing the swells again. The strange vessel was clearly visible, sitting quietly in a natural harbor among several islands, its trees bare of skins, the figures of men silhouetted against the sky. There was a sharp sound like thunder, and some of the men, who had been fishing, scrambled to pull up their lines. As the canoes drifted closer, one of the strangers on the big *wigwaol* shouted in a harsh language that sounded like screeching gulls, making signs that those in the canoes should come aboard.

• • •

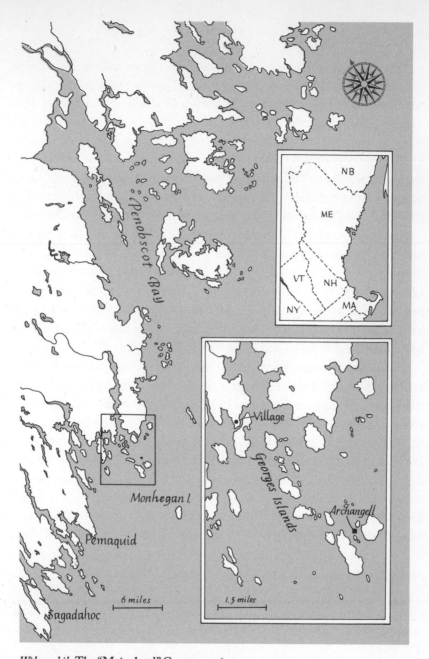

Wôbanakik: The "Maineland" Coast, ca. 1605

By the beginning of the seventeenth century, European explorers such as Samuel de Champlain and George Waymouth were just beginning to map the crenellated shoreline of what is now Maine. Yet Norse vessels had reached the region roughly six hundred years earlier, and Native inhabitants of the Northeast had been in contact with Old World fishermen and whalers for hundreds of years.

The *wigwaol* was, in fact, a squat, square-rigged sailing bark named *Archangell,* its twenty-nine-man crew under the command of Captain George Waymouth, who did not know it was the Grubbing Hoe Moon. By Waymouth's reckoning, it was June 4, 1605,* and his ship was two months out of England on a voyage to look for potential settlements and new fisheries. After several increasingly tense encounters with the local "Salvages," as his men called the Wapánahki, he was preparing to leave—with just one crucial task left unfinished. The arrival of two canoes full of Natives must have seemed providential.

It was Waymouth's second voyage to the New World. In 1602, he'd sailed with two ships, *Discovery* and *Godspeed,* so certain of finding the Northwest Passage to Asia that he carried with him a letter from Queen Elizabeth I to the emperor of China. Waymouth was primed for the trip: his father and grandfather were sailors, and his father, William Waymouth the younger, had taken part in the Newfoundland fishery in the late 1500s before turning to shipbuilding.

Although George Waymouth would have learned much about crossing the treacherous North Atlantic from his father, instead of the hoped-for passage, his ships ran into Resolution Island, at the entrance to Hudson Strait. They turned north along the eastern edge of Baffin Island, following (whether they knew it or not) the route blazed by Martin Frobisher in the 1570s on his own Northwest Passage expeditions. On those rare moments when the fog lifted, the island-strewn coast must have been a daunting sight—flotillas of icebergs, glimpses of foreboding cliffs, thousand-foot hills of bare rock, tundra and snowfields, and higher, glacier-wrapped mountains farther inland.

---

* Dates are, frankly, a bit of a mess during the period covered by this book. Because of the simplistic way in which the Julian calendar, dating back to 46 B.C., dealt with leap days, it was slightly out of synch with the astronomical year, and by the sixteenth century, the eleven-minute annual deficit had compounded dramatically into a serious problem. The Gregorian calendar, instituted by Pope Gregory XIII in 1582, solved this problem but created another. England (and its colonies), not being Catholic, chose not to comply with the "papist" reform until 1752. Then, by an act of Parliament, everyone jumped ahead eleven days in a single night, from September 2 to September 14. The same act also set the first day of the year as January 1, not Lady Day (March 25) as had previously been the case. Except where noted, all months and days in this book are as recorded at the time, and readers wishing to know the modern equivalent should add roughly ten days. Years prior to 1752 have, however, been corrected—that is, a date recorded as March 15, 1620 (Old Style) has been corrected to March 15, 1621 (New Style).

After ten freezing days of fitful exploration, creeping their way north almost by feel in constant fog and mist, the sound of ice scraping against the fragile hulls of their small wooden ships, Waymouth's crews mutinied. The captain backed down and agreed to give up. After poking their noses into Hudson Strait, they limped back across the Atlantic, arriving in England four and a half fruitless months after they'd left, their imperial letter undelivered.

The trip was a failure, but Waymouth wasn't discouraged. Although we know little about the man, including his age during this period, he clearly was not a quitter. He immediately began drumming up support for another go at the Northwest Passage, presenting two hand-illustrated copies of his book *The Jewell of Artes*—about navigation, seamanship, and military tactics—to the newly crowned King James I. A second Northwest expedition wasn't in the cards, but in March 1605 Waymouth was commanding the *Archangell,* sailing out of the Thames for "North Virginia"—the poorly known coast between Chesapeake Bay and Cape Cod, also known as Norumbega—with orders from his commercial backers to seek out prospects for fishing and for settlement by English Catholics.

After ignoring the New World for almost a century—while French, Portuguese, and especially Spanish explorers roamed its coasts—the English were playing catch-up. In the 1580s, several attempts by Sir Walter Raleigh to start a colony on the Outer Banks of North Carolina foundered, and when, after three years of war with Spain, England was finally able to send a rescue party, they found the colony abandoned, with no clue to their fate except for the cryptic word "Croatoan" carved on a post. Whether the members of the "Lost Colony" were killed or assimilated into one of the local Algonquian tribes, history has never determined.

In 1602, Bartholomew Gosnold and about thirty men sailed down the southern New England coast to Cape Cod and Martha's Vineyard, looking for a place to establish a trading post. Although Gosnold and his co-captain, Bartholomew Gilbert, were supposed to leave twenty of their men behind to man a permanent colony, by the time they built a small fort on what is now Cuttyhunk Island, they realized they had only a six-week supply of food, instead of the six months' worth they felt they needed, and retreated to England. They took with them a load of cedar and prized sassafras, the latter thought to be a cure for a variety of maladies, including "the French pox," or syphilis.

The next year, Martin Pring led two small ships, *Speedwell* and *Discoverer,* back to the New England coast, again coming away with a load of young sassafras trees, roots and all. At the same time, Gilbert was exploring farther south, looking for the Lost Colony around Chesapeake Bay, where an encounter with hostile Algonquians would lead to his death.

Given the rudimentary maps and primitive navigation of the day, it's hard to say exactly where Waymouth was planning to make landfall on his 1605 expedition; the Mid-Atlantic coast seems a likely guess. But contrary winds pushed the *Archangell* well to the north, and the crew narrowly escaped dangerous shoals that prevented them from reaching the only land they'd seen since passing the Azores three weeks earlier—a distant cliff that was probably Nantucket Island or Cape Cod.

Five days later, they made a brief landfall on Monhegan Island, "the most fortunate ever discovered," in the words of James Rosier, a thirty-two-year-old who served as expedition chronicler as well as "cape merchant," charged with overseeing its commercial operations and profitability. On Monhegan, they took on desperately needed firewood, while the crew caught "above thirty great Cods and Hadocks, which gave us a taste of the great plenty of fish which we found afterward wheresoever we went upon the coast."

Standing in for the mountainous land they could see far to the west, they found themselves two days later in the natural anchorage formed by several islands, which Waymouth dubbed Pentecost-harbour—for, said Rosier, "we [arrived] there that day out of our last Harbor in England, from whence we set saile upon Easterday." Rosier was the son of an Anglican clergyman, a recent Catholic convert who would, in later years, become an ordained Jesuit, but anyone who risked his life crossing the Atlantic in the frail sailing ships of the day was likely to be devout.

They had come to rest in what today are known as the Georges Islands, which cluster at the mouth of Muscongus Bay along the mid-coast of Maine. With the *Archangell* protected in its harbor, Waymouth's crew worked hard for a week and a half, digging wells for fresh water, cutting new yards for the ship, laying in firewood, and discovering fine clay for bricks. They assembled a long, narrow sailing boat known as a pinnace or shallop, which they'd carried across the Atlantic in pieces down in the ship's hold, and which would allow them to more easily explore the surrounding coast.

When not working, the men gathered blue mussels, prizing the small, misshapen pearls they found inside. Fishing near the shore with a net of twenty fathoms (about 120 feet long), they caught in a single pull "thirty very good and great Lobsters," more cod and haddock, rockfish and flounder. On the shore, they planted a small garden patch with peas and barley and marveled that despite birds that got most of the seed and soil that was "but the crust of the ground," within sixteen days the seedlings had grown eight inches or more. Once the pinnace was finished, they erected a cross on high ground, where it was visible from the sea.

With watering and woodcutting completed, they set out to explore the larger island, which they judged to be four or five miles long and a mile wide. The exploration party was well armed with "fourteen shot and pikes." The "shot" were matchlock muskets, unwieldy firearms that required the soldier to guard a smoldering fuse that set off the shot, and which were so heavy that those shooting them sometimes rested the gun on a wooden stand. The pikes were strong, slender shafts of ash wood ten or more feet long, tipped with razor-sharp spearheads. A pike was a handy weapon against mounted opponents, and more certain than a matchlock, but Waymouth's crew were among the first to learn how clumsy and ill suited pikes were to this new land, where dense forests of oak, pine, birch, beech, and spruce, as well as thickets of raspberries and wild roses, covered the island.

Waymouth was guarding against an ambush, even though they had seen no other people, only the charcoal of old fires on the islands framing Pentecost-harbour. Lying around these remnants of fires were "very great egge shelles bigger than goose egges, fish bones, and, as we judged, the bones of some beast."

Finally, on the morning of May 30, Waymouth and thirteen of his men set off in the pinnace to explore the mouth of a river they'd glimpsed from the ship. Despite the absence of people for the past twelve days, the decision to split the crew must have given Waymouth pause, since it left the *Archangell* significantly undermanned in the event of trouble. Perhaps the weeks of seeing no one along the coast emboldened him.

Only a few hours after the pinnace departed, however, the crew back on the ship spotted three "canoas" coming toward them from what Rosier called "the maine land," a name that ultimately stuck to this rocky coast. The "Salvages" seemed suspicious at first; one man pointed a paddle toward the open ocean and spoke loudly, as if to demand that the Englishmen leave. "But when we shewed them knives and their use by cutting of stickes and other trifles, as combs and glasses, they came close aboard our ship, as desirous to entertaine our friendship," Rosier wrote. The Englishmen gave the visitors tobacco pipes and rings, metal bracelets, and long iridescent peacock feathers, which the Indians stuck in their thick black hair. "We found them then (as after) a people of exceeding good invention, quicke understanding and readie capacitie."

The next day, Waymouth and the pinnace returned, having "in this small time discovered up a great river, trending alongst the maine about forty miles," and concluding from its size and flow that it must come from far inland. Worrying about attacks from shore, they had returned to the *Archangell*, planning to arm the smaller ship's boat, known as a light horseman, against possible dangers on future explorations.

Over the next few days, the number of Wapánahkis visiting the *Archangell* grew. The Indians liked raisins and candy and showed a particular fondness for the boiled dried peas and weevilly ship biscuit that formed a staple of naval cuisine. One can only assume the attraction was sheer novelty rather than the taste, since generations of English sailors cursed "pease" and sea biscuit as nearly inedible. They liked the small beer the crew drank but spat out the grain liquor called aqua vitae. At a demonstration of the *Archangell*'s guns, they fell flat on the deck in alarm. The ship's dogs frightened them and were kept tied whenever the Indians were aboard.

Rosier, in his role as cape merchant, handled the trading. One day he bartered with twenty-eight Indians on the island, swapping "knives, glasses, combes and other trifles" for otter and marten skins and forty beaver pelts. Waymouth indulged in a little grandstanding, using the ship's lodestone to magnetize his sword, then using the sword to lift a knife, "whereat they much marvelled." Rosier, meanwhile, was writing down the Algonquian words for various objects, and when the Wapánahkis realized what he was doing, they lined up, holding fish, fruit, and whatever else they could find, patiently pronouncing the name for each thing while he transcribed it. Lobsters, he wrote, were *shoggah,* mussels *shoorocke,* and an ax *tomaheegon.* To the Wapánahkis, Rosier was making a *wikhegan,* a drawing on birch bark or stone, although to their eyes, these drawings appeared singularly uninformative.

When Rosier pointed to the distant mainland, they thought he was asking the name of that particular spot, which like most Algonquian names was a simple geographic descriptive: *bemoquiducke,* "it sticks far out into the water." Later mariners would call it Pemaquid. When he gestured more widely, signing for the name of the whole land in which they lived, the Indians did not say *wôbanakik,* but *mawooshen,* presumably the specific area along the coast whose villages looked to the Bashabes for leadership.

That evening, Waymouth invited two of the men to dine in his cabin, where they "behaved themselves very civilly, neither laughing nor talking all the time, and at supper fed not like men of rude education." Waymouth wanted the two to sleep onboard, but the other Wapánahkis were cautious and suggested that they take one of the Englishmen with them as a guarantee of their own. The captain was unwilling to order any of his crew to go with the Indians, but a Welsh seaman named Owen Griffin volunteered, disappearing into the dark. He later described watching a two-hour dance, during which "the men all together . . . fall a stamping round the fire with both feet, as hard as they can making the ground shake, with sundry out-cries, and change of voice and sound."

At every opportunity, the Wapánahkis urged the Englishmen to go to "their Bashebas (that is their King),"who lived along the coast to the northeast somewhere and who had an abundance of furs and tobacco. Yet the hairy men, who seemed so anxious to trade here among the islands, making signs for more pelts, more of "their excellent Tabacco," inexplicably showed no interest in making the journey in their huge canoe.

At last, the strangers agreed to go ashore not to the Bashabes, but to the point of land a few miles north where the nearest village lay. Waymouth and fifteen armed men crammed into the light horseman, and although they pulled as hard as they could on the eight sets of oars, they couldn't help but notice that two canoes with just three paddlers each ran circles around them.

As they approached land, the Englishmen saw smoke rising from the trees, suggesting a large encampment. Before they beached the boat, Waymouth called a halt, the light horseman rolling in the swells; he was uneasy. Rosier would go ashore first, he decided, to reconnoiter — but only if one of the Indians, whom they perceived to be a leader, would stay with the English crew.

The sagamore "utterly refused," Rosier wrote, "but would leave a young Salvage." A youngster was no social match for a gentleman like Rosier, so in exchange Waymouth sent the ever-dispensable Griffin, who scrambled into a canoe and disappeared. The crew rowed the light horseman onto a rocky point half a mile away and waited. Everyone was on edge, but Waymouth's boatswain, Thomas King, passed the time by chipping his name, the year, and a small cross into the exposed bedrock above the tide line.

When Griffin returned, his eyes wide, he reported that he had counted 283 Indians, with "every one his bowe and arrows . . . and not any thing to exchange at all." To Waymouth and Rosier, it smelled like a trap, and a sign that for all their apparent goodwill, the "Salvages" were "very trecherous: never attempting mischiefe, untill by some remisnesse, fit opportunity afoordeth them certaine ability to execute the same."

Rather than going ashore, the men rowed back to the ship. Maybe the English were right and the Indians had been planning an attack. Or perhaps Griffin misunderstood their natural caution upon seeing an armed party of strangers rowing toward their village.

Or maybe Waymouth's immediate assumption of treachery was a guilty reflection of his own deceit, because the Englishmen had from the start been planning a surprise of their own. "Wherefore . . . we determined so soone as we could to take some of them," Rosier wrote, "least (being suspistious we had discovered their plots) they should absent themselves from us."

• • •

A thick rope tumbled over the side of the huge *wigwaol,* and two of the Wapánahkis took hold of it and climbed aboard. Ktəhɑnəto and his brother Amóret, however, stayed in their canoes with their two companions, waiting. The hairy men shouted over the side, holding up a steaming gray metal bowl and miming that they should eat. The youngest of the six Wapánahkis, a boy barely into his manhood, clambered up the side of the ship, the muscles flexing in his long limbs, and carried the bowl back down to the bobbing canoes. It contained the hot green berries and something else like baked cornmeal.

As they shared the food among themselves, Ktəhɑnəto asked about the two Wapánahkis on the ship, who had not yet reappeared. The boy said they were inside the *wigwaol,* warming themselves at a fire built among square stones, where the hairy men cooked for themselves, for there seemed to be no women anywhere.

Perhaps, Amóret said, these people have no women. Or perhaps they are hidden, Ktəhɑnəto mused. If the women smelled as bad and looked as ugly as the men, the boy joked, better that they stayed hidden. Then the lad climbed back up the side to return the metal bowl and shouted down that he would stay for a time.

The three remaining Wapánahkis paddled to the largest island, beaching the canoes. They climbed to the high bluff at the northeastern end of the island, within sight of the nearby ship, and kindled a fire, watching for any sign that their companions were ready to leave. Instead, many more of the hairy men came out of the forest, loaded down with bundles of firewood, which they shuttled to the big ship in a small boat. They were poor paddlers, sitting backward, so they had to crane their heads around to see which way they were going; the Wapánahkis could only shake their heads at the absurdity of it. Then seven or eight of the men, with one who acted like a *sɑkəmɑ,* pushed off from the ship and came toward the island, carrying a wooden box and a platter with more food.

The Wapánahkis stood, realizing they hadn't seen any sign of their three friends aboard the big *wigwaol* for some time. Unease was growing within them, and when Ktəhɑnəto and Amóret started down the hill to meet the hairy men, their companion told them he did not want to go and slipped off into the woods.

Did Ktəhɑnəto hang for a moment on his heel, deciding whether to do the same? Was it curiosity that finally won out, or naiveté, or concern for his fellow Wapánahkis, still somewhere inside the unnaturally large ship? Did he, in the long years that followed, regret that he, too, hadn't fled into the woods?

Yet as the boatload of backward paddlers approached and Ktəhanəto's apprehension no doubt grew, he was not making his decision in complete ignorance. It's very likely, in fact, that the young sagamore knew a great deal more about these strangers than the strangers knew about his people.

The Wapánahki and their neighbors to the north, along the Maritime coast, knew from experience that such visitors would be short, ugly men, sallow under their disgusting body hair. He knew that their *wigwaol*, when approached from downwind, would be rank with their unwashed stink. They would jabber incomprehensibly, although the Mi'kmaq had learned to speak a little of their language, and a few words, such as *ania* for brother and *adesquide* for friend, had passed down the coast.

The Wapánahki also knew that they would have goods to trade. For this reason, as word of the giant *wigwaol*'s arrival had spread among the villages of the *wôbanakik* coast, anticipation had initially won out over caution. Occasionally, the Wapánahki could barter with the Mi'kmaq for the hard, black pots like stone they had gotten from the visitors; the Mi'kmaq had regular dealings with these beings and had even learned to sail some of the smaller, winged canoes that used the wind instead of paddles to move. No one, even the elders, could say with certainty whether the strangers were human beings like the Wapánahki, apparitions from the spirit world, or a strange kind of intelligent animal. But the Wapánahki women wanted the black pots, and everyone wanted the shiny gray knives, which were far superior to flint or copper, and the wonderfully sharp *tomhikon* for cutting trees.

Although no one in Ktəhanəto's village had ever seen such people themselves, the Indians knew that these odd men had been coming to the edge of *wôbanakik* for generations. The Mi'kmaq and Maliseet said that some years, there were as many of the seagoing *wigwaol* as there were birds in the sky—dozens of boats with hundreds of smelly men, coming ashore to smoke their fish and trade for furs or meat. The

Mi'kmaq said the visitors were from many different bands, speaking different tongues, but the Innu called them all *mistigoches,* "builders of big boats."

When Ktəhɑnəto was a boy, runners brought news of hairy men taking hundreds of hides from a camp near *panáwahpskek,* whose inhabitants had watched from the forest. More reassuringly, just last autumn, during the Moon of the Eels, a great canoe full of the men had come to *panáwahpskek,* trading with the people there and dealing forthrightly with Ktəhɑnəto's brother the Bashabes, to whom they had given gifts.

But while Ktəhɑnəto may have heard a great deal about such visitors, there was one thing he could not guess. These particular sallow men had not come for cod, or whales, or even furs. They'd come on behalf of wealthy men seeking riches and a new colony. And with the support of their king, they'd come to stymie their nation's imperial rivals, the Spanish and French. The English were badly behind the colonial curve, and to catch up, they needed information. Instead of the ignorant near animals they'd expected to find, the men in the giant *wigwaol* had encountered a race of quick-witted people and immediately grasped that the intelligence and knowledge of a few "Salvages" would be priceless to their venture.

What the Indians had was worth more than beaver pelts or smooth tobacco. They had information, and the Englishmen would get it by guile if they could and violence if they had to. Ktəhɑnəto and his companions were about to be hijacked into history.

By that morning in 1605, the Natives of the Northeast coast had been dealing with Europeans for at least six centuries. Unfortunately, relations had gotten off to the worst possible start, which was perhaps not surprising, given that the first emissaries of European culture we know of were Vikings. Not only that, they were Vikings who had been banished by other Vikings for being too violent—which is really saying something.

Eirik the Red's family had been thrown out of Norway for murder and settled in Iceland, where he was banished again for murdering a neighbor. He sailed away in A.D. 985, discovered Greenland, and founded a new colony there. About that time, a Norse trader named

Bjarni Herjolfsson set sail for Greenland, but he was beset by storms and blown far off course. Bjarni found himself coasting along a wooded shore that was clearly not Greenland. Farther north, the land became mountainous and barren, the waters choked with ice. Bjarni—a Viking cast from a timid mold—made no attempt to land or explore this new world before turning east, bumping at last into Greenland. You'll never believe where I've been, he told Eirik's son, Leif.

Ten years later, Leif Eiriksson bought Bjarni's boat—probably a *knarr,* a stable, wide-beamed cargo ship capable of carrying tremendous loads, rather than the slender, more famous *snekke* warships usually associated with Vikings—and sailed for this unknown place, reversing Bjarni's journey. Leif came home to Greenland with timber, fruit, and stories of Helluland (flat stone land), Markland (forest land), and the impossibly rich Vinland (wine land).

Helluland was almost certainly Baffin Island, and Markland was Labrador, with its thick spruce forests. Vinland was somewhere to the south—perhaps Newfoundland, where the remains of a Viking settlement were discovered in the 1960s, perhaps the Gulf of St. Lawrence or even as far south as New England. This new land was also inhabited by a people the sagas call *skrælings.* Whether these were Inuit from the subarctic, Beothuk on Newfoundland, or more southerly Indians such as the Mi'kmaq, no one knows, but from the get-go the Vikings, starting with Leif's brother Thorvald, casually murdered them. He caught and then executed eight of the nine *skrælings* his crew found sleeping under skin boats on the beach one day. The *skrælings* didn't take this treatment lying down; Thorvald himself was killed later that day in a bloody counterattack.

Although the Vikings tried to colonize Vinland, bringing livestock and women there (at least one child was born in the colony), the *skræling* attacks became more and more frequent. During one assault, Freydis—Leif's sister, a real battle-ax, who according to one saga personally killed five Icelandic women in a plot to increase her profits from the voyage—snatched up a sword from a fallen Viking, bared one breast, and slapped it with the weapon, terrifying the *skrælings* into retreat.

Eventually, though, the *skrælings* drove the Vikings out of Vinland, although evidence suggests that the Norse continued to make peri-

odic voyages there for perhaps several hundred years more, until cooling climatic conditions drove their Greenland colony into extinction
around 1400.* By then, other Europeans were coming regularly to what
was referred to as Hy-Brasil, the Seven Cities, or the Isles of Antilla,
all names for imagined lands west of Ireland. Dreamers assumed Hy-
Brasil was a place of great wealth and opulence; doubters scoffed that
it was just a myth. But the Basques knew it was a very real place—the
land of *bakailao*, or cod.

Not that they were telling anyone. Basque fishermen may have been
making trips to the northeastern coast of North America as early as
the thirteenth or fourteenth century, reaping the unimaginable bounty
of the cod-rich fishing banks off Newfoundland and the Maritimes.
Certainly by the fifteenth century, they were regularly crossing the
North Atlantic for the summer fishing season, landing to salt and dry
their catch, then bringing it back to Catholic Europe, required to eat
fish half the days of the ecclesiastical year.

Good businessmen, the Basques kept their mouths shut about their
sources, but by the 1480s English fishermen from Bristol were seeking
the cod grounds as well, and may have found them. When Giovanni
Caboto (better known as John Cabot) "discovered" his "New Found Ile
Land" in 1497, it was no doubt to the disgust of the Basques, who'd had
a pretty good thing going there for centuries. Jacques Cartier, setting
out in 1534 on behalf of France, relied on directions from Breton fishermen who had been going there for years. When Cartier sailed into
the mouth of the St. Lawrence, he was greeted by so many Mi'kmaqs
and Montagnais (Innu), long accustomed to European visitors and
waving furs to trade from the shore, that his nerve deserted him and
he fired guns to scare them off.

Basque whalers came, too. In 1412 a fleet of 20 whaling ships passed
Iceland, heading west. Beginning in the 1530s, as many as 600 men
a year came to hunt right and bowhead whales, setting up seasonal
camps along the Labrador coast. By the summer of 1578, more than
350 European vessels were fishing off the coast of Newfoundland, with

---

* In 2010, scientists announced that a small group of Icelanders carry an unmistakable genetic
marker found in Asians and American Indians — proof, they said, that Viking explorers brought at
least one Native woman back to Iceland.

another 20 or 30 Spanish whalers working the waters between New-foundland and Labrador. In all, some 20,000 Europeans were em-ployed seasonally in the cod and whale fisheries there. Within two years, the French fleet had grown from 150 to 500 ships.

In 1583, Sir Humphrey Gilbert found the harbor at St. John's, New-foundland, choked with foreign boats—which did not stop him from striding ashore and cutting the thick turf to ceremonially take posses-sion of the land for England, thus formally establishing the English empire. The Basque, Portuguese, and Breton fishermen—never mind the native Beothuk—were unimpressed.

If Ktəhɑnəto had been able to talk to the Indians of the South-east coast, he'd have gotten an earful about Europeans, none of it good. When the Spaniard Ponce de León explored Florida in 1513, the Calusa Indians tried to cut his anchor lines from shielded canoes, while carefully keeping out of range of his ships' cannons and cross-bows, suggesting they'd already learned the hard way to be careful around European weaponry. The hostile reception and the lack of rich gold and silver mines like those found in Mesoamerica kept Spanish colonization at bay for decades.

Not that they didn't try. In 1526, Lucas Vázquez de Ayllón and six hundred colonists sailed from Hispaniola up the North American coast, founding the colony of San Miguel de Gualdape. Just where they tried to settle has been placed variously on the Pee Dee River in South Carolina and Sapelo Island in Georgia. Whatever the location, within three months the colony went bust, Ayllón was dead, and fewer than a third of the colonists were able to limp back to Hispaniola.

French Huguenots tried to settle at Fort Caroline (now Jackson-ville, Florida) in 1564, and that was enough to prod the Spanish into decisive action. They massacred the French, established St. Augus-tine the following year, and salted the coasts of Florida and Georgia with forts to protect their treasure fleets and with missions to convert and control the Indians. The Timucua, who had helped the French colonists, dwindled quickly toward extinction. The Guale, who had already tangled with Ayllón, rose up twice against their invaders, as part of a regional revolt in 1576 and again in 1597 in an especially vio-lent insurrection. Both times, the Spanish retaliated by burning Guale towns wholesale. But the microbial assault from the Europeans was far worse. By 1600, diseases introduced by the Spanish had reduced what may have been a pre-contact population of 1.3 million people in the Southeast to less than a sixth that number.

The centuries of contact between northeastern tribes and Europeans also had left their mark. Three years before Waymouth's voyage, Bartholomew Gosnold was sailing along the Maine coast. To his shock, he encountered a party of six or eight Indians expertly sailing "a Baske-shallop with mast and sails, an iron grapple, and a kettle of copper . . . one of them apparelled with a waistcoat and breeches of black serge, made after our sea-fashion, hose and shoes on his feet." Onboard Gosnold's ship, the Indian commander drew a chalk map of the coast and mentioned the Newfoundland fishing harbor of Placentia, whose name came from *plazenta,* the Basque word for "pleasantness."

"They spoke divers Christian words, and seemed to understand much more than we," one of Gosnold's companions wrote. No doubt the Indians, using the trade pidgin long employed with the Basques, were surprised by the newcomers' obtuseness. By the early 1600s, pidgin Basque was the lingua franca of Northeast trade, and the coastal people of the Maritimes were fluent when meeting their *adesquides.* Mathieu Da Costa—a free black man whose skills as an interpreter commanded a handsome price among Dutch and French traders—was able to make himself understood to the Mi'kmaq and Montagnais in the first years of the seventeenth century, probably using another form of Basque pidgin that had developed on the slave coast of Africa. One early-seventeenth-century visitor to the Maritimes observed that "the language of the coast tribes is half Basque."

Nor were Indian sailors all that rare a sight, manning shallops that might be forty feet long and have two masts. Two Natives joined Samuel de Champlain on one of his voyages, sailing their own shallop from Nova Scotia down the coast of Maine to trade European merchandise they'd acquired from the French. One of them, a Mi'kmaq sachem, informed the explorers that he had already visited France and stayed in the home of the governor of Bayonne some two decades earlier.

The Parisian lawyer Marc Lescarbot, who visited New France in 1606, reported seeing a "chaloupe" manned by two Indians who had painted a moose on the ship's sail, and in 1609 in Penobscot Bay, Henry Hudson encountered "two French Shallops full of the country people" with furs to trade. Hudson, an English sailor in Dutch employ, was not much for niceties. He stormed the Wapánahkis and took one of their shallops by force.

When Rosier and the rest of his crew landed on the island shore, they handed over the platter of peas to the two remaining Indians. As the men followed the Wapánahkis uphill to their fire, they talked quietly among themselves about how to lure back the third Indian who had disappeared into the woods. Rosier opened the wooden box, which was filled with trade items, "thinking thereby to have banisht feare from the other."

When that didn't work, the Englishmen wasted no time. Without warning, they grabbed the two Wapánahkis—and found themselves in the fight of their lives. Though taken by surprise, the Indians fought back with incredible force and agility as the Englishmen dragged them down the hill and into the boat. "It was as much as five or sixe of us could doe to get them into the light horseman. For they were strong and so naked as our best hold was by their long haire on their heads," Rosier recounted. The three who had come aboard the ship willingly were already safely stowed below—whether bound or simply confined, Rosier never said. "Thus we shipped five Salvages, two Canoas, with all their bowes and arrowes," an act that he defended as "being a matter of great importance for the full accomplement of our voyage."

The last Indian, the one who slipped away, "being too suspitiously fearefull of his owne good," was stranded on the island without a canoe, but he managed to stay out of sight for the next three days while Waymouth's crew finished loading water and wood for the return to England and completed meticulous soundings of Pentecost-harbour. During this whole period, Rosier recorded no visits from any Indians.

Then, on June 8, four days after the abductions, two canoes appeared from the east, carrying seven Wapánahkis, including the sagamore who had refused to act as a hostage during the earlier standoff involving Griffin. This was clearly a formal delegation. The emissaries were done up in careful style, some with their faces painted all black or red, others with blue stripes painted over their lips, noses, and chins. One warrior wore a headdress of white bird skins, another a roach of red-dyed deer hair, which "he so much esteemed . . . as he would not for anything exchange" it.

They carried a specific invitation from the Bashabes to come to *panáwahpskek* to trade for furs and tobacco. Waymouth arrogantly brushed them off. If the Bashabes wanted to come to him, fine, but the *Archangell* would not go there. Then, giving the Indians gifts of food and knives, the Englishmen ushered them off the ship, for "we had then no will to stay them long abord, least they should discover the other Salvages which we had stowed below." How Waymouth kept the captives from crying out Rosier does not say, but it's easy to imagine they were gagged.

In fact, Waymouth had every intention of going right to the Bashabes's doorstep. Weighing anchor on June 11, the *Archangell* sailed up the western side of Penobscot Bay, passing one excellent harbor after another. Rosier described the passing landscape with a string of superlatives. He declared the Penobscot "the most rich, beautifull, large & secure habouring river that the world affoordeth." Crewmen who had been with Sir Walter Raleigh when he discovered the Orinoco River of South America said this was better. It surpassed the Rio Grande, the Seine, and the Loire, Rosier said, and although "I will not prefer it before our river of Thames, . . . yet it is no detraction from them to be accounted inferiour to this."

Waymouth led a party on foot several miles inland toward the Camden Hills, passing through open land "having but little wood, and that Oke like stands left in our pastures in England." It was, Rosier said, a "parching hot" day, and the men were broiling inside their Elizabethan armor, but Waymouth eagerly eyed the "notable high timber trees, masts for ships of 400 tun"—the eastern white pines that are to this day known as "Weymouth pines" in England.

Back at the ship, they were met by the same sagamore as previously, now openly begging them to send one man with him to the nearby village of the Bashabes, who would then come to trade "many Furres and Tabacco." Waymouth and Rosier, realizing that their abductions were by now common knowledge, were convinced it was a trick to get an English hostage for exchange, and so they refused. After loading the light horseman with armored men and "shot both to deffend and offend," they rowed twenty miles up the river itself. Rosier described a "great store" of Atlantic salmon leaping from the water and scenes that "did so ravish us with all variety of pleasantnesse."

With that, the *Archangell* turned east for England. Belowdecks, the five Wapánahki captives would not have seen their homeland sink and vanish below the horizon. Back in Mawooshen, the five were being mourned, and word of the abductions and presumed deaths spread far and wide. A few weeks later, farther east along the Maine coast, Samuel de Champlain heard from a man with whom he was trading that the English had "killed five savages . . . under cover of friendship."

During the month that it took the ship to return to England, Rosier spent a great deal of time with the Indians, whose names he spelled phonetically as "Tehánedo" (Ktəhanəto, whose name is also spelled Tahanedo, Bdahanedo, and Dohannida), Amóret, Skicowáros, Maneddo, and Sassacomit. The class-conscious Rosier called Ktəhanəto "a Sagamo or Commander." He considered Amóret, Skicowáros, and Maneddo "Gentlemen," while referring to Sassacomit as a "servant." How accurate these descriptions are isn't known, but if he was right, it's possible four of the men were local leaders. Some scholars suggest that Sassacomit may have been a prisoner from another tribe, taken by the Wapánahki in war and held in servitude.

According to Rosier, the five captives adapted quickly to their situation:

> First, although at the time when we surprised them, they made their best resistance, not knowing our purpose, nor what we were, nor how we meant to use them; yet after perceiving by their kind usage we intended them no harme, they have never since seemed discontented with us, but very tractable, loving & willing by their best meanes to satisfie us in any thing we demand of them, by words or signes for their understanding: neither have they at any time beene at the least discord among themselves; insomuch as we have not seene them angry, but merry; and so kinde, as if you give any thing to one of them, he will distribute part of every one of the rest.

Just how "merry" the five were is impossible to know, but there is evidence that at the very least, the men ultimately made the best of their bad situation. And thanks to the same cold-blooded calculation that led Waymouth to take them captive in the first place, they were treated reasonably—one might even say surprisingly—well once they reached England.

Ktəhαnəto was sent to the London home of Sir John Popham, England's lord chief justice and a keen advocate of colonization of the New World. Most northeastern Indians considered Europeans as a lot to be rather homely, but Popham must have been a real eyeful, as he was described even by fellow Englishmen as "a huge, heavy, ugly man."

Skicowáros, Maneddo, and Sassacomit were taken in by Sir Ferdinando Gorges, a knight in Plymouth who, with Popham, successfully lobbied for a royal charter to colonize the land Waymouth had explored. Amóret's whereabouts are unclear; he may have wound up in the London household of John Slany, a merchant active in Newfoundland ventures.

The five were pumped for information, which Gorges compiled into a book, *The Description of the Countrey of Mawooshen, Discovered by the English, in the Yeere 1602*—a remarkably detailed document that sketches the nine river drainages from what is now Mount Desert Island to the Saco River, listing by name twenty-three villages, head counts of adult men per settlement, and the twenty-one sagamores who oversaw them. If the counts are correct, Mawooshen had a population of about ten thousand, all giving some degree of fealty to the Bashabes, their "chiefe lord."

Historian Alden Vaughan has suggested that the Wapánahki's cooperation may have grown out of their Algonquian understanding of the responsibilities of a captive to his captor. However they viewed themselves, the five were hardly guests in the normal sense—Gorges later referred to at least one of the Indians as his servant—but their treatment over the next several years appears to have been considerably better than that afforded to many earlier captives.

There was, even by 1605, a long tradition of hauling Natives from North America back to Europe, either as mere curiosities and proof that an explorer had reached a distant and exotic land or, like the fifty Indians that the Portuguese explorer Gaspar Côrte-Real rounded up in Newfoundland or Labrador in 1501 and sold into slavery, as assets on the voyage's balance sheet. By 1620, as many as two thousand Natives from the Western Hemisphere had made the trip to Europe, two-thirds of them as Spanish and Portuguese slaves, although this practice fell off dramatically after 1500.

The Indians who crossed the Atlantic were often treated as chattel or performing animals, but they certainly were not mindless cargo. They were people, who reacted to the circumstances in very different ways. Some even went willingly, such as Essomericq, the son of a Guarani chief from Brazil, who around 1503 volunteered to return to France with Binot Paulmier de Gonneville to learn artillery and weapons making. He was baptized during the voyage, was adopted by Gonneville, and eventually married the Frenchman's daughter. Essomericq was among the first of a long series of Indians who—at least in their own eyes—were traveling to Europe as emissaries of a sovereign nation.

Others, ripped out of their daily lives and having no choice in the matter, bore captivity with profound stoicism during the brief and undoubtedly terrifying period before ill use and disease ended their lives. Notable in this regard was an Inuit hunter kidnapped by Martin Frobisher from Baffin Island in 1576. The man's distress was so acute that he bit his tongue in two and died only a few weeks after being delivered to London. Undeterred, Frobisher did the same thing the following year, this time grabbing an Inuit man, who suffered smashed ribs when a Cornish sailor flattened him with a wrestling trick, and an unrelated woman and her infant son. In the month before his punctured lung "putrified" and killed him, the man demonstrated his skills at kayaking and throwing a harpoon for spectators along the Avon River. The woman died four days later, after breaking out in what may have been measles, a disease that would later prove fatal to many North American natives. Her son died shortly thereafter, despite the efforts of a wet nurse.

Still others adapted with alacrity to their new situations. Around 1502, three Natives from "the New Found Island," kidnapped by an unknown ship, were presented to Henry VII, "clothed in beast skins and eat[ing] raw flesh and speak[ing] such that no man could understand them, and in their demeanor like to brute beasts." Two years later, the same anonymous—and doubtless amazed—chronicler met two of the same Natives "apparelled after Englishmen in Westminster Palace, which at that time I could not discern [them] from Englishmen until I was learned what men they were, but as for speech I heard none of them utter one word."

The two sons of a prominent St. Lawrence Iroquois chief visited France in 1534, having been duped aboard Jacques Cartier's ships. They were returned to Canada on Cartier's second voyage in hopes that they would serve as go-betweens, but they maintained an understandably suspicious attitude that the French considered treasonous. Cartier grabbed them a second time in 1536, along with their father and several others, and all eventually died in France.

More often than not, attempts to force Indians into unwanted roles as guides and intermediaries ended badly. In 1561, the Spanish kidnapped a coastal Algonquian along Chesapeake Bay, where they were trying to establish a foothold. Held in Cuba and later in Spain, he was given a Spanish name, Don Luis; learned the language; and was adopted by the viceroy of New Spain. But when he was enlisted as an interpreter for a Jesuit mission on the Virginia coast, he bolted back to his people, who wiped out the settlement not long thereafter.

Just as the Europeans saw the "Salvages" as crucial pawns in their imperial plans, it is clear that many of the Indians who made first contact along the eastern seaboard also sized up the newcomers with a calculating eye to enhancing the positions of their own people. In 1584, two Indians from what is now North Carolina agreed to visit England with Captain Philip Amadas. One of them, Manteo, was described as a *werowance*, a title literally meaning "he is rich" and that referred to a regional chief of considerable power.

The two men spent the next year in England before returning to the Carolina coast. Manteo made a second roundtrip to England in 1586–87. By this point, he was apparently as comfortable in fashionable taffeta as in a loincloth and fur mantle, according to a German who observed him in London. As a werowance, Manteo would have been skilled in diplomacy, which may explain his success in working closely with Sir Walter Raleigh and others planning the Roanoke settlement, which despite the werowance's best efforts would end in failure and mystery as the "Lost Colony."

Why would an Algonquian diplomat help establish an English colony, especially after seeing the seething hordes of English citizens crowding London? Certainly, it wasn't because Raleigh—half-serious, half-mocking—bestowed the title "Lord of Roanoke and Dasamon-gueponke" on Manteo. No doubt Manteo's cooperation was at least partly due to commercial interest. The hairy men from over the sea might smell bad, but they were conduits for new and highly desirable goods, which rapidly found their way along the sophisticated trade networks that stitched together far-flung corners of the continent.

But more than commerce, leaders such as Manteo understood the political and military advantages of partnering with these strangers. Displays such as the firing of the *Archangell*'s guns may have temporarily frightened the unsuspecting Wapánahkis, but the value of having such firepower on one's side against traditional enemies must have been instantly apparent to any Indian seeing such a demonstration.

When the Pilgrims landed at Plymouth in 1620, for example, they were aided by Massasoit of the Wampanoag—not because the sachem (as southern Algonquian leaders were known) was especially kind-hearted, but because disease had left the once-powerful Wampanoag vulnerable to attack by their enemies the Narragansett, and Massasoit saw an opportunity to bolster their prospects through an alliance. (Fifty-six years later, the obvious threat posed by an endless flow of English immigrants would drive some Narragansetts to join a bloody but ultimately futile rebellion led by Massasoit's son Metacom, better known as King Philip.)

Raleigh's success with Manteo may have been at least partly behind Waymouth's intent to kidnap the Wapánahkis, although he and others ignored Raleigh's example of peaceful cooperation. The English also spirited away Natives from across "Virginia," and at times there must have been an Algonquian babel in Ferdinando Gorges's home, where Wapánahkis such as Sassacomit mixed with Wampanoags such as Epenow and Sakaweston, captured in 1611 on Martha's Vineyard and Nantucket, respectively. Epenow was exhibited throughout London. (Historians suspect that this unusually tall Indian was also unusually well-endowed. He was the talk of the town among the ladies of London, and the Indian with a "great tool" described in Shakespeare's *Henry VIII* was probably based on him.)

Epenow was also obviously canny. Once he learned enough English, he sold his captors on the notion of a rich gold mine on his home island, to which he alone could lead them. An expedition set sail in 1614, with particular care taken to make sure their guide didn't escape. Epenow was kept under close guard when they arrived off Martha's Vineyard, where Wampanoags paddled around the ship in excitement. But just as Rosier and the *Archangell*'s crew had been able to openly discuss their abduction plans in front of their Wapánahki victims, knowing they understood no English, so was Epenow presumably able to communicate with his friends and relatives in Wampanoag.

The next day, another large flotilla of canoes swarmed around the ship where Epenow was held, dressed in a long, heavy gown that the English knew would make escape difficult. With the crew distracted, Epenow broke free, shed the cloak, and dove into the water as musket balls splattered around him. Meanwhile, the Indians showered the ship with arrows, wounding many of the crew.

In stark contrast to Epenow's dramatic escape, his fellow captive, Sakaweston, took a very different route—going native, in a sense. Having lived for years in England, he eventually joined the army and went off to fight the Muslims in Bohemia, his fate unrecorded.

Like Epenow, Waymouth's five captives were unusual in that most eventually made it back home, although the route was sometimes torturous and painful. In 1606, Gorges and Popham dispatched the ship *Richard* with enough men to garrison a fort along the Maine coast. They were guided by John Stoneman, who had been Waymouth's pilot, and by Sassacomit and Maneddo, who were to act as go-betweens with the Indians. But the *Richard*'s captain took a risky southerly route through the Caribbean, and in thick fog he found himself surrounded by eight Spanish warships. Although the English offered no resistance, the Spanish boarding party laid into them with cutlasses. "They wounded Assacomoit [Sassacomit] . . . most cruelly in severall places on the bodie, and thrust quite through [his] arme," Stoneman recounted.

All aboard the *Richard* were taken to Spain and imprisoned. Gorges, no doubt seeing a linchpin of his colonial plans snatched from him, expended a great deal of effort in retrieving the Indians from Spanish slavery. He eventually ransomed Sassacomit, who returned not to Mawooshen but to Plymouth. It appears, from fairly oblique references by Gorges and others, that Sassacomit finally returned to his homeland, but not until about 1614. The records are mute about Maneddo, who most likely died in Spanish hands.

Ktəhanəto, probably with his brother Amóret, returned to Mawooshen on a second ship sent out shortly after the *Richard* and that was supposed to rendezvous with it once there. We know little about this voyage except that it safely made landfall, probably in Waymouth's Pentecost-harbour. When the *Richard* failed to appear, the crew explored for a month and then departed, having repatriated Ktəhanəto and Amóret, whose reappearance more than a year after their presumed deaths must have seemed a miracle to the Wapánahki.

The last of the five Waymouth captives, Skicowáros, came home in 1607, when the ships *Gifte of God* and *Mary and John* disgorged more than a hundred Englishmen at the mouth of the Kennebec River to start a colony at Sagadahoc, which stands with Jamestown as the oldest English attempt to settle the New World. As the Englishmen entered Ktəhanəto's village, the sagamore gave a loud cry, and armed men rushed at the intruders before Skicowáros intervened. The two men embraced, then Ktəhanəto made the Englishmen "much welcomme, and entertayned them with much Chierfulness."

Initially at least, the former captives' cooperation was all that Gorges and the other Plymouth Company investors could have hoped for, with Ktəhanəto offering to expedite trade. But relations soon became strained. Instead of playing the loyal company interpreter, Skicowáros—not surprisingly—seemed more interested in remaining with his people than with his erstwhile English employers. In fact, it's unlikely that he even saw himself as fulfilling any formal role. And the English soon grew frustrated by what they viewed as unreliability among the Indians as a whole—disappearing for long periods, skipping appointments, living their own lives instead of being at the colonists' beck and call. What kind of servant was that?

When they met another sagamore, named Sabenoa, who claimed to control the lower Kennebec where the English had settled, the encounter came within a whisker of bloodshed. Rejecting "some Tobacco & Certayne smale skynns" that the Indians offered to trade—a breach of etiquette—the English commander, Raleigh Gilbert, ordered his nineteen men to their shallop, matchlocks at the ready. Firing the cumbersome guns required a lighted match, which one of the soldiers held casually in his hand, "as if he would light a pipe." But Gilbert had underestimated his opponents. The Wapánahkis perceived the Englishmen's weak link, and one of them leapt into the boat, grabbed the match, and tossed it into the water. The English captain, in turn, ordered one of his men to rush the shore and snatch a lighted stick from the Indians' campfire. A tussle broke out as some of the Wapánahkis nocked arrows in their bows and others lunged for the shallop's anchor line, while the soldiers raised their muskets. Only at the last minute did cooler heads prevail.

Within a year, decimated by disease, food shortages, and a brutal winter, and riven by political divisions and poor management that Gorges blamed on "ignorant, timorous and ambitouse persons," the Sagadahoc colony was abandoned. Yet Gorges's efforts weren't completely in vain. In 1614, Ktəhanəto helped Captain John Smith explore the New England coast, assistance that Smith ranked just below the divine: "The maine assistance next [to] God . . . was my acquaintance among the Salvages, especially, with *Dohannida,* one of their greatest Lords, who had lived long in *England.*"

Not every former captive was so forgiving. In 1619, an English crew sponsored by Gorges, on their way to the Chesapeake, landed at Martha's Vineyard, dropping off a Wampanoag named Tisquantum—better known to history as Squanto, a man whose time in England would soon prove crucial to the Pilgrims. Out of the forest walked Epenow, the Wampanoag who had escaped in a hail of gunfire—and who, in fact, the English had assumed they'd killed as he dove into the water. The ship's captain, Thomas Dermer, found him a cordial, let-bygones-be-bygones sort who spoke "indifferent good English" and could afford to laugh at the whole affair. Come back again, Epenow said.

Dermer did. His chores in Virginia finished, he detoured back to Martha's Vineyard in November—and walked into an ambush. Instead of friendly, English-speaking Indians anxious to trade, he found himself under fierce attack. All of the landing party except Dermer were killed. Badly wounded, he reached the boat steps ahead of his pursuers, who would have "cut [off] his head upon ye cudy [cabin] of his boat, had not ye man reskued him with a sword."

Did Epenow set Dermer up for the attack? Or was Dermer attacked by others over whom Epenow had no control? No one knows.

But it's clear that even in its earliest days, the land that had become the First Frontier was a dangerous place, characterized by shifting allegiances and opaque motives, great opportunity and sudden death—a place that would become increasingly treacherous for both Natives and Europeans to navigate as their worlds drew more and more intertwined.

Chapter 2

Before Contact

O N C O L D N I G H T S in winter camp, when the earth slept
and it was fitting to tell stories of great power and magic,
Ktəhanəto and his people listened to the elders talk about
how the Wapánahki came into the world.

There was only forest and sea in those long-ago days. The forest
had no animals, the sea no fish, and nowhere were there any people.
Then, on a summer day, the great chief Kəlóskape, the One Who
Made Himself, and Kəlóskape's twin brother, Malsom, who had the
head of a wolf and darkness in his heart, came down from the sky in
their canoe, down from the rising sun, and at a word from Kəlóskape
the canoe became *oktokomkuk,* the vast granite island north of the land
of the Mi'kmaq.

Kəlóskape shaped some of the rocks into the *mihkomuwehsisok,* the
little people who hide in the shadows and play haunting music on
their bone flutes. One of them was Marten, whom Kəlóskape treated
like a younger brother.

Then Kəlóskape lifted his great bow and shot arrows into the basket trees, the ashes. Each arrow drove deep, piercing the heartwood,
its fletched end quivering in the still air. Again and again Kəlóskape

shot, until from the bark of the ash trees emerged beautiful human beings, whom he called *wαpánahki*, "the people of the dawn." He taught
them to plant and grow, to make lodges and canoes. From the mud,
Kəlóskαpe formed the animals, from the largest *moz* and *potepe* to the
smallest minnow, which he taught the people to hunt. The people
spread out across *wôbanakik* and have lived there ever since, closest of
all human beings to the rising sun.

Science, of course, takes a rather different view. Exactly how the
First Frontier and the rest of the Western Hemisphere was peopled—when, by whom, and by what routes—was long thought to be
a settled question, with scholarship mostly tidying up the details of
a well-documented, straightforward narrative. That story began thirteen thousand years ago, at the close of the last ice age, with Asian
hunters moving out of Siberia, bearing sophisticated flint spears and
chasing big game across a land bridge into North America. Aided by
their technical skill and hunting prowess, their descendants fanned
out in just a thousand years, spreading as far east as Newfoundland
and as far south as Tierra del Fuego, making the once empty land their
own.

But the certainty that once surrounded those assumptions has been
shaken in recent years by discoveries in the field and the lab. Excavations of artifacts, explorations of genes, and techniques to tease apart
the knotted web of language evolution are all painting a far more intricate picture and igniting a ferocious debate. Was there one migration out of Asia, or many spanning tens of thousands of years? Were
the newcomers mighty mammoth hunters following the herds inland
through an ice-free corridor, or master mariners who came millennia
earlier, paddling south by leaps and bounds as they feasted on the
ocean's bounty? Still more controversially, did at least some of them
manage to cross the North Atlantic or even the widest part of the
Pacific, bringing cultures and genes from as far away as Europe?

The story that is emerging, rich and nuanced in its testament
to human achievement, may be no less wondrous than the idea of
Kəlóskαpe's arrows calling forth humans from the trunks of ash trees.
But however people arrived, by whatever means and journeys, the
eventual result was a staggering diversity of human cultures that came

to occupy every corner of the New World, including what would become the First Frontier.

The Seward Peninsula of Alaska comes within a geographic hair-
breadth of Asia today, a dividing line in many ways. The continental
divide between the Arctic and Pacific Oceans bisects the peninsula,
runs over the gray, treeless spine of Cape Mountain, and comes to an
end on the wide sand beach below the massif that curves north and
west along the Bering Strait. Standing there, buffeted by the endless
wind that blows even on the rare cloudless day, you can look west
into Russia, to Big Diomede Island, thirty-five miles offshore, its sheer
cliffs and white snowfields clearly visible, and to what seems to be
another, more distant island to its right — East Cape, actually the tip
of the Siberian mainland. The International Date Line runs down the
middle of the strait, so you are literally looking into tomorrow.

This, archaeologists long agreed, was humanity's portal into the
New World — Beringia, the name for the low, boggy land bridge that
linked Asia and America during the last ice age, when global sea levels were hundreds of feet lower than today.* The term "land bridge,"
which conjures an image of a narrow isthmus, is misleading. Beringia
was more than a thousand miles wide, a mini-continent of grassy tundra that enveloped Asia and Alaska as far south as Kamchatka and the
Aleutian Islands.

Here, at Cape Prince of Wales, history lies very close to the surface.
Every storm that batters the dunes and bluffs unearths evidence of
the past — the gray weathered skulls of walruses and bearded seals;
whale vertebrae as large as footstools, stained peat brown; ivory harpoon points and bone net rims, artifacts of the Inupiat culture. The
descendants of those early Eskimo hunters still live a short distance
away, in the village of Wales.

---

* As early as 1590, the Spanish cleric José de Acosta — seeking to explain how Natives had populated "the Indies," as the New World was then known — speculated that the continents might meet,
or come very close together, in the poorly explored high latitudes. The ancestors of the Indians
might simply have walked to North America, he wrote, "some peopling the lands they found, and
others seeking for newe, in time they came to inhabite and people the Indies, with so many nations,
people, and tongues as we see."

But the Inupiat and their ancestors are relative newcomers to the Bering Strait, having moved into the Alaskan Arctic less than a thousand years ago, long after Beringia was swallowed by the sea. Before them, about six thousand to eight thousand years ago, came people speaking a language now known as Na-Déné or Eyak-Athabaskan, whose descendants now range from the western Arctic to the desert Southwest. And before any of them, archaeologists believe, came the earliest migrants, members of a hunting culture from the arid, ice-free tundra of northeastern Siberia—a region called the mammoth steppe, treeless and brutally cold, which stretched far west into northern Europe. There is very little evidence of these first pioneers, only a few scattered sites, perhaps because their camps were lost when Beringia sank beneath hundreds of feet of frigid water.

Although the Siberians could have walked across the land bridge, for almost ten thousand years they could have gone no farther, because continental ice sheets stretching from the Aleutian Range to the Arctic Ocean would have blocked their passage much beyond where Fairbanks sits today. Only when the ice began to retreat about eleven thousand years ago, opening corridors along the coast and through the Yukon, would the way have been clear.

The first incontrovertible sign of these Paleolithic hunters in the heart of North America was unearthed in the mid-1920s near Folsom, New Mexico, where stone spear points were found mingled with the bones of extinct bison. Of even greater age were the so-called Clovis points, a trove of long, delicately flaked spear tips discovered by a teenager in 1929 near Clovis, New Mexico. By the 1930s, Clovis-style points had been found at a number of other Pleistocene sites across North and Central America, usually in association with the bones of mammoths, long-horned bison, unique species of horses, tapirs, and other extinct Ice Age game.

There was an undeniable glamour to the image of the Clovis people that emerged from these discoveries: peerless hunters, tackling the biggest game imaginable and using weapons that were not only technically advanced but also astonishingly beautiful. Clovis points are up to nine inches long, carefully worked on both sides. They taper like willow leaves and are "fluted," meaning they have a shallow groove

running almost the length of each side, formed when a long flake of
flint was struck off from the point's base; this may have made it easier
to haft the thin blade to a spear handle, or it could have been merely
decorative. Creating a fluted point was the pinnacle of a flint knap-
per's exacting craft, and the Clovis people were masters. Made from
glossy flint or chert, the points range in color from cool blue-gray to
pink to marbled black and buff. At once lethal and lovely, the Clovis
projectiles are works of art.

As science grew in its ability to date archaeological sites, it became clear that the Clovis culture deserved one further honor: that of
founding fathers and mothers. Clovis appeared to represent the dawn
of human occupation in the Americas—a horizon line roughly 13,500
years old, before which there was no evidence of people in the Western
Hemisphere. For more than half a century, every excavation in North
or South America confirmed that preeminent place in American history; no matter where anyone looked, nothing was older than Clovis.

While thirteen thousand years seems like a long time, the Americas
were actually the last region of the world to be colonized by people. Modern humans began moving out of Africa as much as eighty
thousand years ago, expanding east along the Indian Ocean and into
southern Asia. Forty-five thousand years ago, they managed to sail to
Australia and New Guinea, then a single landmass thanks to lower sea
levels, but still separated from Asia by a wide, formidable deep-water
strait. Not long after that, they moved north into Europe, coexisting
with Neanderthals for about fifteen thousand years before that distinct species of hominid became extinct. By eighteen thousand years
ago, with the worst of a cold snap known as the last glacial maximum
passing, modern humans occupied the Siberian mammoth steppe,
with Beringia on the horizon.

The "Clovis First" hypothesis fits neatly into this larger framework,
but there have always been anomalies that Clovis couldn't explain. For
one thing, Clovis points don't really bear much resemblance to the
oldest stone tools found in Alaska, which date from about the same
time as Clovis. Most of the Alaskan tools are of a radically different
style known as microblade technology, in which tiny shards of worked
stone are set into ivory, bone, or wooden shafts—not the single, over-

size points that are the hallmark of Clovis. Nor has anything similar to Clovis, and from an earlier period, shown up on the Siberian side of Beringia to serve as a probable ancestor.

Another anomaly is the Meadowcroft Rockshelter in the hills of southwestern Pennsylvania. Since the early 1970s, scientists have been finding radiocarbon evidence suggesting that humans were stoking fires there at least sixteen thousand years ago, and perhaps as early as nineteen thousand years ago, well before there were supposed to be any people south of Beringia. The experts who first published these results were accused of everything from professional sloppiness to historical wishful thinking, but with more and more sophisticated dating techniques, the evidence has only solidified.

And if Meadowcroft discomfited archaeologists, a site in Chile known as Monte Verde gave them absolute fits. Starting in 1976, the wet, peaty soil along Chinchihuapi Creek, about fifty miles from the Pacific, has given up a variety of remarkably well-preserved artifacts, ranging from the remains of wooden shelter posts to knotted cord, animal bones, and even ancient foodstuffs, such as mastodon meat and quids of seaweed that were chewed like tobacco. Provocatively, radiocarbon dating suggested that the material was 14,500 years old, a finding that, when published, set off a donnybrook among archaeologists, with the "Clovis First" camp attacking the methodology, techniques, and conclusions of the pre-Clovis proponents.

It wasn't until 1997, after a blue-ribbon panel of archaeologists visited the site for themselves and came away convinced of the antiquity of Monte Verde, that the Clovis barrier was officially breached. The years since have seen a flood of new ideas on how the Americas were peopled. Instead of mighty mammoth hunters forging across the prairies, archaeologist Tom Dillehay, who excavated Monte Verde, pictures a maritime culture hopscotching down the Pleistocene coast, using watercraft to move easily across long distances and feasting on the rich array of marine and terrestrial food that was found even at high latitudes during the last ice age.

The lack of evidence for such early boats notwithstanding, the idea has gained significant ground among archaeologists in recent years, backed by discoveries such as 12,500-year-old stone tools and bones in the Channel Islands of California and the 10,300-year-old skeleton of a man in On Your Knees Cave in southeastern Alaska. Abundant marine animal remains from the same period indicate that Alaska's coastal environment was inhabitable during the last glacial maximum, and chemical isotopes preserved in the young man's bones show that he ate a diet heavy in seafood. The stone tools found with him are made of nonlocal rock, suggesting travel or extensive trade networks. Some archaeologists now talk about waterborne colonists from Asia following what they call the "kelp highway," the almost continuous band of fertile, food-rich submarine kelp forest that stretches around the Pacific Rim from Japan to Chile.*

But while the Clovis culture may have been deposed as the oldest human inhabitants in the Americas, there is still broad agreement among scientists that the people who pioneered North America were Asians. American Indians, though usually lacking the epicanthic fold that characterizes the eyes of many modern Asians, share with them other characteristics, including shovel-shaped incisors and the "Mongolian spot," a purplish pigment sometimes seen on the lower backs of infants.

Genetic research has strengthened that assumption. For example, a 2007 study examining the DNA of more than four hundred people from twenty-four Native groups stretching from Canada to South America found that all of them carried a unique genetic marker tying them to the Tundra Nentsi, an aboriginal group from eastern Siberia. What's more, the farther from old Beringia researchers looked, the fewer genetic commonalities with the Siberians they found. The Chipewyan of Canada share more genes with the Nentsi than, say, the Guarani in Brazil, which makes sense if Beringia was the doorway. A DNA analysis of the skeleton from On Your Knees Cave similarly found links with coastal cultures as far south as Ecuador and Chile.

But just because modern Indians are derived from northeast Asian stock, does that mean that the earliest humans in the Americas were,

---

* Any lingering debate over the "Clovis First" hypothesis seemed extinguished in March 2011, when scientists announced the discovery of a site in Texas at which they had excavated a wide array of stone tools in sediment below a layer bearing Clovis tools. What's more, the newly discovered spear points, which date back as far as 15,500 years ago, showed similarities to the Clovis style, suggesting that their makers could have been ancestors of the later Clovis culture.

too? A number of prominent archaeologists are skeptical, because there have been some discoveries that are hard to square with a single migration of northern Asians through Beringia, whatever the antiquity. A comparison of more than fifty ancient skulls from Brazil, dating as far back as eleven thousand years, recently led researchers at the University of São Paulo to conclude that the bones have more in common with Aboriginal Australians and Melanesians than with Asians, "supporting the hypothesis that two distinct biological populations could have colonized the New World." (Although this paper was immediately interpreted to mean that Australian Aborigines had sailed more than eight thousand miles across the Pacific to South America, the Brazilian researchers actually suggested that the first Americans simply shared a common ancestry with Aborigines and Melanesians and that they had traveled to Brazil on foot, crossing via Beringia and spreading widely in the New World before being swamped by later waves of northern Asians.)

But none of these theories of early migration into the Americas made as many headlines as the strange story of Kennewick Man, which started on July 28, 1996, when a couple of college students met on the banks of the Columbia River in Kennewick, Washington, to watch hydroplane races and drink beer. Hoping to avoid paying admission to the event, the two men scrambled through thick brush to a secluded spot where they could see the high-speed boats roar past. One of the pair, a twenty-one-year-old named Will Thomas, noticed what he thought was a bunch of deer bones in the water and an oddly smooth stone by his feet. The bones gave him an idea, and, taking a swig from his beer, he decided to play a prank on his friend.

"Hey, we have a human head!" he yelled, pulling up the stone. To his shock, he found that he was indeed holding a human skull, stained brown and with mud dripping from the eyeholes and braincase. As he gaped, he instinctively held his discovery at arm's length. "I didn't want to get any of that skull gunk on my beer," he later told a reporter.

Neither man knew quite what to do. While they thought about it, the other one tossed a bone he'd pulled out of the mud into the water. (It later proved to be a human femur.) They covered the skull with weeds and went drinking, then came back later to show a skeptical friend what they had found. This time they carried the skull to their truck and left it in a five-gallon bucket while they played Frisbee. Finally, they ran into an off-duty police officer, who quickly called in the county coroner while his fellow officers roped off the area with yellow crime scene tape.

The coroner, recognizing that the skull was not recent, contacted James C. Chatters, a self-employed anthropologist. Chatters looked at the long nose, slanted forehead, and prominent chin of the skull, among other features. He also examined other bones that he and the police subsequently recovered from the river mud (including that carelessly tossed femur) and concluded that the skeleton was that of a white settler buried at a nearby nineteenth-century homestead. Chatters scoffed at the rumor, already floating around Kennewick, that the bones were ancient. Sure, they looked old, Chatters told a reporter. That's what happens to bones — they always look old, even recent animal bones.

But Chatters began to reconsider those flip remarks when he took a CT scan of the pelvis to investigate an odd object buried in it. He was stunned to find the two-inch-long base of a serrated stone spear point, of a style known as Cascade, which dates back almost 8,500 years. Because it was unlikely that anyone was still tossing such ancient weapons at modern settlers, the coroner ordered further tests, including radiocarbon dating.

The results flabbergasted everyone, including Chatters. Kennewick Man, as the remains were known, had died 9,300 years ago, making this one of the oldest and most complete human skeletons found in North America — and one of the most controversial.

The controversy grew from Chatters's conclusion that the skull had Caucasoid features — meaning only that the skull differed to some extent from the broad category known as Mongoloid, to which Asian and American Indian skulls are generally assigned. Caucasoid skull features are found in populations ranging from northern Africa to Europe, and, notably, in much of northern Asia. Nor was Kennewick Man the only ancient human skull from the American West to show such characteristics.

But Chatters's statements were interpreted by most folks to mean that Kennewick Man was a Caucasian—a white European—and the fallout was swift and tangled. Local Indian tribes sued under federal law to have the skeleton returned to them for reburial, and although they eventually lost, it was nine years before anthropologists got their first look at Kennewick Man. In the words of one of the experts conducting the examination, a man's "biography is written in his bones," and Kennewick Man's spoke volumes.

He was probably in his late thirties when he died, robust and tall for those days, standing about five feet nine inches. He still had all his teeth, which were in generally good shape. An isotopic analysis of his bones showed that he ate a diet rich in salmon and other anadromous fish. His heavily muscled right arm, which left distinctive marks on his bones, spoke of a powerful man. The spear thrust that left that stone blade in his pelvis probably occurred when he was in his late teens. "It's no wonder his assailant took after him with a spear," one of the scientists said. "You wouldn't want to tangle hand-to-hand."

What do these Caucasoid skulls mean? It's hard to say. Some, like that of Kennewick Man, share skeletal features with the Ainu, a distinctive, non-Japanese culture of Hokkaido, the Kuril Islands, and Sakhalin in Russia. This finding might bolster the idea of an early, pan-Pacific migration. But other remains resemble those of modern American Indians, which might support the notion of a single wave moving through Beringia.

Science aside, the debate over the origins of Kennewick Man has at times taken on surreal tones. Two years after the skeleton surfaced, and while it was still the center of a legal battle, the bones were moved to a museum in Seattle. During the transfer, members of the Confederated Tribes of the Umatilla Indian Reservation sang traditional songs for the dead. Some distance away, Stephen A. McNallen, a self-styled Norse pagan from Nevada, raised a three-foot-long cow horn filled with mead and entreated the gods of ice and sun to bless the remains of "the Far-traveling One"—the name members of McNallen's Asatru Folk Assembly gave Kennewick Man in their legal filings seeking to have him declared their Scandinavian ancestor.

If that seems far-fetched, it isn't to Dennis Stanford, former chairman of the Smithsonian Institution's anthropology department. He and archaeologist Bruce Bradley of the University of Exeter in England have resurrected an old idea, first proposed in the late nineteenth century, that Clovis's ancestral culture isn't Asian, but European.

Stanford and Bradley, who are themselves expert flint knappers, see remarkable parallels between Clovis technology and the stonework of the Solutrean culture, which lived in southwestern Europe about nineteen thousand years ago, hunting many of the same kinds of big game as Clovis hunters. Although some innovations of one culture could be duplicated by chance in another place, they argue, it's almost impossible for coincidence to explain the near-perfect match between Clovis and Solutrean manufacturing methods and the artifacts they produced. Besides, they note, the earliest Clovis sites occur not in the American West, but in the Southeast, and the three most compelling pre-Clovis sites in North America—Meadowcroft Rockshelter in Pennsylvania, Cactus Hill in Virginia, and Page-Ladson in Florida—are all along the eastern seaboard. "They were from Iberia, not Siberia," Stanford quipped.

But how could European hunters get across the Atlantic? Stanford and Bradley note that during the last glacial maximum, the Eurasian ice sheets traveled as far south as Portugal, pushing humans to a narrow ice-free band along the coasts—and creating a solid bridge between Europe and North America, albeit one made of ice. As Arctic peoples today know, the ice margin is a biologically fecund environment, a place to hunt fish, sea birds, seals, walruses, and other marine mammals. It was seals, Stanford and Bradley believe, that lured the Proto-Europeans across the North Atlantic. "A Solutrean hunter must have been awe-struck when he watched for the first time a pristine seal colony stretching for as far as he could see, basking on an ice floe as it drifted towards the shore," they suggest.

To Bradley and Stanford's critics, the fact that there is no evidence of a seagoing Solutrean culture of kayaks, or similar small nimble watercraft suitable for chasing seals, makes the whole idea ridiculous. They are similarly unmoved by the argument that such objects would rot quickly along the water's edge and that any other evidence was probably submerged by three hundred feet of rising ocean waters when the glaciers melted. Nor is there evidence that either the Clovis or Solutrean people depended on marine resources. In addition, critics say, there is a gap of five thousand to six thousand years between the

disappearance of the Solutrean culture in Europe and the appearance
of the Clovis culture in North America that makes the hypothesis
even more improbable.

Whatever their origins, the earliest humans to reach North America
would have found a land in tremendous flux, including the region that
would become the First Frontier.

At the peak of the last glacial maximum eighteen thousand years
ago, the nearly two-mile-thick Laurentide ice sheet, which covered
more than five million square miles, surged as far south as Pennsylvania. Grassy tundra grew along the glacial fringe and down the higher
elevations of the Appalachians, but most of the East was dominated
by boreal forest, with spruces west of the mountains and an open forest of jack pine to the east. The presence of jack pine, a species that
thrives in dry soil and with frequent fires, suggests that this was a fairly
arid climate, as do old sand dunes and layers of windblown dust that
date to that period.

Thanks to the glaciers, which locked up tremendous amounts of
water, sea levels were as much as three hundred feet lower, and the
exposed continental shelf was a realm of open conifer forests as far
south as Georgia. Florida had almost twice the land area it does today
and was covered by low scrub, arid savanna, and sand dunes, since the
dropping sea level also lowered the water table, turning portions of the
region into a near desert. Hardwood forests, of the sort closely associated with the East today, were likely restricted to a narrow temperate
belt along the Gulf of Mexico and southern Atlantic coastal plain.

As the glaciers began to recede, the retreating ice in effect pulled
communities of plants and animals north behind it, like someone
dragging a banded blanket across the floor. The newly deglaciated land
must have looked like a bombed-out waste, but it was quickly colonized by the windblown seeds of tundra plants such as yellow dryas
and fireweed, turning what had been rock and mud into a palette of
blossoms in short order. With time, these pioneering flowers built a
layer of soil and worked themselves out of a home, as larger shrubs and
trees gained a foothold and crowded them out.

The plants of the tundra and forest would be familiar to a modern naturalist, but the mammals would be cause for no small degree

of alarm. The Ice Age megafauna, as this collection of behemoths is known, formed a bestiary of unparalleled weirdness and diversity, and sharing the landscape with these animals must have been a moving and dangerous experience for the first human arrivals.

There were the expected species, such as caribou, gray wolf, and wolverine in the tundra areas, and elk, white-tailed deer, and cougar in the forested zones. But along with the modern tundra muskox, there were three other species, including the taller, more gracile woodland muskox and the shrub-ox, another forest-dwelling browser. There were steppe bison one-third larger than today's plains bison and another species with horns stretching more than seven feet from tip to tip; giant ground sloths rearing up ten feet on their hind legs, using their massive front claws to pull down branches to feed on leaves; pig-like long-nosed peccaries three feet high; and, around the myriad lakes and bogs gouged out by the glaciers, giant beavers—much like the modern species, except they weighed more than four hundred pounds, making them the size of a black bear.

The land was a Serengeti of exotic wildlife: small, shaggy wild horses; five-hundred-pound American lions; scimitar cats and saber-toothed cats, with their wickedly curved fangs, which hunted nine-ton American mastodons or woolly mammoths measuring ten feet high at the shoulders and carrying curved tusks more than thirteen feet long. There were stag-moose, which had the huge size and long legs of their modern namesakes, but the delicate face of an elk and immense, bizarrely palmated antlers with a thicket of long tines projecting every which way.

The undisputed rulers of the post–Ice Age world, however, were the short-faced bears—measuring almost ten feet long and five and a half feet at the shoulders, weighing more than fifteen hundred pounds, and equipped with lanky legs, a long neck, and a peculiar short bulldog muzzle—the ideal build for running down huge prey such as ground sloths and young mammoths and grabbing them by the throat. There were two species, the giant short-faced bear of the West and the lesser short-faced bear of the eastern seaboard—but even against the somewhat smaller species, a Clovis-tipped spear must have seemed meager protection.

In the North, close to the fast-eroding ice sheet, the world must have looked post-apocalyptic at the end of the Pleistocene. The steadily diminishing Laurentide glacier sat in the midst of its own decay, shrinking rapidly and surrounded by enormous, ever-rising lakes of meltwater, whose drainages were often plugged by ice dams. Glacial Lake Iroquois, for example, grew to three times the size of Lake Ontario, whose basin it occupied, until it breached the ice dam blocking its escape 13,400 years ago. In short order, billions of gallons of water roared down the Champlain and Hudson valleys in a deafening cataract filled with tumbling boulders, dropping the level of the lake more than four hundred feet.

The outflow, which met the fast-rising Atlantic Ocean roughly where the Verrazano-Narrows Bridge stands today, sent a plume of frigid, muddy fresh water far out into the ocean, apparently shutting down the warm flow of the Gulf Stream and plunging the Northern Hemisphere into a sharp cold snap for 150 years.

The almost incomprehensible weight of a two-mile-thick slab of ice had compressed the very bedrock beneath it, and along what is now the St. Lawrence Valley, the glacier had forced the land well below sea level, forming a vast basin. A thousand years after Glacial Lake Iroquois drained, the ice pulled back far enough to allow the Atlantic to rush in, forming what geologists refer to as the Champlain Sea—a cold marine ecosystem extending hundreds of miles inland and encompassing 20,500 square miles. Rimmed by the tundra-clad and ice-capped Adirondacks, Green Mountains, and Laurentian Highlands, this marine world was populated by humpback and fin whales, belugas, cod, seals, lobsters, and clams.

It was into this soggy world of bare-scraped land, violent floods, and cold inland seas that the first humans in the Northeast stepped, presumably moving up from more southerly regions about eleven thousand years ago and probably following the coast and the Hudson River valley. The Paleo-Indians, as these earliest settlers are known, followed the herds of caribou that moved with the seasons from the open tundra to the patchy but ever-encroaching forests of scraggly spruce, poplar, tamarack, and birch.

These were people who, understanding the ebb and flow of their prey, haunted the barren, windswept heights, from which they could watch for the gray, rippling movement of thousands of migrating caribou funneling through the river valleys and along the lakeshores. Leaving little to chance, they pitched their hide tents on the eastern, downwind sides of valleys, so the caribou wouldn't be spooked by their scent, often picking sand dunes for their bivouacs, perhaps because this was the only dry land amid an eternity of boggy muskeg.

The whole band was involved in the hunt, a carefully executed ambuscade. It's easy to imagine the scene, since less than a hundred years ago, subarctic cultures such as the Ihalmiut of Nunavut and the Gwich'in of Alaska and the Yukon still hunted caribou in much the same fashion.

Along the traditional migration routes, the people likely built picket lines of stone cairns, which the modern Inuit call *inuksuit*. To the caribou, these cairns looked like people lining the treeless ridges on either side, channeling the herd closer and closer together, keeping the animals penned up. A few men trailed the herd to keep the nervous animals from bolting back and escaping. Finally, at a signal, women, children, and the elderly leapt from hiding on either side, shouting, flapping hides, and releasing the camp dogs. Now in a blind panic, dozens of caribou raced toward a narrow, preselected defile, where the hunters waited, throwing-spears poised.*

Because good stone was as critical to their survival as fresh meat, when the Paleo-Indians moved into new areas—where the glaciers had scraped away rock, then dropped it as they melted—the people kept a sharp watch for nodules of glossy, fine-grained chert, from which they could knap tools. Once they found the scattered lumps, they could follow the trail of the vanished ice sheet back to the source of the stone. At such outcroppings, they established quarries, where generations of Indians subsequently obtained the raw materials needed to make spear points, scrapers, and drills.

No one knows if these were the direct ancestors of the Wapánahki—in fact, there is good reason to suspect they were not. But all the peoples of the Northeast had long memories, and their myths and

---

* In 2009, researchers found dramatic confirmation of this Paleo-Indian hunting technique. A hundred feet below the surface of Lake Huron, along what nine thousand years ago was a land bridge between Michigan and Ontario, they found camps, stone blinds in which hunters would hide, caribou "drive lanes" walled with stone and stretching for hundreds of yards, and stone *inuksuit*.

legends retain what may be a glimpse of the postglacial world. The Wapánahki of mid-coast Maine, for instance, long preserved a legend of the *wǽskʷekkehs,* or "stiff-legged bears," animals that had enormous teeth and looked like mountains covered in shaggy brown vegetation. Similar monsters, which may be echoes of the mammoths and mastodons, appear in the stories of other Woodland cultures, including the Naskapi of Labrador and the Cree and Ojibwa of the upper Great Lakes.

The Paleo-Indians of the Northeast lived in a fast-changing world. By about ten thousand years ago, the bedrock was rebounding from the weight of the vanished ice, draining much of the Champlain Sea and cutting off its connection to the Atlantic. By one modern estimate, once the seaway to the Atlantic was blocked, rain and snowmelt may have flushed all the salt water out of the Champlain Sea in as little as a decade. This would have caused an ecological catastrophe for the marine life in the lake, and what must have been a rich source of food for the Paleo-Indians would have become far less bountiful, even after freshwater species established themselves.

The land was also changing. About 9,500 years ago, the forests of New England closed in and took on a more temperate character, with a lot of hemlock and oak and less boreal spruce and fir. Most of the big Ice Age mammals—including mammoths, American mastodons, saber-toothed cats, shrub-oxen, steppe bison, and short-faced bears—had become extinct or were, like the American mastodon, rapidly on the way out.

The cause of this mass extinction has been debated for decades. Traditionally, it was ascribed to climate change as the earth emerged from the most recent glacial period. But most of these species had weathered similar flip-flops from glacial to interglacial climates for millions of years with few losses. In addition, scientists noted, most of the extinct species were big—about one hundred pounds or larger; there were few extinctions among smaller mammals.

Perhaps climate wasn't the culprit, but human hunters. Geoscientist Paul S. Martin first proposed what he called the overkill hypothesis (more melodramatically dubbed the "Pleistocene blitzkrieg") in the 1960s. This theory contends that humans, armed with Clovis weapons and facing naive prey with no innate fear of people, swept across the continent, wiping out the Ice Age megafauna. Martin and his supporters note that similar mass extinctions occurred in Australia and Madagascar shortly after humans arrived there, too.

Recently, yet another theory has been proposed, which seeks to explain both the Pleistocene extinctions and the fact that at the same time, the Clovis culture also vanishes from the archaeological record. This coincides with the twelve-hundred-year cold snap known as the Younger Dryas, thought to have been triggered by the breaching of a massive ice dam, sending cold water rushing into the Atlantic from Glacial Lake Iroquois.

But what if the cause of both the freezing climate and the disappearing Clovis culture came from above? Lately, scientists have been arguing fiercely over the suggestion that a cosmic collision occurred about 12,900 years ago, just as the Younger Dryas began. Perhaps a comet or immense meteor slammed into the earth, setting off global fires, filling the atmosphere with soot and dust, and plunging the world into cold. There is some evidence—controversial though it may be—of such an impact.

For hundreds of years thereafter, signs of human occupancy are hard to find in much of North America, especially the Northeast. Those who remained, especially in the Southeast, responded to the warming climate and changing environment with new cultures. At the end of the Pleistocene, humans were living in small, highly nomadic bands that subsisted on hunting and gathering. As the temperature warmed—to averages that were, for a time, higher than today's—those family-based, mobile bands were replaced by more sedentary villages, especially along the coasts and major rivers.

Archaeologists have tracked not only the shifting material culture—for example, smaller projectiles and ax heads made by patient grinding instead of chipping—but also signs of increasing social sophistication. People were living in larger and larger communities, and with so many willing hands, as early as six thousand years ago they began to build immense earthen mounds, some of which eventually grew into enormous complexes. The oldest, as well as one of the largest, is known as Poverty Point, in Louisiana. It features half a dozen concentric ring mounds bracketing a peninsula—in all, more than seven miles of mounds, on which stood homes for up to a thousand people.

The unknown culture that built it had to move almost a million square yards of earth, basketful by basketful.

Along with local social growth, a continental trade network blossomed. At Poverty Point, archaeologists have found points made of flint from Alabama, the Ozarks, and the Midwest; stone bowls from the southern Appalachians; and decorative copper from the Great Lakes. But the ancient sites also reveal signs of increasing social stress—embedded spear points, healed injuries, crushing skull fractures resulting from blunt objects—probably caused by rising human populations and greater competition for resources. War had come to the First Frontier.

Around the end of what archaeologists call the Archaic period and the beginning of the Woodland period, about a thousand years ago, three major cultural developments swept Native communities across eastern North America: agriculture, ceramics, and bow-and-arrow technology. All three are usually referred to as "revolutions," because they reshaped Indian culture in profound ways, but as archaeologists trace their progression through time and across the landscape, their real impact seems to have been more evolutionary than revolutionary.

Human beings are conservative creatures; we tend to stick with what we know. For thousands of years, boiling water entailed dropping a series of heated rocks into a bowl chipped out of stone or a bag made from closely stitched skins. As the first rocks cooled, they were fished out and replaced with hot ones until the liquid boiled—a tedious and time-consuming process. Pottery that could be heated directly on a fire would have been a radical improvement.

About 4,500 years ago, a culture known as the Stallings Island people, who lived along the Savannah River in South Carolina and Georgia, made that inventive leap, creating the first ceramic pots found in North America. But revolutionary as it may have been, the innovation didn't exactly set the world on fire. It took fifteen hundred years for pottery to spread just to the surrounding areas of the coastal plain. By about 1500 B.C., pottery was ubiquitous throughout the Southeast, but it took another two millennia to spread into the Northeast.

Around the same time that early potters were experimenting with clay, pioneering farmers were learning to grow gourds, squash, sunflowers, maygrass, and goosefoot (an amaranth-like grain related to quinoa). Although agriculture didn't replace hunting and gathering, it took on an increasingly important role by the beginning of the Woodland period, about 700 B.C., especially in the Southeast.

One plant whose seeds do not appear in archaeological digs from that time is corn, or maize, which was domesticated in Mexico and did not make its way into the Southeast until about A.D. 175. Given the eventual preeminence of corn in Native life and religion—it was revered as a divinity by everyone from the Cherokee to the Iroquois—you might expect its introduction to have had an immediate and tectonic impact. Instead, like pottery, it took almost a thousand years for corn to become the dominant food crop of eastern Indians living south of New England, where the climate made farming an iffy proposition.

Weaponry also underwent a dramatic change. Most of the "arrowheads" picked up by generations of rural kids from freshly plowed fields were, in fact, made for spears—not as impressive and elegant as the Clovis points, but made to tip slender throwing-spears, or darts. Such darts were often six or eight feet long, fletched with feathers for stability, and hurled with tremendous force and accuracy using a throwing stick, or atlatl, which greatly magnified the force of the throw.

Sometime between two thousand and fifteen hundred years ago, however, atlatls and darts were replaced by bows and arrows, which were smaller, lighter, and more portable. An arrow fired from a powerful bow had more velocity and range than a throwing-spear, and it provided far greater versatility. One could shoot from a crouch, from a tree, or while standing motionless, and a trained archer could fire off several arrows in the time it took to fit and throw a dart.

Here again, the new technology was not universally embraced. Archaeologists have found evidence that in some areas of North America, bows replaced spears almost in an eyeblink. Elsewhere, however, spear throwers lingered for hundreds of years, often used alongside bows before the new technology won out. (An even older technology, the hand-held thrusting spear, lasted right up to the historic era in war and hunting.)

As technology and agriculture changed, so did social and political organization. In parts of the Southeast, especially west of the Appalachians and in the lower Mississippi Valley, social structure became increasingly complex, and with it the physical infrastructure that these hierarchical societies required. Archaeologists have uncovered the remains of huge mounds and plazas, some with no signs of residences; scientists can only assume these were regional ceremonial centers, supported by surrounding communities.

These mounds are small beer compared with those of the great mound-building cultures along the central and lower Mississippi. By A.D. 1100, the city of Cahokia, close to modern St. Louis, boasted a four-tier, flat-topped pyramid, which was made of 22 million cubic feet of soil and had a base measuring almost a thousand feet on each side. With a population of fifteen thousand to thirty thousand, Cahokia would have put medieval London to shame. It was the largest city north of Mexico until 1775, when New York finally surpassed it. Yet Cahokia was just one of a number of large urban and ceremonial centers in the same region.

In other parts of the South, such as the highlands of the southern Appalachians, mound building was a considerably more modest undertaking, and the settlements were far smaller. In the Little Tennessee River valley, around A.D. 500, farmers were tilling rich bottomland to grow corn, sunflowers, bedstraw, and maygrass, but much of their food was still coming from the forests and rivers. And because the woodlands and waterways of the southern Appalachians are among the most biologically diverse areas on the planet, it must have seemed like a feast was laid out for the inhabitants during much of the year.

In summer, wild cherries, blueberries, blackberries, and raspberries were ripe for the picking. In autumn, a smooth stick hurled into the trees brought down a shower of hickories, chestnuts, butternuts, and walnuts. Acorns could be ground and processed into a coarse flour. The rivers held fish and dozens of species of freshwater mussels. Excavations have revealed the bones of elk, deer, bears, turkeys, turtles, and raccoons, among many other species. Closer to the coast, not surprisingly, the ocean's bounty formed the backbone of subsistence, with villages supplementing shellfish, fish, and waterfowl with farming.

Here, as elsewhere, clay pipes are a common artifact found at old settlement sites, for among the plants grown most enthusiastically was tobacco, one of several species domesticated throughout the Americas.

This plant presumably played as important a ritual and religious role in prehistoric life as it did among contact-era Natives. (Ironically, the tobacco grown in the eastern United States today is a different species, *Nicotiana tabacum*, which originated in South America and was brought to Europe from the Caribbean by Christopher Columbus. Introduced to Virginia by English colonists, it quickly replaced the so-called Aztec tobacco, *N. rustica*, also originally from Mesoamerica, which in turn had replaced the native *N. quadrivalvis*—not surprising, given that Aztec tobacco packs up to ten times as much addictive nicotine as the native species.)

Just as tobacco, stone, and other materials moved across North America in trade, so did ideas, religions, and customs. One of the most remarkable examples of a cultural convergence across an immense area of the continent was what archaeologists call the Southeastern Ceremonial Complex or Southern Cult—deceptive names, since artifacts connected to this belief system have been found from Florida to Texas in the South to the western Great Lakes and Ohio Valley in the North. Cahokia and the other mound complexes were an expression of this cult, which sparked intricate artwork depicting a layered cosmology that—as far as anthropologists can decipher—featured an upper world and an underworld, each with a pantheon of supernatural beings, with the daylight world of humans poised in between. A sacred tree linked all three realms.

There are similarities between the imagery in Southeastern Ceremonial Complex art and that in art from Central America—to say nothing of the pyramid-like mounds at Cahokia and elsewhere. That has led some people to speculate that the cult was imported from Mesoamerica, although most archaeologists now dismiss the likelihood of any direct connection, since the Southeastern Ceremonial Complex was thousands of miles from the centers of Aztec and Maya culture, and the similarities may be coincidental.

Trying to understand something as abstract as religion based only on artifacts is an uncertain business at best, as is reconstructing the movements and development of the people who held those beliefs. For example, it appears that most of the Indian cultures of the middle coastal plain remained more or less in place for long periods of time, evolving over thousands of years into the Algonquian-speaking groups that first met Europeans. However, around 500 B.C., new cultures, bringing with them their distinctive pottery styles, pushed into what is now Virginia and the Carolinas. Two thousand years later, these people's descendants would include the Siouan-speaking Catawba and Tuscarora and the Iroquoian-speaking Tutelo and Saponi.

By the fifteenth century, the cultural landscape of what would be the First Frontier was incredibly rich—although much of it remains beyond our understanding, populated by people who vanished in the earliest years of European contact. These people left almost no trace in the historical record—nothing but mute remains stratified in the soil of old village sites and camps. This is particularly true of much of interior Florida, the Southeast, and the Appalachian ridges and plateaus west to the Ohio Valley. It's a telling comment on our ignorance that one of the most authoritative works on the Indians of Florida includes a map whose legend reads, "Peoples about whom something is known."

The coastal inhabitants of Florida—primarily the Calusa on the southwest coast and the Timucua in northern Florida and neighboring Georgia—developed radically different cultures, although in the hot subtropical climate, both wore elaborate breechclouts woven of palm fibers and shawls made either of palm fibers or Spanish moss.

The Calusa were not farmers—with the rich estuaries of the Gulf of Mexico close at hand, they didn't need to be—but they were ambitious builders, creating complexes of mounds and earthworks, as well as ditches and canals that prefigured the drainage schemes that make "Florida swampland" a punch line today. In fact, the spiderweb of canals radiating out from ceremonial centers in southwestern Florida were probably water trails, some wide and deep enough for decked barges made by lashing enormous dugouts together. The canals facilitated trade and political control, held by chiefs who owed tribute to a paramount leader. One such *certepe,* or great lord, welcomed Spanish visitors to a vast thatch-roofed house that could hold two thousand people. The few wooden Calusa objects that survive show that they were superb artists, creating vivid masks, carvings of humanlike animals, and decorated plaques.

The Timucua to the north did not share the Calusa's rigid, tribute-based hierarchy; there were about thirty-five separate chiefdoms, some

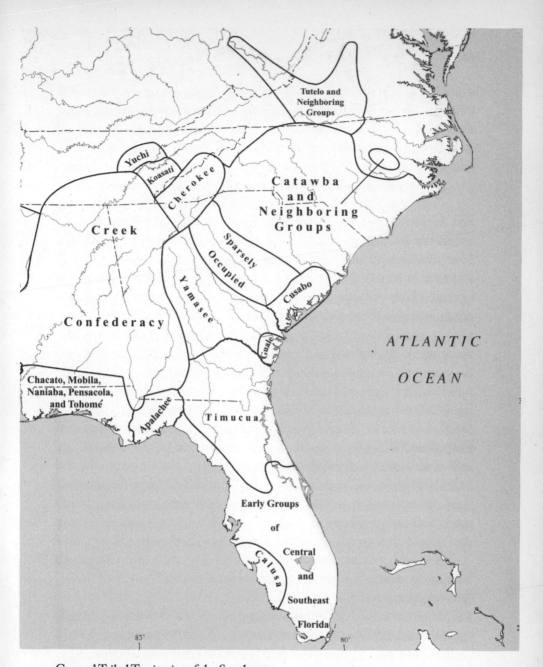

Tutelo and
Neighboring
Groups

Yuchi

Koasati

Cherokee

Catawba
and
Neighboring
Groups

Creek

Sparsely

Occupied

Yamasee

Cusabo

Confederacy

Guale

Chacato, Mobila,
Naniaba, Pensacola,
and Tohomé

Apalachee

Timucua

ATLANTIC

OCEAN

Early Groups

of

Calusa

Central

and

Southeast

Florida

85°

80°

General Tribal Territories of the Southeast

Mapping Native cultural groups and boundaries, especially during the early historical period, is fraught with difficulty. This map depicts generalized Indian territories from the sixteenth to eighteenth centuries. It includes refugee groups such as the Yamasee, which came into existence from the remnants of shattered prehistoric chiefdoms and moved several times between the territory shown here and Spanish Florida. (*Handbook of North American Indians*, vol. 14, Smithsonian Institution Press) loosely allied, stretching from northern Florida up to the Altamaha River in Georgia. Farming was far more important to the Timucua than to the Calusa, and Timucua villages were smaller, as were their houses—circular structures of wood and thatch big enough for a single extended family. French drawings from the mid-1500s show a Timucua man wearing nothing but a woven breechclout, his body heavily tattooed and his hair pulled into a topknot, from which dangles a furry animal tail. A woman, drawn by the same artist, wears a mantle of Spanish moss and earplugs of dyed, inflated fish bladders. Her body is decorated with equally elaborate tattoos.

Just to the north, inhabiting the Low Country and Sea Islands of northern Georgia, were the Guale, neighbors and apparently frequent enemies of the Timucua. Like almost all Indians associated with the Southeastern Ceremonial Complex, the Guale ritually consumed "the black drink," a potent draft brewed from the caffeine-rich leaves and twigs of the yaupon holly, *Ilex vomitoria*. The liquid, drunk from decorated whelk shells and apparently spiked with snakeroot and other ingredients, was a purgative used by high-ranking men in cleansing ceremonies. Taken unadulterated, yaupon tea (also known as cassina) imparts a pleasant buzz and for a time actually gave coffee a run for its money among colonial Europeans.

Calusa? Timucua? These are hardly names that ring a familiar bell with most Americans today. If you know anything about Florida Indians, you are more apt to think of the Seminole—a group actually descended from members of the Creek confederacy in Georgia and Alabama, who didn't migrate to Florida until the early eighteenth century, becoming known by the Creek word *simanoli*, meaning "wild" or "undomesticated."

Elsewhere in the sixteenth-century South, there were, of course, the Cherokee, who were restricted to the rugged, impossibly beautiful highlands of the southern Appalachians. From those mountains across the Carolinas to the ocean lay a welter of culturally linked villages that the Spaniards termed "the Province of Cofitachequi," an amalgam of Catawba-speaking chiefdoms. In 1540, these villages may have been ruled by a paramount chieftainess the Spaniards called "the Lady of Cofitachequi," who was carried about on a ceremonial litter.

In northeastern North Carolina and adjacent Virginia lived several

The manner of their fishing.

A Cannow.

The Manner of Their Fishing

The watercolors of John White, based on multiple voyages to the Outer Banks of what is now North Carolina in the 1580s, provide a unique glimpse of contact-era life among the Algonquians of the Mid-Atlantic coast. Here White shows the use of weirs, spears, and a fire built in the clay-covered hull of a canoe to lure fish to the surface at night. Among the fish are hammerhead sharks, while pelicans and other water birds fly overhead. (Library of Congress LC-USZ62-576) groups speaking closely related dialects of an Iroquoian language. The largest group called themselves *skarò·rə?* "hemp gatherers," referring to the many uses to which they put dogbane, or Indian hemp, a source of fiber and medicine. From *skarò·rə?* also came the Anglicized name Tuscarora.

From the Outer Banks north all the way to the subarctic, the eastern seaboard was almost exclusively the realm of the Algonquians. A diverse mix of cultures with prehistoric roots in the North, these groups were related by language but as often at war as at peace with one another. Algonquian, as a language family, bestrode the continent in the sixteenth century, reflecting millennia of human migration. It included not just the eastern Algonquian-speaking peoples of the Atlantic coast, New England, and the Maritime Provinces, but also the Ojibwa, Sauk, Potawatomi, Illinois, Miami, and Shawnee of the Great Lakes and Midwest; the Cree of the Canadian prairies; the Gros Ventre and Blackfoot of the Rockies; and the Cheyenne and Arapaho of the western Great Plains. Most remarkably, the Yurok and Wiyot of northern California also spoke related tongues.

The southern fringe of Algonquian territory lay in northeastern North Carolina, among groups such as the Roanoke, Croatan, Chawanoke, and Weapemeoc. Ironically, although relatively little is known about the Carolina Algonquians, we have a beguiling glimpse of their daily lives, thanks to a series of delicate watercolors painted by John White in 1585 on Roanoke Island. For many people, White's paintings are the iconic images of the eastern Algonquians, showing their palisaded villages and arch-roofed bark houses; their gardens and cornfields, the latter with bark shelters on raised platforms, from which a guard (usually a child) could scare off flocks of birds with well-aimed stones; dancers moving around posts carved with human faces; and Indians fishing from a dugout canoe, a hammerhead shark and horseshoe crab in the water below and other people in the background spearing fish or trapping them in a weir.

Life was probably much the same for the Algonquians who inhabited the Virginia coastal plain. Although those to the south may have been politically affiliated with their Carolina cousins, many of the villages along the western shore of the Chesapeake Bay were united under powerful chieftains. In fact, by 1607 the Jamestown settlers found one top chief, or *mamanatowick*, a man they called Powhatan (but whose Algonquian name was Wahunsenacwah), controlling a territory of almost eight thousand square miles known as Tsenacommacah. Inland from Tsenacommacah lay land under the control of non-Algonquians, the Manahoac and Monacan, who spoke a Siouan language. To the north were the Piscataway and, on the eastern shore of the upper Chesapeake, were the Nanticoke, both Algonquian peoples.

The Nanticoke traditionally traced their origin to the Lenape, who occupied much of the Middle Atlantic seaboard. Lenape bands speaking Unami dialects lived along Delaware Bay, in eastern Pennsylvania and southern New Jersey. Those speaking a related dialect, Munsee, lived from about the Delaware Water Gap north and east through the lower valley of the Hudson (a tidal estuary they called *muhheakantuck*, "the river that flows two ways"), including the slender lozenge of land known as *manahatta*, "the island of many hills." Later known as the Delaware, the Lenape moved from summer camps near their cornfields or good fishing rivers to more permanent winter settlements, where they sheltered in longhouses roofed with chestnut bark, each up to a hundred feet in length and housing several families.

Indian warfare tended to be a matter of low-intensity raids for revenge or captives—rarely the kind of wide-scale, all-out war that would later emerge in conflict with Europeans. Lenape villages, like those of many Algonquians, were sometimes built in defensible locations such as hilltops and protected by log stockades. But the Susquehannock to the west specialized in building centralized, heavily fortified settlements. An Iroquoian people, the Susquehannock probably didn't move from the headwaters of their namesake river to its lower reaches until the sixteenth century. They were either lured south by the promise of easier trade with the first European visitors or, more likely, pushed by pressure from the *haudenosaunee*, "the people of the longhouse" (the Mohawk, Oneida, Onondaga, Cayuga, and Seneca, better known as the Five Nations of the Iroquois), who controlled much of what is today upstate New York. (Iroquoian-speaking villages enclosed in triple palisades of log walls also dotted the St. Lawrence Valley, but it is unclear how close a cultural connection, if any, these people had to the Five Nations.)

The Iroquois homeland along the Finger Lakes—bounded by

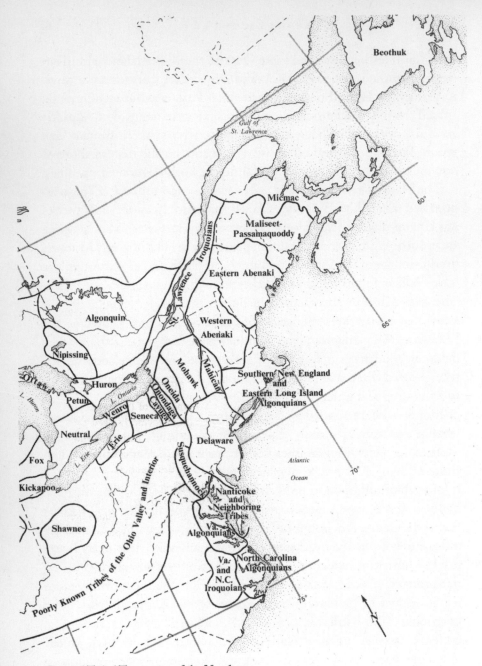

General Tribal Territories of the Northeast
This depiction of tribal boundaries in the Northeast shows roughly how they looked in
the earliest days of the historical period and is in some cases a simplification of cultural
structure. (*Handbook of North American Indians*, vol. 15, Smithsonian Institution Press)

mountains to the south and Lake Ontario to the north, and with rivers flowing to the south and east—lay at the center of continental trade networks that would only increase in importance as the colonial fur trade blossomed. The strategic position of the Iroquois, as well as the politically powerful alliance among the Five Nations, meant that they would exert a profound influence on frontier history for centuries to come.

To the Mohawk—*kanien'kehá:ka*, "the people of the flint," the easternmost of the Five Nations—Lake Champlain was "the doorway to the country," and they were the keepers of the door. But to the Algonquian peoples to the east, such as the Mahican of the upper Hudson, and to the western Abenaki (one branch of the Wapánahki), the lake was a welcome barrier against incessant Mohawk attacks. The Abenaki called it *bitawbagok*, "the lake between."

At the northernmost edge of the eastern seaboard—north of the Maliseet-Passamaquoddy in New Brunswick and the Mi'kmaq in Nova Scotia, north of the Innu (whom the French called Montagnais)—lived the Beothuk, in Newfoundland, an island culture about which little is known. Clad in caribou skins, the Beothuk painted themselves liberally with red ocher mixed with animal fat. Because of this, the first Europeans who met the Beothuk dubbed them "red Indians" and thus tagged an entire continent's population with a misnomer.

The Beothuk traded and even intermarried with the Mi'kmaq and Innu, but their relations with the northernmost people in the East, the Inuit, were generally hostile. The Beothuk called the Inuit "the four-paws," and encounters between them usually ended in a fight. Interestingly, the Inuit were quite recent arrivals to the eastern Arctic and subarctic, which had been occupied for more than a thousand years by a very different Eskimo culture known to archaeologists as the Dorset, who specialized in hunting seals through holes in the pack ice.

But by about A.D. 1000, the Arctic was changing. The Arctic Ocean was warming, the ice was breaking up, and large marine mammals such as the bowhead whale were expanding their range. The Dorset, unable to cope, withered away, while the Thule—ancestors of today's Inuit—found a world perfectly suited to their lifestyle. Communities banded together to hunt whales from large skin boats, and Thule men used bows and arrows (which the Dorset lacked) to kill caribou and bears. From their homeland in western Alaska, the Thule spread east with remarkable speed, either taking over a landscape no longer occupied by the Dorset or assimilating the last survivors of that culture into their own. In just a century or two, they reached Greenland.

The First Frontier was a mind-bogglingly fecund land. An eagle flying anywhere between the northern timberline and the Gulf of Mexico would have seen, reflecting the sky like countless mirrors, the carefully dammed ponds of tens of millions of beavers. There were passenger pigeon flocks, just one of which (more than three centuries later) would be estimated to contain 2.2 billion birds. Sea birds by the hundreds of millions jammed breeding colonies along the northern Atlantic coast, where half a dozen species of great whales breached and blew. In the Gulf of Mexico, green sea turtles swarmed in such numbers that they were later an encumbrance to sailing ships.

Inland lay a mosaic of thousands of unique ecosystems: subarctic tundra and subtropical swamp forests; tidal marshes of cordgrass knitting together land and sea; sphagnum bogs in the North rimmed with tamarack and spruce; longleaf pine forests in the South, where gopher tortoises and armadillos burrowed. Ivory-billed woodpeckers made the ancient cypresses of the southern swamps ring with their hammering, while screeching flocks of orange-and-green parakeets wheeled above sleeping alligators. In the lush forests of the Appalachians, tulip trees twenty-five feet in circumference leafed out in April over carpets of wildflowers. In autumn, American chestnut, butternut, shagbark hickory, and dozens of kinds of oak trees produced an annual bounty on which great flocks of turkeys foraged and black bears grew corpulent.

Fingers of prairie, maintained by fire and among which herds of bison grazed, threaded west into the Allegheny and Cumberland Mountains. An eastern forest race of elk roamed the Appalachians and coastal plain from northern Georgia to at least northern New York. Woodland caribou, pale as the snow, could be found as far south as Vermont and New Hampshire. Moose—gangly-legged, lop-eared, and long-nosed—ranged south to Pennsylvania. Trailing them all

were the hunters—lynx, bobcats, fishers, wolverines, several species
of wolves, and the widespread, adaptable mountain lion, which left its
rounded tracks from the cold forests of New Brunswick to the saw-
grass marshes of the Everglades.

The rivers ran black with fish—Atlantic salmon weighing up
to thirty pounds, choking waterways from the Housatonic north
to Newfoundland, and hundreds of millions of other anadromous
fish, including shad, blueback herring, alewives, rainbow smelt, and
striped bass. Atlantic sturgeon up to fourteen feet long, plated like
alligators and weighing eight hundred pounds, returned each spring
from the ocean to spawn—and to dance, the immense fish leaping
repeatedly from the water as they gathered to mate. Natives wielding
spears hunted them at night by torch light, though it must have been
a chancy undertaking. The leaping sturgeon were so abundant that
Europeans would later consider them a hazard to navigation.

This was the First Frontier in the last days of its isolation. It was
fruitful beyond modern recognition, it was riotous in its natural wealth,
but it was not wilderness. There *was* no wilderness, at least not in the
sense of land unoccupied and unaltered by humans. The Indian cultures of North America had had a profound and far-reaching effect on
their home.

Take agriculture. Given that fields were cleared by girdling trees
and waiting for them to die, then tilled with hoes made from the
shoulder blades of deer, one would expect that Native agriculture was
a small-scale operation, especially when compared with the modern
variety. But by the sixteenth century, farming—especially the cultiva-
tion of maize—had transformed much of the eastern seaboard. In 1539,
Hernando de Soto's conquistadors reported fields in northern Florida
"spread out as far as the eye could see across two leagues of the plain."
In the Iroquois lands of New York, early Europeans reported seeing
both Onondaga and Seneca villages surrounded by roughly six square
miles of cornfields, and one missionary in Ontario in 1632 got lost
repeatedly—not in deep forest, but in endless fields of maize. Because
the soil in such fields eventually wore out, forcing the relocation of
entire villages to new sites to be cleared of forest, the impact on the
landscape over time must have been nearly universal.

Nor was it a trackless emptiness. The land was seamed with foot-

paths linking every corner of the continent, funneling trade and warfare—trail systems such as the one known variously by the eighteenth
century as the Great Catawba War Path, the Iroquois Path, and the
Tennessee Path, which ran from southern New York through Pennsylvania, West Virginia, Kentucky, Tennessee, and the Carolinas, and
which connected with trails stretching into Canada and Florida.
Along such heavily used paths, visitors often encountered bark lean-
tos spaced roughly a day's march apart, rebuilt again and again by generations of travelers.

Europeans often found themselves befuddled not by an absence of
trails, but by too many from which to choose. A local Indian, however,
knew that one path, say, offered the most direct route but entailed
several tiring ascents up steep mountains, while another featured easy
passage through low passes but wasn't a good choice in spring because
of poorly drained ground. Yet another route might be fine in midsum-
mer, when streams were low and crossings were easy, but not after
heavy autumn rains brought up the water level.

The population that laid out and maintained this trail network was
immense. Whereas historians once assumed that North America was
a continent sparsely inhabited and lightly touched by its original residents, they now speak of tens of millions of pre-Columbian Indians,
who had an all-encompassing impact on the landscape and its ecosystems.

Basing their calculations on nose counts given by early settlers,
anthropologists once estimated the pre-contact population of North
America at about a million—although they often dismissed as ex-
aggerations the firsthand accounts of French, English, and Spanish
observers on which they based those estimates. In the early 1900s, for
example, Smithsonian ethnographer James Mooney carefully combed
through the early records to make the first continental approximation
of Indian populations—but time and time again, he rejected his own
sources. Typical was Mooney's reaction to the claim, reported in the
1670s, that the Narragansett in Massachusetts once fielded five thousand warriors; the scientist said that the claim "need not be seriously
considered."

In fact, as Mooney himself knew, those original records largely came after the virulent epidemics of the early seventeenth century, which in most places swept away all but a handful of Indians even before the first European settlements had been founded. Indeed, it's unlikely that Europeans would have gained a foothold had the immense arable land of the Atlantic coast not been conveniently cleared of its inhabitants by disease.

More recent estimates put the precolonial population of North America at 3.8 million people, with as many as a million of them living between the Appalachians and the Atlantic. Some scholars put the population at almost twice that. The Natives' impact on the land was profound, in terms of the amount of land converted to agriculture, the species they hunted and fished, and most particularly the way they shaped their world through the use of fire.

The first European explorers, coming from countries where mature, old-growth forests had been cut down centuries earlier, spoke glowingly of the tall, straight, commercially valuable trees of the New World. But behind these descriptions, almost in the spaces between the words, lies the truth about the eastern woodlands: they were not the universally dense, dark, brooding forests of our historical imagining. Those forests certainly existed, but in many cases the newcomers described delightfully open, parklike woods free of undergrowth, intermixed with wide grasslands and savannas—a mosaic of habitats that, it now seems clear, were created and maintained by fire.

In 1524, on Narragansett Bay, Giovanni da Verrazano remarked on the extraordinary extent of Native agriculture, reporting fields extending for more than seventy-five miles from the bay, "open and free of any obstacles or trees," and the forests beyond so uncluttered that they "could be penetrated by even a large army." Recall James Rosier's observation about the forests of Mawooshen having "but little wood" and looking like the pastures of England. William Strachey, who was with the expedition that returned Skicowáros to Maine in 1607, reported a rich land, "the trees growing there on being goodly and great, most oake and walnut with spacious passadges betweene, and no rubbish," the usual term in those days for underbrush.

Around the same time, Captain John Smith was with the Algonquians in Virginia, where "by reason of their burning them for fire," a man "may gallop a horse amongst the woods any way, but where the creekes or Rivers shall hinder." Leonard Calvert, arriving along the lower Potomac, reported the land "growne over with large timber trees, and not choaked up with any under-shrubs, but so cleare as a coach may without hinderance passe all over the Countrie."

Landing on Martha's Vineyard in 1602 with the Gosnold expedition, the Reverend John Brereton observed forests of "high timbred Oakes," along with beech, elm, cedar, walnut, hazelnut, and black cherry, but noted that "in the thickest part of these woods, you may see a furlong or more round about." Elsewhere, he said, "medowes very large and full of greene grasse; even the most woody places . . . doe grow so distinct and apart, one tree from another, upon greene grassie ground . . . as if Nature would shew her selfe above her power, artificall."

Artificial is exactly what it often was and had been for a very long time. Enough of the Northeast coast was savanna and dwarf oak-pine-blueberry "barren," a fire-dependent habitat, that one of the most common birds there was the heath hen, an eastern race of the grassland grouse known as the greater prairie-chicken, found from Massachusetts to Virginia. Although the heath hen seems to have been better adapted to shrubbier environments like the barrens than its midwestern cousins (it's hard to say for sure; the subspecies became extinct in the 1930s), it clearly needed open land.

In 1670, a visitor named Daniel Denton wrote that both sides of the Raritan River in New Jersey were "adorn'd with spacious Medows, enough to maintain thousands of Cattel," and "where is grass as high as a mans middle, that serves for no other end except to maintain the Elks and Deer . . . then to be burnt every Spring to make way for new."

Indians set fire for a variety of reasons, besides creating a flush of new vegetation that would attract "Elks and Deer." Fire could prepare soil for planting, burn away pests such as ticks and chiggers, and drive game into an ambush. Yet this Native dependence on fire was overlooked by modern historians and scientists for years. Even those who first recognized the importance of Indian-set fires in shaping the landscape did not view such management kindly. "The Indian is by nature an incendiary, and forest burning was the Virginia Indian's besetting sin," wrote Hu Maxwell, a historian with the U.S. Forest Service in 1910. "The tribes were burning everything that would burn . . . [and] if the discovery of America had been postponed five hundred years, Virginia would have been pasture land or desert."

Hardly. Indians had been burning Virginia and much of the rest of North America annually for fifteen thousand years, and it was certainly not a desert. But they had clearly altered their world in ways we are still only beginning to appreciate. The grasslands of the Great Plains, the lupine-studded savannas of northwestern Ohio's "oak openings," the longleaf pine forests of the southern coastal plain, the pine barrens of the Mid-Atlantic and Long Island—all were ecosystems that depended on fire and were shaped over the course of millennia by its frequent and predictable application.

Given that Indians and fire were a constant throughout the long postglacial period, when climate change was already rearranging eastern North America's natural communities, it's fair to say that the land of the First Frontier coevolved with humans and their ready ally, the flame. If there ever was a "balance of nature" in the East, it was a case of natural systems finding some measure of equilibrium amid the pressures of hunting, agriculture, and fire, as practiced by more than a million hungry, busy human beings.

And that was the face of the East at the moment when the first regular contacts began between the New World and the Old: the close-mouth Basques trading iron kettles for furs while their catch dried in the sun; the Bristol merchants sniffing along behind them to find the source of the cod; the trickle of ships that would soon become a colonizing flood.

There is even one small but tantalizing hint to suggest that exploration was not entirely a one-way street. After all, crossing the North Atlantic from east to west, Europe to America, meant fighting headwinds and contrary ocean currents the whole way—hence the usual route far to the south, picking up the northeasterly trade winds to the Caribbean. But if you were heading in the opposite direction, you'd have the prevailing winds at your back and the Gulf Stream flowing toward Europe.

We know that many of the coastal tribes created huge, stable, ocean-going dugout canoes for fishing and whaling. Why did none of them cross the Atlantic?

Perhaps some did. "In the year 1153," according to the sixteenth-century Portuguese historian Antonio Galvano, "it is written, there came to Lubeck, a city of Germany, a vessel bringing certain Indians in a canoe, which is a vessel propelled by oars . . . but this canoe must have come from Florida, the country of cod fish, which lies about the same latitude as Germany."

Galvano may have based his comment on the work of Otto of Frei-sing, who wrote in the twelfth century about the arrival of "Indians" in Lübeck during the reign of Otto's nephew, Emperor Frederick I Barbarossa. But Otto may have been borrowing, too. Stories about Indian merchants washing up in northern Germany bump through the historical record all the way back to A.D. 43, when the Roman geographer Pomponius Mela used the tale as evidence that the Indian Ocean and Baltic Sea were connected.

Galvano didn't cite a source for his claim. When he unearthed this tidbit four hundred years after the fact, he casually made the astonishing assertion that Indians paddled to coastal Germany. To the Portuguese of Galvano's day, "Florida" was a nebulous concept, a name applied to almost any point in the New World. But to an Iberian such as Galvano, who every Friday ate *bacalao* (salt cod), "the country of cod fish" very specifically meant Newfoundland and its surrounding fishing banks.

"The Germans greatly wondered to see such a barge, and such people, not knowing from whence they came, nor understanding their speech, especially because there was then no knowledge of that country," Galvano wrote. He admitted that it may seem incredible that such a small boat could cross such immense seas; "nevertheless it is quite possible that the winds and currents might bring them there."

And that's it—nothing more. On its face, it certainly does seem incredible, not only because of the distances and dangers that had to be overcome, but also because in order to make landfall in Lübeck, such mariners would have had to bypass Iceland, Scotland, the Faeroe Islands, and Norway (if coming from the north) or thread the English Channel without bothering to stop in England or France (if coming more directly from the west). And regardless of the route, they would *then* have to circumnavigate Danish Jutland, poking like a thumb into the North Sea, and avoid the four hundred islands surrounding it before deciding to finally land at Lübeck.

It's preposterous, surely. But what if, by some infinitesimally small chance, Otto and Galvano were right? What if the descendants of the *skrælings* who battled the Norse in Vinland wondered about the land from which those violent intruders had come and decided to seek it out for themselves, stocking a great sea canoe with bladders of water, dried meat, and sealing harpoons? What if men who, like their counterparts in Europe, looked at the horizon and wondered, then decided to stop wondering and take action?

Despite the historical odds, despite common sense, it's thrilling to imagine a big dugout rolling with the swells, wending its way through the flat Danish islands and into the Baltic waters under an overcast sky. With final, tired strokes, the paddlers drive the boat onto the muddy shore and step out—men in leather and furs, their arms and faces tattooed, otter-skin quivers on their backs, their feet clad in sealskin boots, bone and copper in their long black hair.

They can see fields, where strangely clad men and women stop what they are doing and point—oddly pale humans, bleached as though they'd been trapped underground, many of them weirdly and unnaturally hairy.

The two groups face each other, shifting uneasily. Then one of the Indians slowly raises a hand in the universal sign of peace.

The day Americans discovered Europe.

Chapter 3

Stumbling onto a Frontier

I N THE LATE summer of 1582, three powerful Englishmen, united by an interest in the new lands of America, gathered to question a man who, it appeared, had managed one of the greatest treks of that or any other age: a wilderness journey across nearly three thousand miles of unknown land, affording him an unprecedented familiarity with the New World, its inhabitants, and its riches.

The man's name was David Ingram, and he was either one of the greatest accidental explorers of history or a liar of epic proportions. Or very likely both.

Ingram, who was about fifty that summer, was a presumably illiterate sailor from the Thames port of Barking, just east of London. Fifteen years earlier, he'd had the bad luck to sign on with a legendarily ill-starred voyage, one that ended with a ferocious sea battle and his being abandoned on a foreign shore with more than a hundred other starving men. For years, Ingram had been plying tavern audiences with his stories of crossing America on foot, before a miraculous rescue restored him to England. These tales, percolating up through the strata of Elizabethan society, had eventually reached the curious ears of influential men for whom the New World was an abiding passion.

At their command, Ingram was plucked from the alehouses he frequented and brought before them.

Francis Walsingham was principal secretary to Queen Elizabeth I, a member of her Privy Council, and her spymaster. He was also fascinated by the prospects of the fabled Northwest Passage to the Orient. Sir George Peckham saw the New World as a refuge for Catholics, persecuted in Elizabeth's Protestant England, a concept that fit neatly within the colonization schemes of the third gentleman, Sir Humphrey Gilbert, who held a royal patent permitting him to colonize "remote heathen and barbarous landes." Gilbert was a frightening prospect as an overlord. As military governor in Ireland, he had wreaked such terror and devastation putting down a 1569 rebellion that his half brother Sir Walter Raleigh said he "never heard nor read of any man more feared . . . amonge the Irish nation." To Raleigh's mind, that was a compliment. (Later, in 1583, Gilbert would claim Newfoundland for England, despite the presence of a fleet of foreign fishing ships, then vanish with his frigate during a storm.)

There are a great many unknowns cluttering David Ingram's story, so it is best to start with what is indisputable. In 1567, when he was about thirty-five years old, Ingram hired on as a seaman with Sir John Hawkins, who was outfitting a fleet of four ships he owned with his brother. To these ships were added two aging hulks belonging to the queen, *Minion* and *Jesus of Lübeck,* the latter an especially decrepit vessel bought secondhand by Henry VIII. One of Hawkins's smaller ships, *Judith,* was commanded by a young firebrand relative of his named Francis Drake.

The stated purpose of the expedition was to seek a supposed gold mine in Africa, but anyone who knew John Hawkins could guess its real goal. Three years earlier, Queen Elizabeth had granted him a coat of arms, whose heraldic devices included "a demi-Moor proper bound in a cord"—that is, an African slave tied up for sale. Hawkins had made two previous slaving voyages to Africa, selling the captives in the Caribbean (if the Spanish planters declined to buy, he'd shell and burn their plantations until they relented), buying cod in Newfoundland for resale back home, and making a tidy profit in the process. Along the way, Hawkins also introduced the sweet potato and tobacco to England.

This time, things did not go so well. Hawkins's fleet (bolstered by several additional ships he acquired along the way) captured five hundred Africans, but only after a bloody fight in which Hawkins was nearly killed by a poison arrow. By the time they crossed the Atlantic, hammered by a hurricane, the *Jesus* was so decayed that fish were said to swim freely into and out of the hull, and only constant pumping kept the ship afloat.

Hoping to make repairs, the fleet put in near the newly established Spanish fort of San Juan de Ulúa at Veracruz, on the Gulf coast of Mexico, in September 1568. Hawkins's bad luck continued. The following day, the Spanish treasure fleet arrived with the new viceroy, and a few days later the Spaniards, fuming at the English intrusion, attacked. The English sailors managed to sink three Spanish ships, but they lost all but two of their own in the melee. Hawkins's *Minion* was crammed with two hundred of the surviving sailors, while Drake's *Judith* held almost all their supplies. When Hawkins signaled his cousin to come and take some of the men, Drake ignored the command and sailed under cover of darkness for England, taking almost all their food. "The Judith . . . forsooke us in our great myserie," Hawkins bitterly recalled.

For two weeks after the battle, the *Minion* wallowed in the Gulf of Mexico, trapped by headwinds. Although she was loaded down with riches from the sale of the slaves, the starving men onboard were reduced to eating anything they could find, including their pets. "Hides were thought very good meate," Hawkins recalled, "rattes, cattes, mise and dogges none escaped that might bee gotten, parrats and monkayes that were had in great prize."

Hawkins knew he could never get so many men back to England, so he announced that half of them would be set ashore to fend for themselves until he could send a rescue vessel. To hear Hawkins tell it, he was acceding to their wishes: "Our people being forced with hunger desired to be set a land, whereunto I concluded." But Miles Phillips—one of the 114 men rowed through great, storm-fueled waves, beaten over the side of the boat a mile from shore, and left to sink or swim—saw it very differently: "It would have caused any stony heart to have relented to hearr the pitifull mone that many did make, and how loth they were to depart."Two men drowned, and the rest found themselves marooned with no rations and only one useless matchlock musket and two rusty swords among them.*

The castaways were, by their own reckoning and Hawkins's, near the northwestern rim of the Gulf of Mexico, about two hundred miles south of the Rio Grande. The Englishmen divided into two groups. About twenty-five of them, including Phillips, set out to the west, looking for the Spanish, while fifty-two men, including David Ingram, decided to go north. After an Indian attack that killed their leader and two others, half of those men doubled back to rejoin Phillips and the others.

After almost two weeks of wandering with little food or water, picked off by the arrows of hidden Indians, Phillips and his fellow survivors found the Spanish settlement of Tampico. They also found themselves entrapped in the Mexican Inquisition—questioned, tried, and in a few cases burned at the stake for the heresy of Protestantism.

Meanwhile, Ingram's small group simply vanished. Here's where the certainties about his story evaporate, beyond this one final fact: David Ingram, forsaken on the wild shores of the Gulf of Mexico, somehow reappeared in England.

Thirteen years later, under the pointed questions of Walsingham, Peckham, and Gilbert (and carefully recorded by an anonymous scribe), Ingram recounted a tale he'd told many times before. Right off the bat, it differs from what's known about the Hawkins disaster. Instead of being marooned on the Mexican coast, Ingram said that he and the others were abandoned "about 140 leages west and by North from the Cape of Florida"—a position near Apalachicola in what is today the Florida Panhandle.

From there, Ingram said, they walked—in less than eleven months, helped along the way by each Indian village they encountered—almost three thousand miles to Cape Breton, Nova Scotia. There he and two remaining companions were rescued by a French ship named

---

* Hawkins and the *Minion* managed to limp back to Europe, where forty-five of his emaciated survivors died after gorging on too much meat. He finally reached England with only 15 of the 408 men who'd started out with him. Ten years later, by then a vice admiral, he and his once unreliable cousin Sir Francis Drake would lead the English fleet to victory against the Spanish Armada.

Gargarine, repaying their passage with the gift of one of the "great pearles" they'd picked up along the way and by helping the French trade with the locals. In their final weeks, they encountered "the maine sea upon the Northside of America," and the Natives drew for them pictures of ships and flags, "which thing especially proveth the passage of the Northwest." At this point, one can almost hear Walsingham and the rest sigh with satisfaction: here was the Northwest Passage to the Orient at last.

Ingram's account, as condensed and transcribed by the nameless clerk (who refers to "this Ingram" or "the examinate"), runs to only a few thousand words, none of it in the first person. The original, published in 1583, is lost, and the versions that survive differ on some details. Even where they agree, they offer an infuriating mix of enticingly plausible details and transparent whoppers, and they leave unsaid far more than they explain.

Ingram claimed that his group moved almost without pause from the territory of one Indian "king" to the next, the kingdoms spaced about 120 miles apart and each chief willing to help these strangers—at whose "whitenes of their skins" the Natives marveled—along on their journey. He said barely a word about hardships, dangers, attacks, and privations, and he was—most fundamentally and inexplicably—completely silent on the fate of the other men he was with. We know the group was twenty-six strong when they left the Gulf, but by the time they hailed the French ship, there were only three left—Ingram and two men named Richard Browne and Richard Twide. On the fate of the others, Ingram's narrative is mute. (Miles Phillips, one of the *Minion* survivors who endured the Mexican Inquisition, speculated years later that the other men were "yet alive, and marryed in the sayd countrey.")

Doubtless in response to the noblemen's questions, Ingram recounted the animals and plants he saw—shaggy black "buffes" with crooked horns; "a bird called a Flamingo, whose feathers are verie red"; bears, foxes, wolves, and rabbits. He talked about the climate, fierce thunderstorms, and whirlwinds that could have been tornadoes. He discussed the land ("most excellent, fertile and pleasant") and the curious customs of the Indians, whom he described as especially pliant and docile—"how easily they may be governed when they be once conquered."

But most of all, he had a great deal to say, all of it pure hooey, about treasure-laden cities—kings seated on bejeweled thrones, women wearing golden plates around their necks, feasting halls with "pillers of massie silver and chrystall," and nuggets of gold as big as a man's fist simply lying in the streams for the taking. Here is where Ingram's account veers into obvious fantasy, and one reason historians have for so long dismissed it—along with his claims of having seen elephants, odd red sheep, domestic cattle, an immense horselike beast with foot-long tusks growing from its nostrils, and a crested bird as big as an eagle ("verie beautifull to beholde, his feathers are more orient [lustrous] than a peacockes").

"The Relation of David Ingram of Barking, in the Countie of Essex" was published in 1589 by Richard Hakluyt in the first edition of his *Principall Navigations, Voiages and Discoveries of the English Nation,* the great compilation of early exploration accounts. There is no record of Hawkins ever challenging Ingram's claims, and (perhaps conveniently) his two companions, Browne and Twide, were by then both dead. But as more and more Europeans actually visited the eastern seaboard of America—and found no golden cities or crystal palaces, no elephants or red sheep, no fist-size nuggets lying about in every stream—Ingram's reputation took a beating. Hakluyt dropped Ingram's "Relation" from subsequent editions, and the old sailor has been considered more or less a fraud ever since.

Yet there must be a kernel of truth buried beneath all the hogwash. Samuel Purchas, who posthumously published some of Hakluyt's manuscripts, used Ingram as a cautionary tale, noting "the reward for lying being, not to be believed in truths." As Purchas realized, Ingram *somehow* got from the wild coast of Mexico to the halls of London; he couldn't have made up everything. Hidden amid all the lies is a remarkable story of survival.

One possibility, proposed by history writer Charlton Ogburn and others, is that Ingram followed the Gulf coast and crossed what is now the southeastern United States before being picked up by a French ship on the Atlantic coast. This is made even more plausible by the largely tropical or subtropical flavor of the plants, Indian customs, and landscapes Ingram described, although the general lack of French shipping in those waters at that time is a problem. If not the epic, three-thousand-mile march of legend, it would still be an impressive trek of almost fifteen hundred miles.

Ingram may then have observed the eastern shore of the continent from the deck of a northbound ship, since he described a few things that are manifestly boreal in nature, such as his spot-on description of flightless great auks, which were not found south of Maine. He called them "penguins"—the first use of that word—and said the black-and-white birds were "the shape and bigness of a Goose but . . . cannot flie."

As for the rest—the elephants and sheep and gold-bedecked halls—this was Ingram's one chance at glory, spinning his tale to three of the most powerful men in the realm, whom he must have known were intensely interested in the commercial prospects of the New World. One suspects he wasn't going to let the opportunity slip away; thus we have submissive, courteous Natives whose sunlit lands drip with riches.

Besides, as historian David B. Quinn has noted, his story was "the end product of a long run of tavern tales where the original story [was] eventually overlaid by the 'truth' of fiction." In the storyteller's mind, memory and fantasy can fuse. After all those years, Ingram probably believed most of it himself.

Whether Ingram profited from his story is unknown. It seems that at least he was offered a place in Gilbert's expedition to Newfoundland the following year, 1583, but his fate is unrecorded. Gilbert disappeared in more theatrical style, going down with his frigate *Squirrel*, lost in a tempest on the way home. The last his companions saw of him, he was reading calmly on the deck surrounded by monstrous seas, shouting across to their ship, "Wee are as neere to heaven by Sea, as by lande."

By this time, the European race for North America was already nearly a century old. What the Basques, Bretons, and Bristolmen tried to keep secret for so long—the presence of cod-rich waters and islands to the west—had become public knowledge. In 1497, the Venetian sailor Giovanni Caboto, known by his Anglicized name, John Cabot, tried to find a northern shortcut to Asia but was stymied.

"Understanding by reason of the Sphere that, if I should saile by the way of Northwest winde, I should by a shorter tract come into India," Cabot set off with two square-rigged Portuguese caravels. "I began therefore to saile toward the Northwest, not thinking to find any other Land then that of Cathay [China], and from thence to turn toward India, but after certaine dayes, I found the land rann towards the North, which was to me a great displeasure." Cabot sailed farther and farther north, hoping to find a way around this inconvenient landmass, until "dispairing to find the passage, I turned backe againe."

Despite the image of ignorant sailors worried that they would sail off the edge of a flat world, scholars and geographers had guessed since the days of the ancient Greeks that the world was a sphere, and by medieval days, it was widely assumed that one might reach the Orient by sailing west. Columbus and Cabot were merely among the first to actually risk it. And sanguine assumptions about the symmetry of global geography convinced many learned men that there must be a water passage around or through any intervening landmass—the so-called indrawing sea based on classical ideals of geographic harmony. It was the birth of the Northwest Passage, the search for which would consume lives and fortunes for the next three centuries or more.

For much of the early sixteenth century, America wasn't a destination; it was an obstacle, a huge and annoying barrier to the riches of Asia. In 1508, Cabot's son Sebastian probed for a way through, and in 1523 the French—Johnny-come-latelies to American exploration—sent another Italian, Giovanni da Verrazano, to look for the passage. He struck land, probably what is now Cape Fear, North Carolina, and turned north, thinking that he could see the Pacific across a narrow isthmus. (In reality, it was Pamlico Sound, inside the Outer Banks.)

On the Carolina coast, one of Verrazano's sailors—swimming through high surf to trade a few bells and small mirrors with some Indians on the shore—was slammed around by the waves and nearly drowned. The Natives hauled him ashore, dried him out by a fire, and then "hugged him with great affection" before sending him back. Sadly, Verrazano didn't reciprocate the courtesy. Farther north, along the Mid-Atlantic coast, he found a teenage girl and an old woman clutching a boy and two infants, hiding in fear. And with good reason. "We took the little boy from the old woman to carry with us to France," Verrazano reported, "and would have taken the girl also, who was very beautiful and very tall, but it was impossible because of the loud shrieks she uttered as we attempted to lead her away."

At the mouth of the Hudson, in Narragansett Bay, Verrazano kept meeting friendly, welcoming Natives. But this bonhomie gave way as he rounded Cape Cod and headed north into Maine or the Maritimes, where the Indians seem to have grown wise to the ways of Europeans. The Natives shot at Verrazano's men whenever they tried to land and would trade only by lowering ropes down cliffs to the explorer's boats. They had no use for baubles, wanting steel knives and fishhooks instead, and when the trading was over, "the men at our departure made the most brutal signs of disdain and contempt possible."

No doubt the contempt grew. A few months after Verrazano's visit, a Portuguese captain named Estevão Gomes, sailing under Spanish colors, explored the East coast, from Cuba and Florida clear up to Newfoundland, looking for the Northwest Passage. To his disgust, the land proved immense and impenetrable, and in a final bid to turn a profit, he kidnapped dozens of Indians on the Maine or Nova Scotia coast and took them back to Spain.

Spanish maps based on Gomes's voyage accurately show a landmass with no indrawing sea, but few were ready to give up on the Northwest Passage. Englishman John Rut sailed from Bristol in 1527 in one of Henry VIII's ships (reassigned from its usual duties hauling royal wine from Bordeaux) and another vessel, subsequently lost at sea, looking for "the land of the Great Khan." In Newfoundland, they found the usual cod-driven hubbub—"eleven saile of Normans and one Britain [Breton] and two Portugall barkes, and all a-fishing"—and probed well to the north, encountering icebergs before turning back.

Jacques Cartier, a Breton who grew up in the fishing town of Saint-Malo, where the smell of Newfoundland cod hung heavy, thought he'd discovered the Northwest Passage in August 1535—though, more accurately, he was simply following the directions of two young Stadacona

Iroquois brothers, Domagaia and Taignoagny, whom he'd kidnapped the previous year. The sons of the sachem Donnacona, they'd made the best of their ten-month ordeal, learning to wear French clothes and speak enough of the language that Cartier had great hopes for them as interpreters and go-betweens.

To Cartier's shock, Domagaia and Taignoagny wanted no part of his scheme. Reunited with their father, they largely held themselves aloof from the Frenchmen—except, having learned the real value of European goods, to serve as a brake on trading, "giving them to understand that what we bartered to them was of no value, and that for what they brought us, they could as easily get hatchets as knives." Cartier and his men, having barely survived a frighteningly cold winter, kidnapped Domagaia, Taignoagny, and Donnacona, along with a handful of others, all of whom died in France before the Frenchman could make his third and final expedition in 1541. Questioned by the Natives at that time, Cartier explained the absence of the missing Indians by claiming they were happily married in France.

The Spanish, meanwhile, were looking to expand their newly established empire in the Caribbean and Mesoamerica into the north. Taking his cue from the stories of an Indian boy captured somewhere along the Southeast coast, taught Spanish, and taken back to Spain, the nobleman Lucas Vázquez de Ayllón envisioned the land of "Chicora" to be an earthly paradise, and in 1526 he set out to colonize it, sight unseen. With six ships crammed with livestock, soldiers, women, children, and African and Indian slaves, he sailed to the coast of South Carolina.

He found Chicora to be much less paradisiacal than he'd imagined. Shifting south to Georgia (possibly Sapelo Island), Ayllón established San Miguel de Gualdape, where the colonists confronted the stark reality of winter along the Southeast coast. Beset by disease, attacks from the Guale (whom they planned to enslave), and starvation, the colony lasted just three months, costing Ayllón and hundreds of his followers their lives.

No Spaniard, however, plunged as deep into the eastern American frontier as Hernando de Soto. If David Ingram was telling the truth about the unfailing courtesy and generosity of the Indians he encountered on his epic journey, it attests to their forgiving nature, since Ingram must have crossed de Soto's trail of almost unmatched destruction and cruelty.

Starting at Tampa Bay in 1539, de Soto and 620 men—armored, mounted, accompanied by surly mastiffs on leashes, and driving herds of pigs before them—moved slowly north and west through what is now Georgia, the Carolinas, Tennessee, and Alabama, then on to the Mississippi, where de Soto died in 1542, his body consigned to the river. The ever-dwindling number of survivors plunged on into Louisiana and Texas, then backtracked to the Mississippi, built boats, and sailed down the Gulf to Spanish Mexico in 1543, barely alive and most of them grateful to be shed of the place.

Although de Soto and his men were often welcomed when they arrived in a new chiefdom, the mood quickly soured when the Indians discovered what they were dealing with. The Spaniards demanded vast stores of Indian food, grabbed women and slaves, and searched frantically for gold and riches like those that had made de Soto rich in Peru. If the locals put up a fight, the Spaniards would burn captives, kidnap leaders, and torch entire communities.

Perhaps it's not surprising, then, that de Soto's men were also the first Europeans to personally experience that iconically American practice—scalping. The Apalachee Indians, with whom the conquistadors fought running skirmishes through their first winter, regularly ambushed the Spaniards. In one case, a dead soldier's companion said, "Hardly had he fallen when they cut off his head, or rather I should say all the scalp in a circle, and carried it away as a testimony of what they had done."

Taking body parts as war trophies has been part of virtually every human culture that ever raised a weapon in anger. But scalping—flaying the top of the head and preserving the skin and hair—is somewhat rarer in history. The ancient Scythians scalped their enemies on the steppes of Eurasia, but the practice is most closely associated with North America, where scalping among Natives was a culturally complex means of demonstrating bravery, gaining prestige, and appeasing the dead. In some cultures, scalps could even be ceremoniously "adopted" as a replacement for deceased relatives.

As with de Soto's chroniclers, the first Europeans to record scalping struggled with descriptions and terms in ways that make clear it was something new to their experience.* During his time with the Iroquois at Stadacona in 1535, Jacques Cartier was shown "the skins of five men's heads, stretched on hoops, like parchment," and was told they belonged to Mi'kmaqs, with whom the Stadacona Iroquois fought. Jacques Le Moyne de Morgues, part of the French force that established Fort Caroline near what is now Jacksonville, Florida, in 1564, saw Timucua warriors using sharp reeds to scalp fallen enemies. And Samuel de Champlain noted scalps and the act of scalping on a number of occasions among the northern Algonquians with whom he was allied.

By the late sixteenth century, the race for North America appeared to have been won by Spain. Having conquered most of South and Central America and the Caribbean, the Spanish tried to export a proven colonial model to the Southeast. Known as *repartimiento,* it was a system of forced labor in which Indian communities had to provide a set number of workers—the *reparto,* or "distribution"—to their overlords. It was not technically slavery, but to the Indians it must have felt like it, as their bones attest.

---

* Starting in the nineteenth century, but especially since the 1960s, a myth has taken root that scalping was a European invention, to which Natives were introduced after contact. Although there is a certain appeal to this view of history—after all, the list of wrongs committed against American Indians by European invaders is long and bitter—it is demonstrably false. Scalping had an ancient and incontrovertible history throughout North America.

Scalping leaves diagnostic cut marks in the skull, and such wounds have been found by archaeologists at sites dating back thousands of years in places such as Kentucky, Tennessee, Ohio, and North Dakota. More recently, but still well before European contact, almost five hundred people were massacred at Crow Creek, South Dakota, during an attack around 1325, most of them mutilated and virtually all scalped after death. (In fact, a few of the victims had been scalped in some earlier attack and survived.)

Scalps were hardly the only body parts taken as trophies. Entire heads were often removed, and early observers noted that scalps were sometimes preferred only when distance precluded taking the whole thing. The lodge of a Huron war captain, in fact, was known as *otinontsiskiaj ondaon,* "the house of cut-off heads."

Scalping, especially among Algonquians and other tribes in the Northeast, needs to be seen within the widespread cultural framework of ritual torture (and sometimes ritual cannibalism), enslavement, and the customary adoption of women, children, and occasionally men as replacements for lost relatives. Through torture, an enemy's spiritual strength could be assumed by his captors—a force also tapped through the possession of a scalp, which was far more than a mere trophy.

While the sheer horror of scalping initially appalled European immigrants, the practice was eventually embraced by the French and English as colonial military and political strategy. Scalp bounties were a perennially popular way of trying to fight fire with fire. While bounties were initially paid to Indian allies, frontiersmen of all races eventually lifted hair. According to one historian, "Scalping [became] as Anglo-American as shillings and succotash."

Archaeologists excavating remains at Mission Santa Catalina de Guale on St. Catherines Island, Georgia, have found that, compared with the bones of their pre-contact ancestors, the Guale who lived there carried more bulk (probably because they ate less seafood and more carbohydrate-rich corn, a dietary change that also brought on anemia), had longer and heavier leg bones, and bore the marks of more powerful arm muscles from field work and long journeys carrying crushing loads. Most telling, nearly two-thirds of the Indians suffered from osteoarthritis, a disease common in those beaten down by grunt labor but exceedingly rare among the Guale before contact. Little wonder that the Guale and neighboring tribes tried several times to throw off the Spanish yoke, beginning in 1576.

The other colonial powers, unable to challenge Spain directly, nevertheless tried to slip in through the cracks in La Florida. The French efforts began in earnest in 1562, when Jean Ribault left a garrison at what he grandly named Charles Forte, on what is now Port Royal Sound in South Carolina.

The woods there were a shipwright's delight—sprawling live oaks for the knees and timbers, and tall, straight pines for the masts. "We set ourselves fishing," Ribault wrote, and "two draughts of the net were sufficient to feed all the company of our two ships for a whole day." Turkeys and quail were everywhere, and the land had a sweetness, a "fragrant odor that only made the place to seem exceeding pleasant." Food and timber aside, the idea was to harry Spanish shipping, prospect for commodities and riches, and provide a haven for French Protestants. The twenty-seven men erected a three-sided stockade and settled in to wait for Ribault to return with more supplies, something he'd promised to do within six months.

Ribault, though, found France consumed by religious war. He slipped away to England, where he convinced Elizabeth I to back his colony. She in turn enlisted a genuine rogue, Thomas Stucley, who may have been her own half brother, rumored as he was to be an illegitimate son of Henry VIII.

A soldier and privateer who was a master at playing all possible ends against the middle, Stucley "during his lifetime served England, France, Spain and the Pope, betraying each in turn except the latter." A contemporary simply described him as "the rakehell." A confessed

They Reach Port Royal (1591)
This engraving by Theodor de Bry, based on a watercolor by Jacques Le Moyne de Morgues, commemorates the 1562 arrival of Le Moyne and other French Huguenots under Jean Ribault at Port Royal in modern South Carolina. With their seagoing ships anchored in the sound, Ribault's men explore the surrounding rivers in a smaller vessel. On land are fanciful depictions of Indian villages, hunting parties, and local plants and animals. (Library of Congress LC-USZ62-380)

double agent, he'd already turned coat several times and been imprisoned in the Tower of London, but Elizabeth saw in him and Ribault an opportunity to needle Spain by taking control of Charles Forte.

Stucley—feelers secretly extended to Spain—delayed his departure as long as possible, then finally set off in 1563. Instead of sailing for La Florida, however, he turned pirate again (possibly with the queen's consent; she made a great profit by English buccaneers preying on French shipping) before being captured and clapped in irons.

The French garrison at Charles Forte, meanwhile, was dying a miserable death. Out of food (their supply house lost in a fire), torn by mutiny, and seemingly forgotten, they gambled on a desperate throw of the dice. They cobbled together a ship from local timber, sewing sails from their remaining clothing, and tried to sail home. Becalmed and reduced to cannibalism, they were eventually rescued by an English vessel.

When the French tried again at Fort Caroline, the Spanish—fed up with the intrusions into what they considered their colonial realm—simply slaughtered the garrison (including Ribault, who had just arrived with eight hundred men) and took the site. But even the Spanish weren't having much success with the northern reaches of La Florida. In 1570, they sent a Jesuit mission to Ajacán, at the mouth of Bahía de Santa María—what the English would call Chesapeake Bay. The Spanish considered this the northern edge of their domain and a likely entrance to the Northwest Passage.

The Jesuits had with them an Algonquian they'd kidnapped as a teenager around 1560 and baptized as Don Luis de Velasco. They assumed he was safely acculturated, since he'd lived for years in Havana and Spain and had even been adopted by the viceroy. Little is known about Don Luis. He was said to be the son of an Algonquian chief and was almost certainly from one of the tribes that would later make up the Powhatan confederacy. After a decade with the Spanish, he told the priests he was eager to return home and help them win souls there. Instead, he shifted almost immediately back to his people and soon led an attack that wiped out the mission in January 1571.

It's been suggested that Don Luis, recognizing the cultural danger that European colonialism posed, acted to safeguard his people from it. If so, the danger did not remain at bay for long. The Spanish returned, hanging fourteen captives in revenge (having baptized them first, just to make it right with God). It also seems likely that the Powhatan would have learned of the experiences of other Algonquians to the south, where other *tassantassas*—strangers—had settled briefly on Roanoke Island in the Outer Banks.

Thomas Hariot, a mathematician who went on that 1585 expedition to Roanoke as surveyor—and who had spent the preceding two years in England learning the local dialect from the visiting werowance Manteo—admitted that "some of our companie . . . shewed themselves too fierce, in slaying some of the people, in some towns, upon causes that on our own part, might easily enough have been borne withall." In one such incident, the English burned an entire village and its winter corn supply when a silver cup turned up missing.

Unlike Don Luis, who knew better, the Roanoke tribes considered the newcomers deities. But whether they were gods or men, the Natives saw trouble in the European arrival. "Some would likewise seeme to prophesie that there were more of our generation yet to come, to kill theirs and take their places," Hariot wrote.

The failure of the Roanoke colony was only a speed bump in English colonization. In 1607, it was followed by competing settlements at Jamestown, just south of where the Jesuits had been killed a generation earlier, and the Popham colony at Sagadahoc, Maine. Popham was abandoned a year later, and the Jamestown colonists barely hung on by their fingernails, more than half perishing in their first miserable year of starvation and disease.

The marshy land and brackish tidal water of Jamestown Island were partially to blame. The 104 colonists had ignored pointed instructions from their backers, the Council for Virginia, that they not "plant [the colony] in a low and moist place, because it will prove unhealthful." But regardless of where they settled, few of the colonists had any skills that lent themselves to a wilderness venture. They were "poore Gentlemen, Tradesmen, Serving-men, libertines, and such like, ten times more fit to spoyle a Common-wealth, than . . . begin one," groused Captain John Smith, the young but experienced soldier who became their leader. The colonists were more interested in hunting for gold than building shelter or digging a well. "Ten good workemen would have done more substantiall worke in a day, than ten of them in a weeke."

The council had also warned the colonists to "have great care not to offend the naturals," but their immediate concern was not the local Indians but the Spanish, lest the settlement suffer the same fate as the French at Fort Caroline. This is one reason the marshy but easily defensible Jamestown Island was ultimately chosen. The Spanish remained a theoretical threat, but relations between the English and Algonquians were strained from the start and took no more than two weeks to explode into open warfare.

Just upriver from the settlement lived the Paspahegh, who were tributary to Powhatan, the *mamanatowick,* or primary chief, who controlled the eight-thousand-square-mile confederacy known as Tsenacommacah. (Had the small English outpost realized the name meant "densely inhabited land," it might have given them pause. Powhatan ruled an estimated fourteen thousand Indians of about thirty tribes, most of whom he'd conquered, but the colonists were blissfully ignorant of these facts.) The Paspahegh's werowance at first made a show of force, with a hundred armed men, but later sent a gift of a deer and agreed to give the English the island—or at least that's how the English chose to interpret his signs and gestures.

Unfortunately, that meeting ended in a minor brawl over a hatchet, and shortly thereafter the Paspahegh and neighboring tribes attacked the largely undefended settlement, killing a man and a boy. With most of the guns still packed away, and without any defensive walls, the colonists managed to hang on for an hour before the ship's cannons opened fire from the riverfront, bringing down a tree limb among the warriors and scaring them into flight.

It was the beginning of a long seesaw between détente and conflict, as Powhatan tried to find ways to cooperate with, co-opt, or expel the *tassantassas*—a process during which the Algonquians would prove as resistant to the increasingly bullying commercial focus of the English as they'd been to the clumsy religiosity of the Jesuits. Although Powhatan is remembered for steering his powerful confederacy through these early, tumultuous years of contact, much of that resistance would flow from a single man, one of Powhatan's younger brothers, the werowance Opechancanough.

Given his later, fearsome reputation, some historians have suggested Opechancanough was the kidnapped teenager Don Luis, who had led the uprising against the Jesuits in 1571, or that he took part in the attack on the mission. There is nothing to support this speculation, and given his dealings with the English, he hardly needed a Spanish backstory to explain his hostility toward the colonists. The werowance of the twelve-hundred-strong Pamunkey, who lived up the James River from the colony, Opechancanough was a tall, vigorous man with a long scalp lock, worn on one side of his shaved head in the Algonquian fashion. A natural leader, he found himself more and more at odds with the English as time passed.

Opechancanough's village was typical of those across Tsenacommacah—a loose scattering of oval houses known as *yihakin,* framed with bent saplings; covered with bark or, more often, mats made of woven reeds; and sheltering an extended family around a central fire. Werowances and paramount chiefs like Powhatan lived in longhouses; at Powhatan's capital, Werowocomoco, on the modern York River, the leader's house and ceremonial areas were set off from the rest of the village by two D-shaped ditches, each hundreds of yards long, but there was no defensive palisade of logs.

The rivers and tidal creeks provided the easiest and surest transportation by canoe. These were not the lithe bark canoes found to the north, but heavy, much more cumbersome dugouts made from tall, straight pines. The villages were surrounded by rich agricultural land where corn, squash, and other crops were grown. Children armed with piles of stones were set on guard against crows and other raiders, while the women hoed weeds and harvested.

The Jamestown colonists, however, spared little time for farming, what with hunting for gold and jewels, and had little success when they tried. Although their free-ranging pigs and chickens were increasing (the former becoming a menace to Indian crops that stone-throwing children couldn't deter), they found their stock of corn "halfe rotten, and the rest . . . consumed with so many thousands of Rats"—a plague that, like the feral pigs, they'd brought with them. Forever short on food, they depended on surplus corn purchased from the Indians. Smith, the de facto governor of the colony and increasingly fluent in Algonquian, traveled among the villages of Tsenacommacah, bartering for a few dozen bushels at a time. Acting as though he was doing the Natives a favor by trading trinkets for corn, he was unwilling to let "the Salvages" see how weak the colony, torn by internal strife and politics, really was.

In midwinter 1609, Smith and fifteen others sailed to Opechancanough's village to trade for corn. Striking a bargain, the werowance told them to return the next day, when they found large baskets of corn waiting for them—along with seven hundred Powhatan warriors surrounding the chief's house and fields.

There was no love lost between Smith and Opechancanough. During the colony's first winter, the werowance and a party of Pamunkeys had captured Smith far up the Chickahominy River, killing his companions and putting an arrow through the soldier's leg. It was from this captivity, and his supposed imminent execution, that

Smith was "rescued" by Powhatan's daughter Matoaka, barely twelve, whose nickname was Pocahontas. Mostly, though, Smith depended on bluff—and the very real threat of English muskets—to cow the Algonquians. He'd done this not long before, demanding tribute of four hundred bushels of corn from a small tribe called the Nansemond after an attempted ambush of the English had failed, and burning houses until they complied.

Smith now tried bluff again. First, he challenged Opechancanough to a hand-to-hand fight. Then he "snatched the King by his long locke in the middest of his men, with his Pistolle readie bent against his brest." Smith claimed that the chief was "neare dead with feare"—an unlikely condition for a war leader in a culture that prized personal bravery—but Smith was forever blowing his own horn. "You promised to fraught [load] my Ship ere I departed," he told the Pamunkey, "and so you shall, or I meane to load her with your dead carcasses." He got the grain. "Men may thinke it strange there should be such a stirre for a little corne," Smith wrote, "but had it beene gold with more ease wee might have got it; and had it wanted, the whole Colony had starved."

For the first two years of Jamestown's existence, neither the Indians nor the colonists seemed to know exactly how to act toward each other, uncertain of intentions or capabilities. Powhatan tested the English defenses with open attacks like those in the initial days and weeks after the ships arrived. He tried to capture or kill the *tassantassas* leaders with stealth, as Opechancanough had done, and to suborn disaffected members of the colony into turning against their fellows—not a difficult task, given Werowocomoco's prosperity and the periodic famine at Jamestown. At times, he offered what seemed like sincere assistance, and he and his people certainly welcomed the chance to acquire English goods, especially weapons.

Neither side understood fully what the other was doing. Smith's ritual "execution" involving Pocahontas may have been an Algonquian adoption ceremony in which Powhatan granted the Englishman status as a werowance, bound to serve the *mamanatowick*. If so, Smith no more comprehended that honor and responsibility than did Powhatan when, later the same year, the English crowned the chief with much pomp and declared him a vassal of their king.

But the old *mamanatowick* finally saw the colony for what it was—a

C. Smith Taketh the King of Pamaunkee Prisoner

There was no love lost between Captain John Smith and Opechancanough, the younger, physically imposing brother of Powhatan, paramount leader of the eight-thousand-square-mile chiefdom in which Jamestown was founded. In one dramatic encounter in 1609, Smith and fifteen men, desperate for corn, confronted Opechancanough and seven hundred warriors. Grabbing the Pamunkey werowance by his long scalp lock and jamming a pistol into his chest, Smith held the man hostage until the food was loaded onto Smith's ship. That humiliating act only fueled Opechancanough's hatred of the English, which, after he assumed control of the chiefdom, led to two bloody attacks on the Virginia colonists. (Library of Congress LC-USZ62-51971) growing menace in the heart of his empire. "Some doubt I have of your comming hither," Powhatan told Smith in 1609, "for many doe informe me, your comming hither is not for trade, but to invade my people, and possesse my Country." This was confirmed when a fresh influx of colonists arrived, including a bevy of aristocrats who wrested the leadership from the lowborn Smith, forcing him to return to England. Then they set about confirming Powhatan's worst fears.

Smith had been no shrinking violet when it came to dealing with the Indians, but the newcomers, having spread to other encampments up and down the James River, simply terrorized the villages close to where they settled, stealing food, beating people, burning houses, and even desecrating the dead. The English were no longer just strangers, *tassantassas*. They had proved to be *marrapoughs*, "enemies."

The Algonquian confederacy struck back, killing hundreds of colonists until the survivors cowered behind the triangular palisades of James Fort, too scared to venture outside even to bury their dead. It was England's first experience with war in the New World, and it was terrifying. John Ratcliffe, one of the original captains, had been captured while trying to barter for corn and skinned alive with mussel shells, then burned. The men with him were left where they fell, full of arrows, their mouths stuffed with corn bread: *You want food*, Powhatan seemed to be saying. *Here's food.* Another party, led by one of the newly arrived young fops, acquired a significant quantity of corn by murdering several members of the only tribe not actively hostile toward the colonists, then sailed away with it, going directly home to England.

Hemmed in by an Indian siege, more than half the two hundred colonists died during the winter of 1609–10. They cannibalized the houses for firewood, and some literally cannibalized the dead for food, including the corpse of an Algonquian dug up and eaten by "the poorer sort." One Englishman "did kill his wife, powdered [salted] her, and had eaten part of her before it was knowne."

"Now whether shee was better roasted, boyled or carbonado'd, I know not, but of such a dish as powdered wife I never heard of," Smith wrote with ghastly humor that immediately turned bleak. "This was that time, which still to this day we called the starving time; it were too vile to say, and scarce to be beleeved, what we endured."

When spring came, the emaciated survivors buried their cannons beneath the filthy ground of James Fort, packed themselves onboard ship, and prepared to sail for home. Powhatan had won. Just as the Spanish had been driven out of Tsenacommacah a generation earlier, so now had the English. But if the *mamanatowick* breathed a sigh of relief, it was premature. At almost the very moment that the colonists were setting sail from the James River, a relief fleet under Sir Thomas West, Lord De La Warr, hove into sight with supplies, soldiers, and more colonists. Jamestown was reestablished, and the bloody fighting would go on for four more years, until an uneasy peace was sealed by the marriage of John Rolfe, one of the newcomers in De La Warr's fleet, and Pocahontas, who had converted to Christianity after being kidnapped a year earlier.

Rolfe and Pocahontas were married in April 1614, just as John Smith was returning to the New World, having been offered command of a whaling and fishing expedition based far to the north on Monhegan Island, in what was still called Norumbega.* That summer, the fishing well in hand, he took eight men—including George Waymouth's former captive, the Wapánahki sagamore Ktəhɑnəto—and roamed the coast from Maine south to Cape Cod in a small boat, trading for pelts. Smith fell in love with the land he named "New-England." "Of all the foure parts of the world I have yet seene . . . I would rather live here then any where," he wrote.

But everywhere he went (except the "high craggy clifty Rockes and stony Iles" east of Penobscot Bay), Smith saw a densely occupied country. Penobscot Bay was "well inhabited with many people," and the coast of Massachusetts was likewise "well inhabited with a goodly, strong and well proportioned people." The "country of the Massachusits," Smith said, was the "Paradice of all those parts . . . the sea Coast as you passe shewes you all along large Corne fields, and great troupes of well proportioned people." The area that is now Cape Ann and Gloucester Harbor struck him as an especially magnificent location, "not much inferior . . . (for) any thing I could perceive *but the multitude of people*."

---

* The master of a second ship accompanying Smith's fishing expedition, Thomas Hunt, enraged Smith by slipping off and kidnapping twenty-four Nauset and Patuxet Indians, whom he sold into Spanish slavery. One was Tisquantum, or Squanto, who spent the next five years in captivity, returning just in time to aid the Pilgrims.

Again and again, the earliest European explorers found the same discouraging situation: a lush and verdant land that cried out for civilized use but that was already solidly inhabited by healthy, well-armed Natives, who, though generally friendly, could brush away any attempt to settle their territory. It must have seemed a cruel joke for Providence to have presented this New World to Christian explorers, having already populated it with another race.

And then, it seemed, Providence changed the rules.

In the early seventeenth century, Western medicine still blamed the outbreak of disease on various "miasmas"—fetid air, the effluvia of rotting vegetation or bodies, the "falling damps" of swamps, and noxious elements released by earthquakes or floods. People did realize that some diseases could, inexplicably, pass from person to person, and even the least educated could see that some diseases conferred immunity on their survivors, while others did not. But how diseases emerged and how they moved were facts still unknown. Antony van Leeuwenhoek's discovery of bacteria and other microbes was more than eighty years away, and the germ theory of disease transmission would not finally supplant miasmas until the mid-nineteenth century. Yet these were far from academic questions for the Native inhabitants of North America.

Bitter feuds have been fought in the pages of scholarly journals on the size of pre-contact Native populations, on the origins and nature of the epidemics that followed European arrival, and on the timing and severity of the nearly hemispheric depopulation that they caused. Arguments have been made for a North America (north of the Rio Grande) that held as few as 900,000 or as many as 18 million people in 1491. Those arguing for the higher figure put the human population of the entire Western Hemisphere at as many as 112 million, far more than were then living in Europe.

For all the controversy, two points are clear. First, by the time all but the very earliest European colonists arrived, Indian populations along the Atlantic seaboard were a fraction of what they had been just a few decades before. Second, without this monstrous loss of life (which probably exceeded 90 percent of many Native groups), Europeans would have had little chance of establishing more than a beachhead in the New World.

Even at the time, the colonists recognized how lucky they were, although it was generally acknowledged as the hand of God moving to clear the pagans from a new Eden, not a chance throw of the dice. Even in the nineteenth and twentieth centuries, after concepts of immunity and genetics were better understood, the epidemics were still taken as a sign of some inherent weakness in the Native constitution.

No one knows how soon after contact epidemic disease hit the New World. It's unlikely that the Vikings allowed Natives to linger close enough to spread germs, but the Spanish inadvertently unleashed waves of infection almost from the moment they settled the Caribbean and South and Central America. Coupled with the collapse of indigenous societies, starvation, and slavery—all of which magnified the effects of disease—epidemics took a horrific toll.

On the island of Hispaniola, the native Taino may have numbered between 300,000 and several million before Columbus landed there in 1492. A year later, days after fifteen hundred Spaniards and herds of pigs, horses, and other domestic animals arrived, Columbus and almost all of his men fell ill with what may have been an unusually potent strain of swine influenza. Up to half the Spaniards died, but once the disease spread to the Taino, the carnage was orders of magnitude greater. After that, one disease after another, including a major smallpox epidemic in 1518 that left only a few thousand survivors, whittled down the population. By the end of the sixteenth century, the Taino were effectively extinct.

The same catastrophe played out again and again on a staggering scale. By 1568, just three-quarters of a century after Columbus made landfall, 90 percent of the Amerindian population was dead, from the Mexican cities of the Aztecs (infected in 1520, according to one story, by an African slave with smallpox in the train of Hernán Cortés) to the Inca Empire in the Andes, where a smallpox epidemic in the 1520s may have killed fully half the people in a single blow. The decimation of the Indian population has been called the greatest loss of human life in history.

Such virulent epidemics, accompanied by extraordinary mortality, are referred to as "virgin soil" outbreaks, since the Indians had never been exposed to these pathogens and had no acquired immunity to them. Although some early scientists, such as the French Comte de

Buffon, saw this vulnerability as a sign of racial decrepitude (a prejudice that was more subtly advanced as late as the twentieth century), the Indians were not inherently more susceptible to disease; their immune systems were just as robust as any European's. But in order to beat back an infection, the immune system must be primed by earlier exposure, often in childhood. Europeans and their livestock brought novel microbes against which New World inhabitants didn't stand a chance, including many viral diseases that are far more dangerous to adults than to children.

The diseases of Europe were, by and large, so-called crowd diseases—maladies such as smallpox, typhus, and measles that require large populations living in close proximity. Some, like cholera, spread through feces and contaminated water. Others, such as typhus and plague, are passed through external parasites such as lice and fleas. Many crowd diseases emerged from domestic livestock. Measles appears to have derived from the virus that causes the cattle disease rinderpest, while one strain of pertussis (whooping cough) apparently arose in pigs. The variola virus that causes smallpox seems to have evolved from either the cowpox virus (which despite its name is primarily a rodent disease) or from a similar pathogen that causes camelpox. Malaria evolved in birds, while influenza moves among human, pig, and avian hosts, shuffling its genes each time to produce exotic and sometimes deadly strains.

While the New World had some large urban centers in Mesoamerica and the Andes, most of its population was more diffuse than Europe's, limiting the development of crowd diseases. Measles, for example, requires a minimum population of 200,000 to remain viable. Native Americans also had far fewer domestic animals—and thus fewer avenues for viruses and bacteria to travel to human hosts.

So when boatloads of Europeans began anchoring along the American coast, they brought a microbial stew against which Native immune systems had no immediate defense. And with only a few exceptions, the traffic was one-way, because there were few endemic pathogens of similar virulence to attack the newcomers. The Americas had plenty of diseases of their own—from parasitic infestations such as hookworms, tapeworms, and roundworms to pneumonia, and possibly chickenpox—but little to match the sheer killing power of the European pathogens.

One exceptionally nasty bug that did travel east across the Atlantic was a variety of the bacterium that causes the chronic skin disease known as yaws. It found its own virgin soil by hitchhiking with Columbus and his crew back to Europe, where it morphed into syphilis, setting off an epidemic in 1495. In its first incarnation, syphilis was utterly terrifying, marked by intense pain and acorn-size pustules oozing smelly green matter. It was called the "great pox" to differentiate it from the deadly but less awful smallpox. Like many diseases, syphilis has become steadily less potent with time, since venereal diseases spread more readily when their carriers do not look like something from a horror movie; carriers with less virulent strains proved with time to be the more effective vectors.

In this global game of microbial exchange, it now appears that tuberculosis was working both sides of the street. Long assumed to have originated in the Old World, it has since been found in the lungs of a thousand-year-old Peruvian mummy, and recent DNA evidence suggests that TB first infected human ancestors millions of years ago in Africa. This is also true of the prevalent species of the human head louse, which has been found on Peruvian mummies as well.

Any isolated human population would be at similar risk as the Native Americans were. In 1707, smallpox broke out in Iceland, where the disease had been absent for generations; more than a third of the population died. In 1846, measles appeared for the first time in sixty-five years on the isolated Faeroe Islands in the North Sea, sparking an epidemic that infected more than three-quarters of the nearly eight thousand islanders and killed more than one hundred, most of them adults. And rat-borne human plague, which moved out of Asia in the sixth century and flared up repeatedly thereafter—most notoriously in the fourteenth century as the Black Death—killed between 30 and 60 percent of the population of Europe in just two years.

Had Indian populations, however reduced, been facing a single disease, they probably would have recovered, since survivors would have acquired immunity. But the germs came in swarms. Influenza would lay a village low, then pneumonia would pick off the weakened survivors. Smallpox and measles would strike simultaneously, the latter leaving its victims highly susceptible to tuberculosis. And while lucky survivors would acquire immunity to smallpox or measles, the flu would constantly shift strains, so that each year brought a new variety against which they were rarely protected.

In the urbanized cultures of central Mexico, introduced diseases could quickly spread to pandemic levels, but there is no evidence that they spread much beyond what is now the northern border of Mexico, perhaps because the populations were widely scattered enough to allow each outbreak to wither and die. Instead, diseases had to wait for the northward advance of English, French, and Spanish colonial efforts in the mid-sixteenth century. When they came, the results in Florida and along the Southeast coast were typically appalling.

In 1585–86, when the English tried to settle Roanoke Island, their arrival was followed by what Thomas Hariot called a "rare and strange accident": "Within a few dayes after our departure from every . . . Towne, the people began to die very fast, and many in short space, in some Townes about twentie, in some fourtie, and in one sixe score, which in truthe was very manie in respect of their numbers," Hariot wrote. "The disease also was so strange, that they neither knewe what it was, nor how to cure it, the like by report of the oldest men in the Countrey never happened before, time out of minde."

"This happened in no place . . . but where we had bin," Hariot concluded. Although he was a mathematician and astronomer, whom Sir Walter Raleigh had sent to America as the expedition's science adviser, Hariot failed to connect the dots—probably because his own company experienced no comparable illness. In "all that space of [the Indians'] sickenesse, there was no man of ours knowne to die, or that was specially sicke," Hariot said. He chalked this up to celestial justice, since in every village where the contagion struck, the Algonquians had—to Hariot's mind—used some "subtile devise" against the colonists.

Despite the enormity of loss, many European writers acknowledged the ghastly death rates among Natives only in passing, and often only in terms of how the epidemics affected their own priorities, such as filling labor quotas or making religious converts. Exactly what diseases lay at the root of these outbreaks is rarely clear to the modern reader, and it's become a minor cottage industry among historical epidemiologists to try to diagnose sixteenth- and seventeenth-century epidemics from the meager clues.

But whichever specific microbes wreaked such havoc, the effects were ghastly. Where de Soto in 1541 found a packed and thriving urban environment along the central Mississippi Valley—the heart of the mound-building Southern Ceremonial Complex—French explorers 150 years later found a landscape that was, for all practical purposes, void of people. Similar collapses occurred among the Coosa chiefdom in Georgia. According to some archaeologists, after de Soto passed through their territory, the Coosa's numbers rapidly fell by more than 95 percent. Throughout the Southeast, de Soto's arrival effectively marked the end of the mound-building society that had flourished for more than four hundred years.

In the North, however, the relative scarcity of epidemics until the early seventeenth century has puzzled and intrigued anthropologists and medical experts alike. Even leaving aside the Norse, the Indians of the Northeast had protracted contact with Basques and other fishermen, starting at least as early as the fifteenth century. These encounters might have sparked epidemics that no European was present to chronicle, but the evidence suggests that disease outbreaks in the sixteenth century must, at worst, have been occasional and locally contained, especially when compared with the overwhelming waves of infection and death sweeping across the Spanish colonies.

Ironically, the first recorded instance of what seemed to be an epidemic illness in the Northeast struck Europeans, not Indians. Jacques Cartier's men, settled in for the winter of 1535–36 along the St. Lawrence River near present-day Quebec City, heard that an illness had befallen the Indians at nearby Stadacona and forbade them to come near the French fort or ships. Soon, however, "the strangest sort [of sickness] that ever was eyther heard of or seen" began to spread among the French, who found their strength gone, their legs swollen, their skin speckled with bloody spots, their gums rotten and bleeding, and their teeth falling out. Cartier was so alarmed that he ordered the body of a newly dead man "ripped"—crudely dissected—"to see if by any means possible we might know what it was."

The impromptu autopsy did nothing to solve the mystery. Twenty-five of Cartier's company died, and all but three or four were so weakened they could barely move through the icebound ships to care for themselves. But then, said Cartier, "God of his infinite goodenesse and mercie . . . revealed a singular and excellent remedie." Actually, however, it was the Indians, who knew exactly what was ailing the French, who came to the rescue.

The disorder was winter scurvy, the inevitable consequence of a meat-heavy diet with little or no vitamin C. While no one knew the physical cause of the malady, the Indians understood from long experience how to treat it. Despite ample reasons to let the French suffer, Domagaia—one of the two brothers Cartier had kidnapped and taken to France—dispatched women to collect the bark and needles of a local conifer, which was used to make a tea that cured the surviving Frenchmen in short order. "That Cartier was blind to this generosity is perhaps seen in enthusiastic thanks to God, rather than to [Domagaia], for the miraculous cure," one historian has noted.

Epidemics, though delayed, eventually reached the Mid-Atlantic region. The short-lived Jesuit mission on Chesapeake Bay observed a disease of unknown origin that the Algonquians there said had been raging among them for six years. Along the New England coast, the "multitude of people" John Smith and others had described—the countless villages; the fields of corn, beans, and squash; the flotillas of canoes—fell victim to a great dying between 1616 and 1622. That pandemic cleared the Northeast just as the first boatloads of English and French settlers were arriving. For religious people, it was impossible not to draw the obvious conclusion.

A few years earlier, as a company of French sailors were trading for beaver pelts on what is now Peddocks Island in Boston Harbor, Indians suddenly attacked, burning their ship and killing all but five of the men. One of the enslaved survivors learned Massachusett well enough to berate his captors in their own tongue for "their bloudy deede," saying that God would revenge himself upon them. "The Salvages (it seemes boasting of their strength) replyed and sayd, that they were so many that God could not kill them," wrote Thomas Morton in 1637, recounting what was by then a common tale among the settlers in and around Plymouth Colony.

Contrary wise, in short time after the hand of God fell heavily upon them, with such a mortall stroak that they died on heapes as they lay in their houses; and the living, that were able to shift for themselves, would runne away and let them dy, and let there Carkases ly above the ground without burial. For in a place where many inhabited, there hath been but one left a live to tell what became of the rest . . . And the bones and skulls upon the severall places of their habitations made such a spectacle after my coming into those partes, that, as I travailed in that Forest near the Massacussets, it seemed to mee a new found Golgatha.

"By this meanes," Morton concluded, "the place is made so much more fitt for the English Nation to inhabit in, and erect in it Temples to the glory of God."

The disease that caused the New England pandemic has never been satisfactorily identified. It wasn't smallpox, which would hit a few years later, and references by English and French observers to "plague" do not necessarily mean it was bubonic plague; the word was often used as a general term for contagion. Eyewitness accounts of jaundice among its victims suggest it may have been liver failure resulting from hepatitis, which can be quite virulent in immunologically naive communities.

Historians have suggested that what finally unleashed epidemic disease of all sorts on the Northeast was the arrival of families, especially children. The sea voyages of the fifteenth and sixteenth centuries were long enough (six weeks or more) for most viruses to burn themselves out among a small crew. The smallpox virus could more easily survive the passage, but the crewmen would likely not have served as carriers; having been exposed to it, as to measles and similar diseases, in childhood, they would have been immune. The Spanish, bringing hundreds of soldiers at a time (de Soto's band, for example, numbered more than six hundred men) and ships whose holds were crammed with African slaves, provided the diseases easy passage. Such access was lacking in the Northeast until the early seventeenth century, when French and English families began arriving.

For the French, the sweeping population clearances were a challenge to their twin goals of a stable fur trade and religious conversion, although there was a run on deathbed baptisms, willing or no. The outbreaks also robbed the French of their trading partners. Messamouet, for example, who had visited France and sailed his shallop with Samuel de Champlain a few years earlier, disappears from the historical record after an epidemic struck the Mi'kmaq in 1610, killing the majority of the people in many villages.

But for the agrarian English colonists, the pandemic left a fertile land primed for the hoe. In 1619, Thomas Dermer sailed down the Maine and Massachusetts coast and found "some antient Plantations, not long since populous now utterly void." Four years later, Christopher Levett enthused over the prospects for an English settlement in southern Maine that he would name York, saying there was "good ground, and much of it already cleared, fit for planting of corne and other fruits, having heretofore ben planted by the Salvages who are all dead."

Farther north, the villages of the Wapánahki had been devastated as well. In 1611, Pierre Biard, a Jesuit priest at the colony of Port Royal, found much disease along Penobscot Bay, in a village whose major sagamore was "Betsabés, a man of great discretion and prudence." This was the same grand sachem whom George Waymouth and James Rosier called the Bashabes six years earlier, and with whom Champlain had traded before that. The Indians, Biard later wrote, "are astonished and often complain that, since the French mingle and carry on trade with them, they are dying fast, and the population is thinning out. For they assert that, before this association and intercourse, all their countries were very populous . . . Thereupon they often puzzle their brains, and sometime think that the French poison them."

In a peculiar way, the fortunes of war and accidents of geography could act as lifesavers. Thanks to their relative isolation, villages on Noepe (Martha's Vineyard) and Canopache (Nantucket) were initially spared, as was most of Cape Cod, while mainland coastal villages were decimated. As Dermer skirted the shore in May 1619, he found a few survivors—"a remnant remaines, but not free of sicknesse," he wrote. "Their disease [is] the Plague, for wee might perceive the sores of some that had escaped, who described the spots of such as usually die."

Dermer's guide was Tisquantum (Squanto), returning home after being kidnapped by the English five years earlier. Having survived abduction, enslavement by the Spanish, and subsequent release, he was anxious to see his village of Patuxet, home to as many as two thousand Wampanoag. Instead, Dermer and Tisquantum found the place empty, full of bones and weeds. The coastal Wampanoag had been all but obliterated, and Dermer finally reached the survivors two days' journey inland.

One of the sachems with whom Dermer eventually met was probably Massasoit, who faced an enormous problem. Not only had the epidemic gutted the Wampanoag, but it had spared their enemies to the west, the Narragansett, who, it seemed, had deflected the spirits that brought on illness through a burning ritual in which valuable possessions and even homes were sacrificed. More likely, the state of war between the two groups ensured that none of the infected Wampanoags came in contact with the Narragansett. It was into this power vacuum that the Pilgrims would step two years later, to be welcomed by Massasoit, who had empty land to share, had powerful enemies to hold at bay, and was badly in need of allies.

Eventually, the pathogens reached everywhere. English immigrants, their bloodstreams awash with plasmodium parasites, were bitten by mosquitoes as they worked in the fields of their new homes, and malaria began to spread through coastal Indian communities from Massachusetts to Virginia. The Swedes, settling along Delaware Bay in the 1630s, were confronted by Lenape sachems who accused them of bringing a manitou, or evil spirit. Don't be ridiculous, replied a young man named Peter Lindeström. "Our ship had [not] brought along any evil . . . as many of our people were dying and the sickness had come among them," he said. "[We] told them that sickness had formerly often been among them, through which whole tribes had died out."

The Lenape told Dutch colonists that before smallpox came, "they were ten times as numerous as they are now, and that their population had been melted down by this disease, whereof nine-tenths of them had died." Along the lower Delaware River, a Lenape in 1670 lamented, "In my grandfather's time the small-pox came; in my father's time the small-pox came; and now in my time the small-pox is come."

Thanks to extensive Indian trade networks, the diseases ran well beyond the handful of European outposts. In the summer of 1634, twenty-five French children arrived at Quebec City, and not long after, an outbreak of what was described as "a sort of measles, and an op-pression of the stomach" erupted among Montagnais (Innu) and other traders coming to the French settlements. They carried the contagion farther back into the Great Lakes, to the twenty or so large villages of the Wendat-Tionontate, the confederacy better known as the Huron-Petun, along the eastern shore of Lake Huron. "The people of the countries through which they pass are all sick, and are dying in great numbers," wrote one Jesuit. As many as 2,500 people, a tenth of the Wendat-Tionontate, may have perished in this single outbreak; within five years, between one-half and two-thirds of the Huron-Petun were dead.

No one really knows what cultures were lost in the virgin soil out-breaks. By the time Europeans finally breached the wall of the Alle-gheny Mountains to reach the Ohio Valley in the early eighteenth century, they found an almost completely empty land, into which were starting to move refugees from crumbling tribes farther to the east and south. All that is known about the original inhabitants—known today simply as "the Monongahela people," for lack of a better name—comes from the excavations of their stockaded villages, built on defensible high ground. But not all enemies are visible, nor can all be stopped by log walls.

The new illnesses did not simply cause physical suffering. They brought down entire social orders, killing elders in whom authority was vested and knowledge preserved, breaking up families and com-munities, disrupting hunting and planting (and thus generating or exacerbating famine), sparking warfare and tides of refugees, and rob-bing religious leaders of public confidence.

But Indians also fought back, with the weapons they had. The Cherokee blamed smallpox on evil spirits known as *kosvkvskini* and battled them with a seven-day ritual they developed known as the *ito-hvnv*, or "smallpox dance," in which the entire village remained in the council house, taking special medicine and listening to prayers. While this may not have slowed the disease—in fact, corralling everyone in a council house may have aided its spread—it gave the Cherokee a cul-tural bulwark against what must have been a terrifying loss of control over their lives.

Disease was not the only agent of change. The Europeans also brought trade goods, such as metal hatchets, knives, and pots; wool blankets; cotton cloth; and glass beads. Traditional trade networks were in flux as furs flowed from the interior to the coast, where many tribes (the Mi'kmaq in particular) set themselves up as middlemen. With the traders came missionaries, bringing a new god and a new religion, exploiting the upheaval for which they themselves were partially responsible.

The French first brought the Récollets, a Franciscan order, to minister to the Indians in New France beginning in 1615. The Récollets believed that conversion was the end of a long process of acculturation, in which the Indians would learn to speak and live as Frenchmen. In this they were similar to English missionaries in New England, who insisted that converts abandon their culture, their clothes, their long hair, their names, and often even their homes to live in "Praying towns." Neither the Récollets nor the English (at least initially) enjoyed any great measure of success.

In 1625, though, the Jesuits—an order founded by a former military man and with none of the passivity of the Récollets—arrived in New France, embarking on what would be a stunningly successful campaign of conversion. This was especially striking given their earlier failure in Virginia, but the Jesuits had learned from their mistakes on the Chesapeake. The "Black Robes" abandoned the idea of civilizing the Indians in the European model before trying to win their souls. Instead, they lived among the Mi'kmaq, Huron, Montagnais, Abenaki (as the Wapánahki came to be known), and other tribes. Instead of depending on Native go-betweens such as Don Luis de Velasco, the priests themselves learned the Indian languages and ways, suffering the same privations and approaching the question of conversion with a shrewd mix of fiery religious fervor, utter fearlessness, and deep pragmatism.

The Jesuits tried to discredit Native shamans and religious leaders, while allowing Indian traditions to color this transplanted Catholicism. Beyond gathering souls for the church, the Jesuits provided a thin but crucial buttress for the French presence (and policy) across the far-flung empire that was New France. Like many of their Protes-

tant counterparts in New England, they had little theological problem
with warfare against apostates and heretics. Some of the Jesuit fathers
not only incited attacks against the enemies of France, but also participated in the raids themselves.

In many ways, the missionaries found the most chaotic fields to be
the most productive; those whose lives were upended by disease, war,
and cultural unrest were especially open to proselytizing. Many among
the Indians, however, blamed the Black Robes themselves for the evil
changes that had come upon them.

They "attribute to the Faith all the wars, diseases, and calami-
ties of the country," Father Paul Le Jeune said, and "they assert that
their change of Religion has caused their change of fortune; and that
their Baptism was at once followed by every possible misfortune. The
Dutch, they say, have preserved the Iroquois by allowing them to live
in their own fashion, just as the black Gowns have ruined the Huron
by preaching faith to them."

While they might not have understood the germ theory of disease
transmission, Indians from the Huron in the north to the Timucua in
the south recognized that contagion seemed to follow all these newcomers from over the sea. So why not simply abandon their villages,
the perceived sorcery of the priests, and the assumed vengeance of the
newcomers' god? Aside from the very human desire to stay close to
places and people they knew, by the early 1600s many Indian societies had already become so tightly woven into trade arrangements and
military alliances—as well as so closely bound by religious ties—that
it would have been difficult or impossible for them to just pull up
stakes and move.

The seasonal round of movements, from winter camps to summer
villages, from the deep interior to the coast, began to come to an end.
Both for the traders and the missionaries, it was better that the Indians live permanently in one place, where schools and churches (and
trading posts and military oversight) could be established.

Nor were the Europeans, even from the earliest moments of contact,
at all shy about casting their lot militarily with their new neighbors. As
early as 1609, Champlain and two of his men joined a mixed force of
Montagnais, Huron, and Algonquians to defeat several hundred Mohawk, with Champlain personally killing two Mohawk leaders with a

single blast from his harquebus. It cemented early Algonquian loyalty to the French but also cast the die for French-Iroquois hostility that would last for more than a century.

The Mohawk, meanwhile, were receiving support from their Dutch trading partners in their war with the Susquehannock. In turn, the Susquehannock welcomed English colonists from Virginia to their palisaded towns along the lower Susquehanna River—colonists who were already trading with Powhatan's confederacy, the Susquehannock's enemies.

European traders immediately noticed that the Indians sought some items in their inventory more eagerly than others. Anything blue or red was snapped up—cloth, ribbons, beads—while white or other light-colored material was usually scorned. Small mirrors, metal bells,

Champlain Fights the Mohawk

From the very beginning, Native leaders and colonists found themselves enmeshed in alliances and tangled in feuds—between imperial rivals or tribal enemies—whose implications they rarely understood. In 1609, Samuel de Champlain and several hundred Montagnais, Huron, and Algonquian allies fought a pitched battle against Mohawk warriors along what the Frenchman dubbed Lake Champlain. The wheel-lock harquebuses carried by Champlain and his two French companions proved decisive, killing several Mohawk leaders and demoralizing their battle-hardened warriors, but the victory sparked generations of conflict between New France and the Iroquois League. (Library of Congress) and cheap jewelry were prized, often over utilitarian objects such as knives or axes. "Generally covetous of Copper, Beads and such like trash," John Smith dismissively said of Virginia's Natives, but this may not have been simple (or simply due to) taste. European objects, at least initially, were thought to embody the supernatural power of these new beings. Even years later, the language of the Montagnais-Naskapi in Quebec and Labrador reflected the connection between *məntu'*—a sense of the magical and inexplicable—and these new objects: *məntu'wian* for cloth, and *məntu' mines'it* for beadwork.

One thing the Indians especially wanted were European weapons. Powhatan, squeezing the hungry colonists at Jamestown during one of their early famines, traded twenty turkeys for twenty swords, and he rarely missed an opportunity to bargain for weapons thereafter, although the English tried to limit those acquisitions. (Opechancanough's Pamunkey, having acquired a bag of gunpowder in those initial years of contact, tried to plant the dark grains as though they were seeds.)

The real prizes were firearms, which, despite the cumbersome nature of matchlock muskets, gave the Europeans an overwhelming advantage of distance and accuracy, especially when musketeers were assembled in ranks. Gaining guns became a fixation of both Powhatan and Opechancanough, when he took the reins of the confederacy after Powhatan's death in 1618.

Opechancanough's ascendancy marked a striking change of fortune. In 1609, he'd been humiliated in front of his people when John Smith jammed a pistol into the werowance's ribs and stole his corn—an indignity reinforced later by a beating that Smith gave Opechancanough's son. Thereafter, the Indian leader endured a long period of diminished prestige among the werowances of Tsenacommacah, even losing one of his favorite wives to a political rival. But he slowly rebuilt his power base, and the aging Powhatan relinquished to his younger brother more and more of his authority, until Powhatan's death brought to Opechancanough the mantle of *mamanatowick*.

By then, the English presence had grown dramatically. None of the original attempts to make Jamestown profitable for its investors—including experiments with minerals, silk, tar and pitch, and glassmaking, among other endeavors—had yielded much profit. So in 1616, the Virginia Company in England opened the colony to privately backed "particular plantations." Typical were Warresquioake, about sixteen miles downriver from Jamestown, where sixty freemen and African slaves worked the fields, and Martin's Hundred, on the north shore about seven miles below Jamestown Island.

Settled in 1620 by more than two hundred immigrants, Martin's Hundred encompassed 21,000 acres, with its daily life centered on Wolstenholme Towne, a small cluster of wood-framed, wattle-and-daub houses with thatched roofs on a bluff above the James River. The layout of the village was one any English colonist, from Virginia to Plymouth to Popham in Maine, would have instantly recognized. In fact, English and Scottish settlers displacing the northern Irish in Ulster were using the same town plan, which featured a palisaded fort called a bawne.

From the Gaelic word *bàdhùn*, meaning "cattle fortress," a bawne was originally a tall fence around a cottage's farmyard, which served as a nighttime corral in times of peace and as a defensive barrier in times of war. By the early seventeenth century, the concept had grown into a central fort, and many of the investors underwriting plantations issued very similar instructions to emigrants, whether bound for Virginia, New England, or northern Ireland, on how to plan, build, and defend their communities.

The main bawne at Martin's Hundred was a lopsided log fort with two "flankers" (watchtowers protruding from the walls that gave the defenders a clear line of fire) that protected the leader's fairly rustic house and served as a refuge in case of attack. A cannon mounted in one of the flankers covered a small complex of company-owned barns, sheds, and workrooms, which, like many of the private homes, had bawne fences of their own.

Tobacco was driving the expansion of Martin's Hundred and the rest of the Virginia colony. The Indian crop had become all the rage in England, and after so many commercial failures, it was responsible for the budding success of Jamestown and its surrounding settlements. It was also consuming more and more Algonquian land, at a time when virgin soil epidemics were ravaging the people of Tsenacommacah and reducing their numbers. The balance along the Virginia tidewater was rapidly shifting in favor of the white *tassantassas*, and Opechancanough decided to move before it was too late.

For years, English missionaries had tried to convert the Powhatan people, with marginal success. Opechancanough, playing a cunning game of political chess, finally agreed to let young Algonquian men learn the new religion—if they were also schooled in musketry. The Indian stockpile of guns had slowly but steadily grown, and with it the sophistication to use the new technology.

For their part, the colonists were smug and complacent—"settled in such a firm peace, as most men there thought sure and unvoidable," Smith's history recounts. "The poore weake Salvages being in every way bettered by us, and safely sheltered and defended, whereby wee might freely follow our businesse." An emissary visiting Opechanca-nough in the middle of March was told by the grand chief that "he held the peace so firm, the sky should fall or he dissolved it."

The sky fell a week later, on March 22, 1622—Good Friday morn-ing, a day Opechancanough knew had special significance for the *tas-santassas*. Up and down the tidewater, unarmed Indians appeared out-side the bawne walls of towns and the dooryards of isolated houses, holding up game. There was no sign of danger, for the Indians bore no weapons.

"They came unarmed into our houses, without Bowes or arrowes, or other weapons, with Deere, Turkies, Fish, Furres, and other provi-sions, to sell, and trucke with us, for glasse, beades, and other trifles," wrote Edward Waterhouse, recounting the coordinated, simultaneous assault on dozens of plantations and villages along both sides of the James River.

Yea in some places, [they] sate downe at Breakfast with our peo-ple at their tables, whom immediately with their owne tooles and weapons, eyther laid downe, or standing in their houses, they basely and barbarously murthered, not sparing eyther age or sexe, man, woman or childe . . . And by this meanes that fatall Friday morning there fell under the bloody and barbarous hands of that perfidous and inhumane people, contrary to all lawes of God and men, three hundred forty seven men, women, and children, most by their owne weapons.

The surprise was not universal—one farmer, tipped off by an Indian with whom he had grown friendly, rowed his small boat furiously across the river in the middle of the night, alerting Jamestown and some neighboring plantations—but most outlying settlements had no warning. When the carnage was over, nearly 350 people—a third of the English colonists—were dead. Fifty-two were killed at Warresquioake alone, although a few men were able to barricade themselves and their families inside, defending themselves with musket fire. Among the handful of survivors was "Antonio the Negro," a newly arrived slave, who would go on to gain his freedom, move to the Eastern Shore, and become one of the largest black landholders there.

At Martin's Hundred, virtually the entire settlement was burned, and about 80 of the 140 people who lived there were killed. Victims who had been bludgeoned to death with their own garden spades were mutilated, scalped, and left among the smoking ruins. Many of the survivors were taken captive.

As they had been a dozen years earlier, the English were pushed to the wall. But when Opechancanough tried to evict them completely, he failed. The colonists regrouped, abandoning the smaller outlying farms and consolidating the survivors into five or six strongly defended plantations. (The authorities had to threaten to burn the property of one Mistress Proctor before the "proper, civill, modest Gentlewoman" would consent to move.) And then they struck back with a ferocity that may have surprised Opechancanough and his werowances.

War raged for three more years, including a massive two-day battle in which the Indian musketeers proved themselves the equal of their English counterparts. The English also tried a grand deceit of their own, inviting the Indians to a peace parley, then serving poisoned liquor that killed hundreds. Eventually, both sides grew fatigued with the fight and, after years of sporadic violence, made peace in 1632.

The great chiefdom of Tsenacommacah, which once commanded eight thousand square miles, was by then sadly reduced. The flow of colonists increased each year, and the land they now claimed along

all the tidewater rivers was expanding at a growing pace. Opechancanough made one last attempt to reclaim the Chesapeake tidelands for
his people—another surprise assault in 1644, when the old leader was
probably in his late nineties, so infirm his servants had to hold open
his sagging eyelids.

The death toll was even higher this time—some five hundred settlers—but the colonies were so populous and firmly entrenched that
nothing could move them. In the wake of the attacks, Opechancanough, barely able to walk, was captured by Sir William Berkeley, the
governor of Virginia. Berkeley planned to ship the old werowance off
to England, but before he had a chance, a soldier guarding Opechancanough shot him in the back, and the last of the great Algonquian
mamanatowicks was gone.

PART II

"LET US NOT LIVE TO
BEE ENSLAVED"

Chapter 4

"Why Should You Be So Furious?"

FROM THE FIRST attack on the freshly landed Jamestown
settlers to Opechancanough's murder and the eventual collapse of Powhatan's once great chiefdom, the early history of
the Jamestown colony seems like the quintessence of frontier relations
between Native tribes and encroaching Europeans, no matter through
what prism one views it.

For generations, the massacre of 1622 was considered the epitome of
Indian deception and cruelty, and the eventual triumph of the English
settlers over these forces of savagery was thought to be foreordained
by the divine. "In consideration of Gods most mercifull deliverance of
so many in this Cuntrie of Virginia from the treachery of the Indians,"
proclaimed Governor Francis Wyatt shortly after the attack, "the 22th
day of March . . . forever hereafter . . . be in this Cuntrie celebrated
Holy." More recently, historians have reexamined the strategy of Powhatan and Opechancanough from the perspective of an active Algonquian resistance against an often brutal invasion and blatant theft of
land, concluding that Opechancanough in particular was "the archetype of later and better-known patriot chiefs."

But in many ways, Jamestown was an exception, especially in regard to the immediate and rapidly escalating violence between the

two extraordinarily different cultures. A period—often a surprisingly long period—of mutual accommodation and cooperation was the rule along much of the First Frontier. Though ginger and tentative, and fraught with miscalculations, these early relations provide a hint of what might have been, had not greed, human nature, and the over-whelming number of European immigrants eventually become too much for the tenuous peace to bear.

Initially, such coexistence was certainly abetted by a fundamental misunderstanding over land. To Europeans, land was a commodity to be owned outright by one person or entity, and ownership was clear-cut and guaranteed by law. Among the Indians, to the extent that land was "owned" at all, it was held communally and represented a complex web of vital resources both physical and spiritual, bound tightly with the seasonal round of life and in which was deeply embedded intricate social bonds. While boundaries (not only between neighboring tribes but also between clans, bands, and extended families within the same community) might be sharply drawn and at times breached only at great peril, the notion that a single person might buy land and hold it in perpetuity was a profoundly foreign concept.

So when Peter Minuit and a boatload of Swedes dropped anchor along the lower Delaware River in 1638 and negotiated through a translator with the Lenape, it is quite certain that the five sachems with whom they met had no idea they had "ceded, transported, and transferred the said land"—a huge swath of Delaware Bay some fifty miles south from the mouth of the Schuylkill River—"with all its ju-risdiction, sovereignty, and rights to the Swedish Florida Company," much less that they had been "paid and fully compensated for it by good and proper merchandise."

Minuit had made a similar deal in 1626 for Manhattan Island, and then, as now along the Delaware, the Lenape sachems with whom he met saw the goods he offered not as payment, but as rare and valuable gifts solemnifying an important new relationship. Gift giving was a practice of immense importance in Algonquian culture, and by grant-ing the use of Lenape territory, the sachems were, in their eyes, making a gift of significant reciprocity. They fully expected to continue to use the land themselves, as they always had, exchanging still more gifts as their relationship with these strange newcomers grew. To the Indians, this was good manners. To the Swedes, Finns, and Dutch with whom the Lenape dealt, it was good diplomacy. That these miscomprehensions meshed was a matter of dumb luck, because at least at first, neither side realized how the other viewed matters.

Along much of the southern New England coast, the situation was considerably different. The great epidemics of 1616 to 1619 had emptied much of the land, and English colonists felt free to settle where they liked. Even where there were surviving Indian communities, the Puritans invoked biblical law, under which land that was not actively used—that is, tilled and farmed—was available for the taking.

"What warrant have we to take the land, which is and hath been of long tyme possessed of others the sons of Adam?" John Winthrop asked in 1629, then answered his own question. "That which is common to all is proper to none . . . Why may not Christians have liberty to go and dwell amongst them in their waste lands and woods (leaving such places as they have manured for their corne) as lawfully as Abraham did among the Sodomites?"

The Wapánahki in Maine, though hit hard by disease, generally remained in control of their coast, and many of the English immigrants they encountered were not sin-obsessed Puritans, but entrepreneurs looking for commercial opportunities in furs and fishing. Here the English and the Wapánahki encountered the same initial misunderstandings about land ownership, but as surviving deeds from the 1620s on show, the Natives quickly developed a working knowledge of what "buying land" really entailed.

There's an assumption today that all such land deals were swindles, designed to separate Indians from the land for a pittance, but the Maine deeds suggest that a more nuanced dynamic was at work, with both sides bridging the cultural divide to reach what each saw as an equitable agreement. Yes, the prices paid—in wampum, bolts of fabrics, clothing, tools, food, and household items—seem ridiculously low by modern standards, but such assumptions about worthless beads and trinkets (as they seem to us) overlook the very real appeal of unusual goods from distant lands, to say nothing of their utilitarian value. (If you doubt the virtue of a steel hatchet, try using a stone ax to chop down a tree.) And some deeds spelled out what is implied in many others—annual payments that make the transaction appear more in the nature of a lease than an outright purchase.

With time, Indian deeds also tended to spell out explicitly what rights the Native owners were retaining for themselves—the right to hunt moose or deer, to fish for salmon or sturgeon, to gather certain roots or shellfish, to camp there seasonally. "Nothing in this Deed be Construed to deprive us ye Saggamores [our] Successessors or People from Improving our Ancient Planting grounds nor from Hunting In any of said Lands, nor from fishing or fowling for our own Provission Soe Long as noe Damage Shall be to ye English fisherys," reads one deed for land along the Kennebec River being sold by Warumbee and five of his fellow sagamores. In many respects, the rights carved out by Indian deeds may have been the same uses to which they had always put the land. Having been burned by the all-inclusive concept of English ownership, they were simply spelling it out.

In less than a generation, up and down the eastern seaboard, the bonds between Europeans and Indians had grown remarkably strong, bolstered by trade, marriage, and military alliances. That was not always the case between the colonists themselves, however, even those from the same country.

Jamestown was an Anglican settlement, nominally so in its earliest years, more rigorously after Lord De La Warr's arrival, when he ordered all colonists to attend the twice-daily worship services. The appearance in 1634 of settlers sponsored by the overtly Catholic Cecilius Calvert, Baron Baltimore, just up the Chesapeake on the Potomac River, was seen as a threat to Protestant control of the tidewater, as well as the growing profits that Jamestown's backers were beginning to enjoy. Worse, the land included in the new colony of Maryland had been carved out of Virginia two years earlier by Charles I.

Perhaps because the Virginians started a rumor that the newcomers were the hated Spanish, the Piscataway, who controlled the lower Potomac, met Calvert's brother Leonard and the Maryland colonists with hordes of armed men, who "made a generall alarm, as if they intended to summon all the Indians of America against us." Somewhat reassured, the Piscataway *tayac* (grand chief) granted them permission to live at an abandoned town site at the mouth of the river, which Calvert named St. Mary's City.

In a sense, both the Marylanders and the Piscataway needed friends. The Piscataway were being squeezed between Powhatan's confederacy to the south and the Susquehannock to the north. The Virginians, too, had launched reprisals after the 1622 massacre, even though the Piscataway weren't involved. The Piscataway were also being cut out as middlemen in the fur trade between the English and the powerful Massawomeck, an interior tribe from the headwaters of the Potomac with rich stores of beavers.

Developing a good fur trade was also important to Calvert, and his colonists needed Indian corn to tide them over. As St. Mary's grew, the connections between the Piscataway and the colonists deepened. English Jesuits converted many of the Indians, including the new *tayac*, strengthening the ties between the two communities. Treaties obligated the Piscataway to aid the English in war, and vice versa. Indians and Marylanders built forts together, hunted together, and drank together.

They continued bartering corn and deerskins for English goods. One of the traders, a man named Mathias de Sousa, came to St. Mary's with the original settlers as an indentured servant of a Jesuit priest. Almost certainly of mixed Portuguese and African descent, de Sousa eventually became a free man under Maryland law, and in 1641 his name appeared on the rolls of the general assembly—the first man with African roots to have a voice in American government.

For several decades, the Piscataway and Maryland colonists existed in a kind of equilibrium. Smallpox and other diseases reduced the Indians' ranks, the encroachments of hostile tribes increased, and more and more land and control were ceded to the English, ultimately including the right to chose a new *tayac*. Yet the Piscataway, if no longer a dominant political force in the upper Chesapeake, still held on to their identity. Despite early successes, the Jesuits made little headway with them and eventually abandoned their mission. The English colonial proprietors maintained a largely hands-off policy toward their allies, who by the late seventeenth century still lived in their palisaded village, ruled by a *tayac* of their own choosing (if then rubber-stamped by Maryland authorities). They could still field almost a hundred warriors, roughly the same strength that John Smith had reported almost a century earlier, although some of them were certainly refugees from other, fractured tribes.

This delicate balance ended in the 1690s, when pressure from encroaching farmers reached the tipping point, at the same time that the English Crown assumed direct control of the colony. Francis Nicholson, former governor of New York, took the reins, implementing policies explicitly designed to drive the Piscataway from land deemed too valuable to waste on Indians when it could be used for growing tobacco. Nicholson's policies worked. The last *tayac* gathered his thinning flock and left, eventually settling with their one-time enemies the Susquehannock.

The early tensions between fellow Englishmen in Maryland and Virginia was entirely typical of the times. While France's colonial template was based on fur posts and wilderness Jesuits, the Dutch focused on trade and trade alone, and Spain created a unified network of mission towns supported by the *repartimiento* system, England took a far more ad hoc, idiosyncratic approach to colonization. The settlements it was sprinkling from Jamestown to Newfoundland reflected almost the entire spectrum of English spiritual views, from the rawly secular to Calvert's Catholicism to the astringent Puritanism of the Pilgrims, grafted onto competing commercial interests that made furs, cod, or tobacco at least as important as anything else, including eternal salvation. The mixture was often unstable and could be combustible.

Most modern Americans assume that the colonists came seeking religious freedom, but even the most pious of colonists depended on commercial backers. When the Pilgrims landed at the deserted Wampanoag village of Patuxet in 1620 and renamed it New Plimmoth, their leaders were motivated by a desire to flee the Church of England, but the underwriters who paid for the colony—the Merchant Adventurers of London—were inspired by profits, and they were displeased by the lack of gain in the colony's first year. Their testy letter of complaint arrived in 1621 with a second load of colonists, many of them young, non-Pilgrim men who came to make their fortune and who immediately chafed under the community's literally puritanical religious rule.

(The Pilgrim leaders grudgingly permitted the newcomers to observe Christmas—a holiday the Pilgrims considered essentially pagan—but when the celebrants organized a ball game, a disgusted Governor William Bradford snatched away their balls and bats.)

Especially in New England, competing commercial backers led to competing colonies, and internal disputes about religion and the rights of man led to splintering, schism, and still more settlements. In 1629, land between the Charles and Merrimack rivers, including existing fishing encampments at Cape Ann and Salem, was granted to the Massachusetts Bay Colony, which brought almost a thousand new immigrants to Boston in 1630.

The Massachusetts colonists were mostly Puritans, like the Pilgrims, dedicated to shedding what they saw as secular trappings in the Anglican Church, but they were not separatists, instead believing that the church could be reformed from within—anathema to Pilgrims. And they were shrewd businessmen, although you'd never know it by Governor John Winthrop, who claimed that other English settlements had failed because "there mayne end and purpose was carnall and not religious . . . aymed chiefly at profit."

In fact, Winthrop and others had raised a grubstake of almost £200,000—about $40 million in today's currency—an investment on which the colony's backers expected to turn a profit, from furs and cod fishing, in this new and holy experiment. Business and righteousness would go hand in hand, unlike in previous colonial enterprises, Winthrop said, which "used too unfitt instruments, a multitude of rude and ungoverned persons, the very scums of the land."

But even within the Puritan community, there were opposing factions. In 1636, the Reverend Thomas Hooker, angered by Massachusetts's refusal to grant full suffrage to all men regardless of church membership, took a hundred followers and trekked west for weeks—driving a large herd of cattle, goats, and hogs along the way—to the Connecticut River valley, where they founded Hartford on land ostensibly under Dutch control. By contrast, a boatload of Puritans under the Reverend John Davenport declined to settle in what they saw as the lax spiritual environment of the Bay Colony and sailed down the coast to alight at what they called New Haven, a colony founded on an Old Testament reading of scriptural law

—perhaps the most purely theocratic colony in the New World.

Massachusetts just kept rubbing people the wrong way. Roger Williams was perhaps Davenport's opposite—a separatist Puritan who came to believe in religious tolerance, the rights of Indians, and what he called a "wall of separation" between the church and secular life. Williams bounced back and forth between the Bay Colony and Plymouth before being expelled entirely. In 1636, he founded Providence Plantations, the start of what would eventually become the colony of Rhode Island, a welcoming home for Quakers, Jews, and those scorned in most other colonies.

The Dutch, meanwhile, had spread out along the Noort Rivier (Hudson River), establishing a fort near what is now Albany and a settlement at New Amsterdam on *manahatta*, the Lenape's "island of many hills." They also erected *handelshuisen* (trading posts) east to the Versche Rivier (Connecticut River) and south to the Zuyd (Delaware River). Peter Minuit's purchase of Manhattan for sixty guilders has become shorthand for Europeans swindling Natives out of their land, but it actually reflects an earnest policy by the Dutch to deal fairly with the Indians. They sought to negotiate in (mostly) good faith, stipulate punishment for colonists who abused Natives, and even set wages—though at half the pay given Europeans—when Indians were hired to work.

Although France, Spain, and England are usually thought of as the three-way imperial rivalry in North America, in the early to mid-seventeenth century, the United Provinces of the Netherlands represented a global power to match or exceed any of the others, standing as the world's dominant trading nation, with colonies and outposts from North America to Brazil, Indonesia, India, China, and Japan, all connected by a potent navy. Only Spain—Holland's one-time imperial master, whose yoke the Dutch had only recently thrown off—came close to this immense reach. By comparison, England and France were bush leaguers.

And unlike those nations, the Dutch were interested in one thing and one thing only in North America: business. In fact, they did not speak of their footholds in the New World as colonies, at least not at first. Unlike the English, they had no desire to create growing agricultural communities for large numbers of immigrants; initially, they brought just enough farmers to feed their operations. Nor did they have any great desire to bring Indians to the church, as did the French. The Spanish model, with its casual brutality and enslavement, was an active abomination to Hollanders, having just escaped Spanish rule.

Instead, the Dutch negotiated for the right to build forts and *handelshuisen* along the coast and major rivers, treating their Native counterparts as fellow entrepreneurs. Recognizing Native priority was a decision that also paid shrewd commercial dividends, since the Dutch could shrug as they sidestepped claims from other European powers, even as they considered their own right to do business in the New World—founded on the exploration and mapping of Henry Hudson and other explorers in Dutch employ—as ironclad.

The Dutch holdings were peopled by a multiplicity of ethnicities and religions drawn to the Netherlands' history of tolerance—which went only so far, since the communities included a significant number of African slaves. Backed by the powerful West India Company, the Dutch had a distinct advantage over the foundering, squabbling English colonists: stable supplies of quality trade goods with which to bind the Mohawk, Lenape, Susquehannock, Mohegan, Pequot, and other nations to them in commerce, all based on the barter of beaver pelts.

The European fashion for broad-brimmed felt hats drove this insatiable demand for beavers, whose plush undercoat made superb felt once the long brown guard hairs were removed. Beavers could be easily trapped in the winter near their lodges and bank dens, lured by bait sticks smeared with a vile-smelling secretion called castoreum, with which they mark the boundaries of a colony. Otter, muskrat, lynx, fox, and other furs also were traded, but beaver pelts were the great prize.

No matter where it arose, the fur trade caused profound changes in Indian life. The demand for European goods caused communities to alter their traditional seasonal movements, created middlemen, and intensified rivalries among tribes for access to goods and control of the flow of furs. It transformed a Native lifestyle that had been focused on subsistence and created a new economy based on an international trade triangle: beaver pelts from the hinterlands, manufactured goods from Europe, and what may have been North America's first mass-produced commodity, wampum.

Traders purchased beaver pelts with muskets, kettles, steel knives, dried peas, and liquor, among other goods, but wampum was "the source and the mother of the beaver trade," New Netherland governor Peter Stuyvesant told his backers, who were aghast at its price. "Without wampum, we cannot obtain beaver from the savages."

Wampum—the word comes from the Algonquian *wampompeag* —served as both a form of currency and a medium of diplomacy among many Indian nations, with spiritual aspects that transcended both uses. The small beads came in two colors—white, made from whelk shells, and black (actually dark purple) made from the hard shells of quahog clams. Both kinds were produced by tribes along the Atlantic coast and traded as far west as the Great Plains.

Although crude forms of wampum were important among Northeast Indians long before contact, the introduction of hand-held metal drills in the late 1500s standardized the manufacturing process, allowing craftsmen to turn out remarkably uniform, neatly cylindrical beads about half an inch long. It was still a laborious process. A skilled wampum maker could produce about forty white beads a day from whelk shells but barely half that number of black beads from the much harder clamshells. A large, especially ornate wampum belt might contain seven or eight thousand beads—months' worth of work.

But wampum's importance went far beyond its monetary value. Strings or belts of wampum were used to punctuate important speeches, reinforce calls to war, emphasize overtures for peace, and ratify alliances. They might fulfill tributary demands by conquerors, memorialize treaties, serve as a badge of recognition for a messenger, or compensate the family of a victim of violence.

Although the English colonies sat on a rich trove of wampum raw material, it wasn't until 1627 that they learned about its value from the secretary of New Netherland, Isaak de Rasieres, during his visit to Plymouth. De Rasieres had hoped the English would use their shell wealth to buy furs from the north, leaving Dutch areas to the west alone, and at first his plan worked. Indians in Maine who had scorned English offers of corn for furs snatched up all the wampum they could get. But that simply increased English demand for the beads, in pursuit of which they pushed west along the southern New England coast into Dutch territory and sailed across Long Island Sound to Peconic Bay, where the Indians made some of the best wampum in the region.

Soon the trade triangle fell into place. English merchants imported cheap goods from abroad, including the heavy woolen cloth known as duffel, popular among the Indians. With these goods, they purchased wampum from the Wampanoag, Pequot, Narragansett, and other coastal groups, then resold it to the Dutch at a considerable markup. They also traded it directly for furs with interior tribes along the middle Connecticut River, who needed it for tributary payments to the Mohawk. The furs, drawn from increasingly distant parts of the interior, were shipped back to Europe and sold there at a tremendous profit.

The fur business was booming. Between 1624 and 1632, the number of beaver pelts handled by Dutch traders increased tenfold, to more than fifteen thousand a year. But as beavers were trapped out in the areas closest to the trading forts and grew harder and harder to get—and as the supply of wampum burgeoned—the buying power of a six-foot bead string, known as a fathom, fell from two and a half pelts to half a pelt. Further complicating the exchange rate was the growing use of wampum as currency among the Dutch and English colonists themselves.

Furs and wampum brought the Dutch and English into increasing conflict along the southern New England coast, and they also set up an incendiary situation involving the region's Native communities, which were in tremendous flux. The conflict reached a flash point in the 1630s along a broad river running straight south from the remote fur regions down to Long Island Sound. The Dutch called this waterway the Versche Rivier (Fresh River). The English and French—mangling the Algonquian word *quinetucket,* or "long tidal river"—called it the Connecticut.

The lower Connecticut and smaller rivers to the east, such as the Mystic and Thames, were home to the Pequot, who had been expanding their power in the years before the Dutch arrived. They inhabited about fifteen villages, each of which included roughly thirty wigwams (framed with bent saplings and covered with bark or mats of woven cattails), surrounded by several hundred acres of cornfields cleared from the rich bottomland forest.

Two of the Pequot towns, home to paramount sachems, were built on high ground some distance from the water and were surrounded by palisades for additional protection. The main village was Weinshauks, also known as Fort Hill, at what the English called Pequot Harbour,

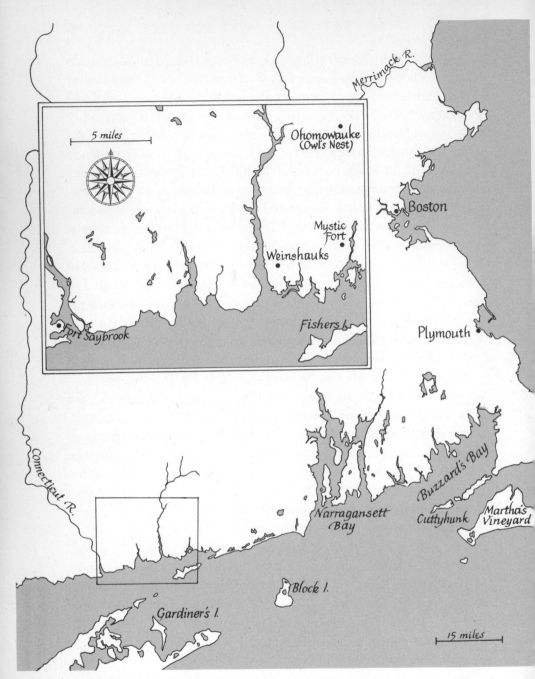

Southern New England, ca. 1637

Controlling the lower reaches of the Connecticut River, which provided access to interior fur supplies, and subjugating many of their Native neighbors, the Pequot held the balance of power in southern New England in the 1630s.

the mouth of the Thames River. Mystic Fort, about five miles to the east, occupied a hill just west of where the Mystic River emptied into a deeply sheltered bay.

Covering two acres and encompassing about seventy wigwams, the formidable Mystic stockade, according to one English witness, was built "close together as they can, [of] young trees and half trees, as thick as a man's thigh or the calf of his leg. Ten or twelve foot high they are above the ground, and within rammed three foot deep with undermining, the earth being cast up for their better shelter against enemy dischargements. Betwixt these palisadoes are divers loopholes through which they let fly their winged messengers."

There was no gate as a European fort would have. Rather, the palisade walls formed two greatly overlapping half circles, with the long, narrow gaps between them creating entrances on opposite sides. These could be blocked off at night with barriers of brush—hardly as sturdy as a heavy wooden door, but enough to slow attackers, who would be forced to enter single file while the defending Pequots poured arrows on them.

Under their primary sachem, Tatobem, the Pequot had solidified their position as middlemen in both the fur and wampum trades, forcing such neighboring groups as the Mohegan into tributary status and essentially ignoring Dutch insistence that they permit free trade. The Narragansett, who likewise craved an exclusive partnership with the Dutch, chafed at Pequot dominance, while some among the Mohegan—so deeply entwined with the Pequot that they were, in some ways, one people—reached out to the English to bolster them in their own internal struggles.

Prodded by the Mohegan, and receiving pleas from a leader of the River Indians, a group that lived under the Pequot's thumb at the mouth of the Connecticut, the English decided to test Dutch claims to the river and Pequot tolerance, grabbing a slice of the trading pie for themselves. It was the beginning of a game of brinkmanship that would have enormous and tragic consequences.

In the summer of 1633, the Dutch built a stockaded *handelshuis*, known as the House of Good Hope, at what is now Hartford. A few months later, the Plymouth colonists trumped them by brazenly sailing a large, newly built bark up the river, right beneath the two Dutch "murtherers," small cannons that could be loaded with nails, scrap metal, or anything else handy and lethal. Despite shouted threats from the garrison, however, the cannons were not fired. Landing about six miles upriver, the Englishmen hurriedly unloaded the precut frame and siding for a house, which, in the words of William Bradford, they "clapt up . . . quickly, and landed their provissions, and . . . afterwards palisadoed their house aboute, and fortified themselves better." Safe within its stockade, the outpost was secure enough that when the Dutch commander arrived with seventy men to evict the squatters, he judged an attack on the fort impractical and left.

Even worse for the Dutch, the Pequot were ignoring their agreement to allow other tribes unimpeded access to the trading posts. Pequot warriors, determined to maintain their monopoly, killed a group of Narragansetts in the fall of 1633. The move infuriated the Narragansett and Dutch alike, but the latter made the first move. Deciding that harsh measures were in order, the Dutch commander lured Tatobem onto a ship to trade, seized him, and demanded a bushel of wampum in ransom. The Pequot paid it, but the Dutch, driving home their point, executed Tatobem anyway. When a Pequot delegation came to Good Hope, the Dutch leveled one of their "murtherers" at them and blew another sachem to bits.

Tatobem's murder rocked the Pequot, but they had to tread a fine line. As the Dutch guessed, the Pequot's need to maintain trade relations checked how strongly and directly they could respond, both to the grave Dutch insults and to the English incursion. What's more, the Pequot leadership was in turmoil, and for the first time, smallpox was sweeping southern New England, compounding the confusion with a ghastly epidemic that killed as much as 80 percent of the Indian population. Several survivors scrambled to replace Tatobem, including his Mohegan son-in-law Uncas.* The victor, a sachem named Sassacus, is usually described as Tatobem's son, but little is actually known about him. The choice of a new chief sachem depended as much on personal

---

*James Fenimore Cooper forever muddied the historical waters by appropriating the name of the ambitious Mohegan sachem for the title character in *The Last of the Mohicans,* changing his tribal identity to Mahican (a group that lived in the upper Hudson Valley) and transplanting him to the middle of the eighteenth century.

qualities and support within the community as blood relations. Unfortunately, Sassacus was unable to hold together the fractious Pequot and their tributary villages. Some of the dissenters, including Uncas and a Pequot sachem named Wequash, who had likewise failed to succeed Tatobem, began turning to the Narragansett for protection.

And there was still the issue of Tatobem's death to address. The Pequot reaction came in a form any Indian would have anticipated and heartily approved of—a revenge attack on the Dutch by the family of the victim.

As archaeological evidence makes plain, warfare was a constant in pre-contact eastern North America, but the reasons for conflict, and the approach and scale of war among Indians, were dramatically different than in Europe. Instead of employing all-out assaults by massed armies for the capture of territory, Indian warfare tended to consist of limited, hit-and-run engagements by small parties. These strikes were designed to take captives, exact revenge, and garner prestige; in most cases, war was an almost endless cycle of tit for tat between traditional enemies. This approach seemed pointless to Europeans; Indians "might fight seven years and not kill seven men," wrote one contemptuous Englishman.

Killing wasn't the point, however. The societal trappings varied, but among the Five Nations and related groups in the Northeast, for instance, such offensives were known as "mourning-war," whose main intent was to capture enemies who could either be adopted into the village as replacements for dead relatives or be ritually tortured to death to assuage the family's grief. Mourning-warfare might entail raids by a few men (usually related through marriage to the dead person's female survivors) or large confrontations involving hundreds of warriors. However great or small the force, the aim was not to kill but to capture; casualties were usually minimal. The 1609 battle in which Samuel de Champlain and his Algonquian allies fought against the Mohawk may have been such a mourning-war clash. If so, the deaths of several Mohawk chiefs at Champlain's hand must have been a terrible loss to the Iroquois.

Ritual torture, which so shocked the first Europeans to witness it, was part of this same elaborate framework of war. Arriving at the victors' village, a captive would be forced to run the gauntlet—two lines of people armed with sticks, knives, and hatchets, who would rain blows and abuse on the running man as he tried to reach a predetermined point of safety. Depending on his behavior in the gauntlet, a prisoner might be killed on the spot, reserved for more formal execution by torture later, or held as a slave. A number of Native languages—Oneida, Ottawa, and Ojibwa among them—had words, like the Mohawk term *enaskwa,* that meant both a human captive and a domestic animal.

Captives could also serve as potent pawns in diplomacy—traded to other tribes or offered as gifts that implicitly spoke of the giver's power and eminence. Women and children were often adopted into the community, helping to restore the population of a village reduced by disease, war, or famine. Frequently, this meant assuming the name and social obligations of someone recently dead. Among the Iroquois, this was accomplished through a condolence rite known as "requickening." A dead person of high rank was usually replaced from within the community, but someone of lesser status might be replaced by a war captive.

There was no question of trying to replace Tatobem—his death would be revenged against the Hollanders in a more direct manner. Biding his time, Sassacus waited until a European trader showed up along the Connecticut, then he personally led the assault on the man. Unfortunately, just as most Indians looked and sounded alike to Europeans, most Europeans were indistinguishable to many Indians. Sassacus chose the wrong nationality on which to avenge Tatobem's murder.

The irony was compounded by the fact that this was probably the first time in his dissolute life that the primary victim—an English rogue named John Stone—got a comeuppance he *didn't* richly deserve. A member of the Virginia colony, where his behavior scarcely raised an eyebrow, Captain Stone was immensely unpopular among the Puritans and Pilgrims, who abhorred his drinking, his carousing, his piracy and smuggling, and the dark stories of his cannibalism after a Caribbean shipwreck. Stone's appetite for liquor and larceny seem to have been about equal. At one point, Stone managed to drink his good friend, the Dutch governor at Manhattan, under the table, then ask the sozzled official to "give" him a Plymouth trading pinnace full of goods. Stone was forcing the crew at gunpoint to set sail for Virginia when the Dutch retook the ship.

Hauled up on piracy charges, he tried to stab the Plymouth governor, Thomas Prence, and in a scandalous piece of wordplay, he called a prominent judge "a just ass." He was caught in flagrante with a married woman and charged with adultery—a hanging offense in the Bay Colony. His pinnace was seized but he escaped, although soldiers eventually flushed him out of a cornfield and took him into custody. Because Stone had powerful friends in England, the colonists agreed to slap him with a heavy fine and banish him on pain of death. Everyone (except, perhaps, a few married women) was relieved to see Stone sail away in the summer of 1634.

Heading back to the more freewheeling climate of Virginia, Stone decided to detour up the Connecticut River. In typical highhanded fashion, he and his men seized two Indians as "guides." Tying their hands behind their backs, Stone didn't realize the abduction was being witnessed by other Natives.

That evening, part of the crew was on shore with the captives, while Stone and a few others remained on the ship. There was no guard, and everyone was drinking. Sassacus himself boarded the pinnace with his men. Stone welcomed him in a boozy haze and poured the sachem a glass. "Captain Stone, having drunk more than did him good, fell backwards on the bed asleep," a Pequot emissary later told the English. Sassacus pulled out a hatchet and tomahawked the unconscious man to death.

At his signal, the Pequots turned on the remaining Englishmen, killing those on shore and freeing the two Indians, while trapping the rest of the ship's crew in the cook room. Whether by design or accident, the gunpowder store caught fire, and the Pequots dove for cover moments before the pinnace exploded. The few English survivors, burned and blinded, were quickly dispatched.

From the Pequot's perspective, this was a completely appropriate, well-measured response to the Dutch murder of a grand sachem—except that the victims weren't Dutch. Even though they felt Stone asked for it when he kidnapped his "guides," the Pequot seem to have been genuinely contrite when they realized their mistake. That October, they offered the Massachusetts colonists wampum and furs in compensation—again, a traditional Algonquian approach to settling what might be termed a wrongful death complaint.

Estranged from the Dutch and Narragansett, the Pequot had now angered the English, who despite their ill will toward Stone were happy to use his death as an excuse to further their ambitions in the Connecticut Valley. They demanded the heads of his killers, the right to freely settle Pequot land along the river, an agreement to trade only with the English, and a lavish payment—four hundred fathoms of wampum, forty beaver pelts, and thirty otter skins, the equivalent of half a year's tax revenue for the colony. It looked as if John Stone would prove immeasurably more valuable in death than the rascal had been in life.

The Pequot might have agreed to the steep terms, but they wanted something the English would not provide—military protection. In Algonquian tradition, a payment of tribute entailed reciprocity, with the victors agreeing to safeguard the newly weakened people. This was not an idle concern; the Narragansett had dispatched two dozen men to capture or kill the Pequot ambassadors meeting with the Puritans in Boston. But the English agreed only "as friends to trade with them, but not to defend them, etc." The Pequot sachems, not wanting a hugely expensive deal that would still leave them without a military ally, declined to ratify the treaty.

For two years, animosity simmered. The English, their numbers swelled by thousands of fresh immigrants, planted new settlements well up the Connecticut Valley, forty or fifty miles upriver from the ocean. At the river's mouth, they established Fort Saybrook, hiring a tall, thirty-seven-year-old engineer named Lion Gardener to build their first real fortification in the region.

Gardener was a professional soldier, a veteran of fighting for the Prince of Orange against the Spanish. He expected to find three hundred men waiting for him when he arrived at Saybrook in the early winter of 1636. Instead, he found two, plus some traders relaying gifts of pelts and wampum from the Pequot.

From the moment of his disappointing arrival, Lieutenant Gardener was a plainspoken realist. When some of the Englishmen at

Saybrook, frothing about Stone's killing, said they would rather have the Pequots' lives instead of their presents, Gardener noted that they were soon heading back to Massachusetts. "If you make war with these Pequits, [you] will not come hither again, for I know you will keep yourselves safe, as you think, in the Bay, but myself, with these few, you will leave at the stake to be roasted."

Not only that, but the tiny community for which he became responsible—two dozen men, women, and children—had only two months' supply of food and indefensible cornfields miles from the settlement. He told his employers in Massachusetts that the most pressing danger was "Capt. Hunger" and that any building had to wait until spring. Gardener, who could see the writing on the wall, pleaded that those in authority "do their utmost endeavour to persuade the Bay-men to desist from war a year or two, till we could be better provided for it."

By the spring of 1636, progress was being made. The fort's palisade was finished in March, and a month later, as he was busy laying out the adjacent town, Gardener marked the birth of his son, David, the first English child born in the new Connecticut colonies.

Tensions continued to rise. Trade wasn't going well, and a dustup between Englishmen and Indians on Long Island that resulted in white deaths was blamed on the Pequot, despite evidence to the contrary. On the Connecticut frontier, able-bodied men were required to keep a gun, ball, and powder at the ready. Uncas, the Mohegan sachem, tried several times to exploit the developing power vacuum as Pequot control weakened, and he now told his English friends that the Pequot were preparing a preemptive attack on the colonists.

Uncas may have been playing the English against his Pequot rivals. If anything, the Pequot probably feared an unprovoked attack as well, since some of the Englishmen had been making "indiscreet speaches" about their plans for the Indians, according to Jonathan Brewster, the son of a prominent Plymouth colonist who was trading on the river. Not everyone believed Uncas's warning, but Brewster told Connecticut Governor John Winthrop Jr. that he had come to trust the Mohegan, "whom I have found faithfull to the English."

Winthrop was dispatched to issue an ultimatum to the Pequot: deliver Stone's killers or face the consequences. "If they should not give you satisfaction," Bay Colony governor Henry Vane and his deputy,

John Winthrop Sr., instructed, "declare to them that we hold ourselves free from any league or peace with them, and shall revenge the blood of our countrimen as occasion will serve."

In late July, the Massachusetts trader John Gallop steered his bark into Manisses (Block Island), where he recognized a pinnace owned by John Oldham riding at anchor two miles from shore. Oldham was the Bay Colony's commercial agent, an active trader among the coastal tribes and, like Stone, not a man with a sterling reputation among the colonists. Gallop hailed the vessel, whose deck was crowded with Indians, but got no response. In fact, the Indians slipped the anchor and set sail downwind toward the mainland. Gallop, suspecting that Oldham had been attacked, steered his ship to cut them off.

The odds were stacked against him. Gallop was alone except for another man and two small boys, and they were lightly armed with just two muskets, two pistols, and some duck shot. There appeared to be fourteen Indians on the other vessel, with muskets, pikes, and swords, but as Gallop closed on the pinnace, he and his tiny crew fired, forcing the Indians belowdecks. Swinging back again, the gale filling his sails, Gallop gathered speed and rammed the smaller ship, almost overturning it; six of the panicked Natives leapt out to their deaths. Gallop again bore down on the pinnace, crashing his anchor into the bow, where it lodged, pinning the ships together. The Englishmen fired their guns right through the thinly hulled pinnace, flushing four more Indians into the ocean. Then they boarded Oldham's vessel, capturing two Indians, whom they tied up. "Being well acquainted with their skill to untie themselves, if two of them be together," Gallop shoved one of the bound Natives overboard to drown. The final two, armed with swords, remained barricaded belowdecks in a small room.

Gallop found Oldham below as well—dead, stripped naked, his head split open, and his hands and legs almost cut off. Gallop and his companions slid Oldham's body over the side of the pinnace and tried to tow the disabled ship back to the mainland, but the wind rose again, and they had to cut it free to drift off, the last two Indians still hiding in the hold.

Oldham's killers weren't Pequots—the Manissean of Block Island were tributaries to the Narragansett, whose sachem tried to forestall English revenge by taking two hundred of his men to the island and recovering most of Oldham's goods, along with two English boys who had been with him. Indeed, Roger Williams later became convinced that Oldham's murder had been plotted by a group of Narragansett sachems. But the murder tipped the Bay Colony leadership over the edge where Indian relations of all sorts were concerned.

Magistrate John Endecott, backed by nearly a hundred volunteers, sailed to Block Island in September, prepared to murder all the Manissean men and capture the women and children. The Indians, retreating to swamps, easily evaded the clumsy soldiers, who contented themselves with burning wigwams, fields, and great stores of corn. Endecott turned next to the Pequot, who were supposedly harboring a couple of the Manisseans who had attacked Oldham, but the old grudge involving Stone's death was dredged up again. Endecott was to demand that the Pequot turn over Stone's and Oldham's killers, along with "one thousand fathoms of wampom for damages, etc., and some of their children as hostages." If they refused the exorbitant demand, Endecott was authorized to use force.

If Massachusetts really wanted peace, the colony could hardly have chosen a less suitable emissary. John Endecott was an unusually strident brand of Puritan, whom one historian called "a strange mixture of rashness, pious zeal, genial manners, hot temper, and harsh bigotry." He was the worst choice for delicate diplomacy.

In this, the so-called Bay-men were taking a cocky step—essentially claiming the right to unilaterally enforce their own policies across the boundaries of other English colonies and a tangle of conflicting Indian jurisdictions. The Bay Colony's move actually met with approval in some Native councils, among sachems such as the Mohegan Uncas and the Narragansett leader Miantonomi, who were happy to see the power of the once strong Pequot eroded.

It was not popular, however, with those who actually lived on what would be the frontlines of a war. With Block Island in smoking ruins, the Bay-men sailed next to Fort Saybrook. Lieutenant Gardener, who was still far from prepared for fighting, was at first incredulous, until he was shown Endecott's written orders. Then he became furious, later recalling, "For, said I, you have come hither to raise these wasps about my ears, and then will take wing and flee."

Guessing that diplomacy would fail and fearing the loss of the fort's cornfields, he begged Endecott to forgo the pretty baskets and other loot his men had stolen from Block Island and bring back the most crucial booty of all — food. "Seeing you will go," he told the commander with grim resignation, "I pray you, if you don't load your Barks with Pequits, load them with corn."

The five English ships sailed off to the east. As they hugged the coastline heading for Weinshauks, the fortified Pequot town on the Thames River, the Indians ran along the shore, shouting at the armed men with the universal greeting along the New England coast: "What cheere, Englishmen, what cheere?"

The soldiers watched in stony silence. "They not thinking we intended warre went on cheerefully until we came to Pequat river," Captain John Underhill wrote. Here the Indians set up a new and more worried chorus: "What English man, what cheere, what cheere, are you hoggerie, will you cram us?"

This meant, Underhill said, are you angry, will you kill us, and do you come to fight? Which is accurate enough, but the nuances of the pidgin word "cram" may have eluded him, since its derivation was Dutch, not English. It probably came from the word *kwalm,* meaning "dense smoke." The Pequot were asking, Have you come to kill us and burn us out?

Endecott wasted little time on niceties. At Weinshauks, his soldiers disembarked, formed ranks, and climbed the hill to the Pequot village. Furious negotiations followed, as the Indians tried to persuade the English to lay down their arms and parley. It smelled like a trap to Endecott's men and may well have been one. "Seeing that they did in this interim convey away their wives and children, and bury their chiefest goods . . . we rather chose to beat up the drum and bid them battle," Underhill wrote. The English columns opened fire, and the rest of the day was Block Island all over again — the Pequots fled with their families while the attackers burned everything they could reach. Many Pequots were injured, and at least one was killed. With acrid smoke hanging thickly over the estuary, Endecott sailed off with "a pretty quantity of corn" in his holds.

Gardener needed the extra stores, for the Indians' reaction was quick and bloody. Having thrown his weight around, Endecott may have considered the matter closed, expecting the Pequot to humbly accept their beating and honor English demands. But from the Indians' perspective, the English had declared war.

The Pequot "slew all they found in their way," killing settlers along the Connecticut frontier, including men Gardener had sent to bring in as much of the corn harvest as they could. The fort's granary was burned, and a sortie to bring in badly needed hay ended with three men killed and a fourth tortured to death. The fort was besieged by land, and although ships could still land there, food was running short, as Gardener had long feared.

Gardener himself had a close brush with some Pequots that winter, ambushed while he and his men were bringing in logs. Gardener kept his men in a half-moon formation, slowly retreating under heavy fire as the wounded limped to keep up, "defending ourselves with our naked swords." Arrows bounced off Gardener's buff coat, a heavy leather armor common in the seventeenth century, but one penetrated deep into his thigh. The Pequots, seeing him hit with so many arrows, were later shocked to find him still alive.

Not everyone was as lucky. Gardener eventually recovered the body of one dead Englishman, a Pequot arrowhead halfway through a rib. He carefully cleaned the bone and sent it, arrowhead still embedded, as a stark token to the Bay Colony, where some armchair experts had earlier "said the arrows of the Indians were of no force."

Not long after, a group of Pequots called for a parley with Gardener, asking him, "Have you fought enough?"

"We said we knew not," Gardener recalled. "Then they asked if we did use to kill women and children? We said that they should see that hereafter."

Gardener's reply silenced the Pequots for a moment. Women and children were captured but almost never killed in intertribal raids. Perhaps everyone sensed that the rules of frontier warfare were in flux.

No matter. "We are Pequits and have killed Englishmen," the warriors replied, "and can kill them as mosquetoes."

But in truth, the Pequot were running out of friends. Uncas and the Mohegan had cast their lot with the English, and Miantonomi and the Narragansett formally agreed to help the Bay Colony, sealing the pact with the gift of a Pequot hand.

Lion Gardener's small garrison held on through the winter, but the attacks—which had been focused on the fort—spread across the region in the spring. The settlement at Wethersfield, the most southerly of the English plantations and forty miles from the meager protection of Fort Saybrook, was attacked. For the first time, the victims included two women and a child, and two teenage girls were kidnapped. (Recognizing the military imbalance they faced, the Pequot were disappointed to find that the young women knew little about how to make gunpowder.)

Of the 250 English colonists living in the Connecticut River valley, 30 died in Pequot attacks. The Bay Colony leaders had already decided that the war, "having bene undertaken upon just ground, should be seriously prosecuted," but Connecticut moved even more quickly, drafting 90 men from its three plantations and dispatching them under the command of Captain John Mason.

Another veteran of fighting in the Low Countries, Mason had immigrated in 1632 to Massachusetts, where he harried pirates off the New England coast, then became a captain in the Dorchester militia. Mason moved to Connecticut in 1635 and was the logical choice to lead the colony's efforts against the Pequot—although not all of his colleagues shared Mason's splendid opinion of his own fighting abilities.

The Connecticut ships anchored at Saybrook, and when Gardener learned of their mission, he and Underhill were distinctly underwhelmed. Neither thought much of Mason, Gardener believing the Connecticut men "were not fitted for such a design." Nor did Gardener especially trust Uncas and his seventy Mohegans.

Mason was under orders to land his men on the Thames River and assault the nearby stockade at Weinshauks, but he worried that doing so would be disastrous. The girls kidnapped from Wethersfield, having been rescued by a Dutch trader, warned that the Pequot possessed a surprising number of guns. Instead of a frontal assault, Mason proposed they flank the Pequot by sailing far to the east, gathering allies among the Narragansett (who had been skirmishing for months with the Pequot), and coming at the fortified towns from behind.

Pinned down at Saybrook by contrary winds, Mason, Underhill, and the ever-cautious Gardener argued for several days. Written orders, which demanded a frontal assault, were not to be disregarded lightly, and finally Mason asked their chaplain to "commend our Condition to the Lord"—that is, to pray on the subject. By dawn, they had the chaplain's acquiescence, and after Gardener insisted that twenty of Mason's poorest soldiers be replaced by men from Fort Saybrook led by Underhill, the expedition set off.

Saybrook was under constant watch; the Pequot saw the pinnaces sailing east, passing Weinshauks, and mistook their departure for a retreat. For several days, there was relief and celebration in the Pequot villages, dancing and singing well into the night.

Mason's flotilla, meanwhile, fought poor winds, taking three days to sail the roughly fifty miles to the west shore of Narragansett Bay, where they met Miantonomi, who agreed to give the English and Uncas's Mohegan warriors safe passage. But the Narragansett sachem warned Mason that his numbers were too small. The Pequot, he said, were "very great Captains and Men skilful in War."

The weather was hot for early summer, and the English soldiers were laboring under their buff coats, leather neck stocks, and metal helmets. Lugging heavy matchlock muskets, they tramped eighteen or twenty miles west to a Niantic town, close to the Pequot border. The Niantic had been tributaries to the Pequot and were aloof enough to the English that Mason—unsure of their loyalties, and worried that word of the soldiers' approach would leak out—ringed the town with his forces, warning the Niantic not to leave their stockade "upon peril of their Lives."

In the morning, though, a number of Miantonomi's men arrived to join Mason, and soon the Englishmen were surrounded by a great circle of eager Narragansetts and Niantics, pledging to fight and boasting of the Pequots they would kill. The expedition marched off with nearly five hundred Indians leading the way.

The sun climbed high in a clear sky, even hotter than the previous day. Short on food and water, and baking inside their gear, the colonials began to faint from the heat. Mason used the fumes from an empty liquor bottle to revive some of them. Worse, as they forded the Pawcatuck River and passed through newly planted Pequot cornfields, he found that many of the Narragansett and Niantic warriors were slipping away. Those who remained told Mason that the two Pequot forts were "almost impregnable" and that they couldn't hope to reach

Weinshauks, where Sassacus lived, much before midnight. His men beaten down by the heat and short rations, Mason abandoned his plan to strike both towns at once and decided to focus on the nearest, Mystic Fort, which was home to another major sachem, Mamoho.

The English commander depended on the firsthand experience of Mohegans such as Uncas and the renegade Pequot Wequash. Also, Roger Williams had some months earlier forwarded intelligence from the Narragansett sachem Miantonomi to the Puritan leadership. If the Pequot saw English ships approaching, the sachem said, they would send their women and children to a swamp three or four miles away called *ohomowauke*, "the owl's nest." For this reason, he suggested exactly the kind of rear attack that Mason was undertaking. Miantonomi had also told Williams that the colonials should strike at night, "by which advantage the English, being armed, may enter the houses and do what execution they please."

Miantonomi did make one request, however, which Williams also conveyed: "That it would be pleasing to all natives, that women and children be spared, &c."

The English and Native forces marched now in silence, moving until an hour after nightfall and bivouacking less than two miles north of the Pequot village. The colonial pickets, listening in the warm moonlight, could hear singing in the town as the Pequot continued to celebrate what they assumed was the English withdrawal. It also appears the Pequot had earlier pulled most of their fighting men out of Mystic, concentrating them at Weinshauks, where the stroke was expected to fall.

Shortly before daybreak, the allied forces, led by Uncas and Wequash, marched south to the hill where Mystic Fort lay. The ranks split—Mason to the east, Underhill to the south—ringing the fort with soldiers backed by three hundred Mohegan, Narragansett, and Niantic warriors.

A dog began to bark, and from inside the fort, they could hear a man yelling "*Owanux! Owanux!*" (Englishmen! Englishmen!). The colonials fired a volley through the stockade walls—their muskets belching flames and smoke in the twilight—then hurried to clear the chest-high brush barricades that blocked the two entrances. Arrows began to flash through the air, pinging off helmets, as the soldiers poured into the village, slashing with their swords at anything that moved.

Inside the compound was chaos, screams, and blood. Although Uncas and his Mohegans had been given yellow headbands before the attack, few of the Narragansetts wore anything to distinguish them from the Pequots, and many were hurt or killed by mistake.

Mason later wrote that he and his commanders had "concluded to destroy them by the Sword and save the Plunder." But after the initial charge, in which about thirty Pequots were killed, the Indians' counterattack was simply too ferocious. "Seeing the fort was too hot for us, we devised a way how we might save ourselves and prejudice them," Underhill wrote. While Mason began to torch wigwams at one

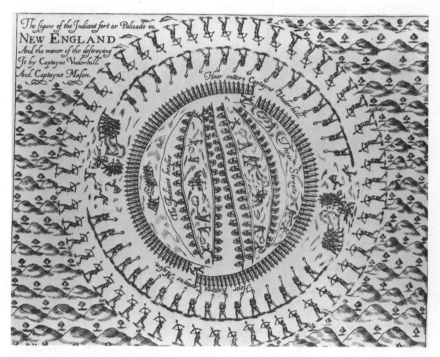

The Indians' Fort or Palizado
An illustration from Captain John Underhill's "Newes from America" (1638) shows Mystic Fort under assault by English and Indian forces. Inside the overlapping walls of the palisade, made of logs "as thick as a man's thigh . . . [and] ten or twelve foot high," about seventy wigwams go up in flames, while a wall of attackers prevents anyone from escaping. (Library of Congress LC-USZ62-32055)

end, Underhill shook out a line of gunpowder among the houses at the other, then set it alight.

"The fires of both meeting in the centre of the fort, blazed most terribly, and burnt all in the space of half an hour," Underhill recounted.

> Many were burnt in the fort, both men, women, and children. Others forced out, and came in troops to the Indians, twenty and thirty at a time, which our soldiers entertained with the point of the sword. Down fell men, women, and children; those that scaped us, fell into the hands of the Indians that were in the rear of us. It is reported by themselves, that there were about four hundred souls in this fort, and not above five of them escaped into our hands. Great and doleful was the bloody sight to the view of young soldiers that never had been in war, to see so many souls lie gasping on the ground, so thick, in some places, that you could hardly pass along.

Mason wrote that the fire "did swiftly over-run the Fort, to the extream Amazement of the Enemy," who despite the overwhelming odds tried to mount a counterassault: "Some of them climbing to the Top of the Pallizado; others of them running into the very Flames; many of them gathering to windward, lay pelting us with their Arrows; and we repaid them with our small Shot: Others of the Stoutest issued forth, as we did guess, to the Number of Forty, who perished by the Sword." Conjuring the language of the Psalms, Mason exulted that "God was above them, who laughed at his Enemies . . . making them as a fiery Oven."

From the first warning dog bark until the last blow fell, the massacre took barely an hour. Only two English soldiers died.

The Pequot were hardly broken, however. It was just five or six miles to Weinshauks, and soon the main body of Pequot warriors was attacking. The English, burdened by more than twenty wounded, some grievously, couldn't retreat. They'd made arrangements to meet their ships at the mouth of the Thames, just beyond Weinshauks, so they advanced through the heart of the Pequot homeland, falling into ambush after ambush. But as slow and clumsy as the primitive muskets were, the steady gunfire—combined with rear-guard skirmishing by Uncas's Mohegans—was enough to keep the Pequots largely at bay as the exhausted Englishmen continued to march southwest.

At the mouth of the river, they met a late-arriving company of Massachusetts soldiers under Captain Daniel Patrick, who was primed for an assault on Weinshauks. Mason overruled him. Underhill loaded the wounded onto the boats bound for Saybrook, while Mason and Patrick, guided by the Mohegans, took their combined forces another twenty miles west across Pequot and western Niantic land, burning what they could between the Thames and Connecticut rivers.

Mason, in his account of the fight, put the death toll at between six hundred and seven hundred people and said that only fourteen were captured or escaped. It was, by any measure, a slaughter to eclipse anything previously seen in New England, and the news shocked and sickened more than just "young soldiers" exposed to the butchery. The Mohegan and Narragansett warriors "cried Mach it, Mach it," Underhill recalled, "that is, It is naught, it is naught, because it is too furious, and slays too many men."

The Puritans hardly thought so. Mason was promoted to major, and the small army's actions were heralded in Plymouth and the Bay Colony. "It was a fearfull sight to see them thus frying in the fyre . . . and horrible was the stinck and sente ther of," wrote William Bradford, "but the victory seemed a sweete sacrifice, and they gave the prays therof to God, who wrought so wonderfully for them." Even Lion Gardener, who considered the war a mistake from the beginning, called the massacre at Mystic a "victory to the glory of God, and honour of our nation."

To Underhill, the logic was simple. "It may be demanded, Why should you be so furious? (as some have said)," he wrote. "Sometimes the Scripture declareth women and children must perish with their parents. Sometimes the case alters; but we will not dispute it now. We had sufficient light from the word of God for our proceedings."

The Pequot tried to regroup, but suddenly every hand was against them. Former tributaries such as the Mohegan, River Indians, and Montauk joined the hunt, sending scalps and heads back to the colonies. Some escaped to the tangled thickets of *ohomowauke*, the owl's nest swamp, where the rhododendrons were in bloom. Most were

killed or taken, and the unusual, blood-red hearts of the rhododen-
dron blossoms in *ohomowauke* were, for generations to come among
the Pequot, a reminder of their ancestors' deaths.

Harried by pursuing Mohegans and Englishmen, Sassacus and
the surviving sachems moved the Pequot west along the coast toward
Quinnipiac. A month and a half after the attack on Mystic, however,
eighty men and two hundred women and children were cornered by a
combined force of English soldiers and Mohegan, Narragansett, and
Long Island Indians in another swamp, called *munnacommock,* near
present-day Fairfield, Connecticut.

As dusk fell on July 13, Thomas Stanton, a colonist who spoke some
Pequot, negotiated the surrender of the Pequot women and children
and the local Indians with whom they had been sheltering. There was
no quarter offered for the Pequot men, who remained pinned down
through the night. At daybreak, they made a desperate counterattack, creating a diversion by which many escaped. Unable to flee, the
wounded were summarily executed by the English.

Sassacus and about twenty bodyguards sought refuge far to the
west, but the Mohawk, to curry favor with the English, made them
a gift of the sachem's head and hands. About three hundred Pequots,
almost all of them women and children, were reduced to slavery. Seventeen, mostly boys, were shipped to Bermuda but wound up instead
on Providence Island, a short-lived Puritan colony off the coast of
Nicaragua.

The rest, whom John Winthrop indicated were "women and maid
children," apparently remained in New England, parceled out among
the colonists. The Reverend Hugh Peter, a preacher at Salem, wrote
to Winthrop asking for his share. "Wee have heard of a dividence of
women and children in the bay and would bee glad of a share viz: a
young woman or girle and a boy if you thinke good." Winthrop was
happy to oblige.

A few Pequots managed to come through this ordeal and regain
their freedom, including the wife of the sachem Mononotto, of whom
Winthrop said he took particular care because she'd shielded the two
teenage girls kidnapped from Wethersfield. But few others enjoyed
such protection. One Pequot woman, for example, was branded in

"punishment" for having been raped. Branding, usually on the shoulder, was also a common sanction for runaways who were recaptured.

Roger Williams, who alone among the colonial leaders had taken the time to learn Algonquian ways, was deeply troubled by the treatment of the captives. "Since the Most High delights in mercy, & great revenge hath bene allready taken," he urged leniency for the Pequot, suggesting that they be kept subject to English oversight but "which they will more easily doe in case they may be sufferd to incorporate with the natives in either places." To make them useful, he recommended that the Pequot be charged an annual tribute in wolf scalps, "an incomparable way to save much cattell alive in the land."

Facing annihilation, the sachems leading the roughly two thousand Pequot survivors had little choice but to accede to a September 1638 treaty at Hartford between the English and their Indian allies. On the face of it, the treaty wiped away the Pequot as a nation. They "shall no more be called Peaquots but Narragansetts and Mohegans," their people parceled out between those two tribes. Nor would the victors "suffer them to live in the country that was formerly theirs but is now the Englishes by conquest." As for the Mohegan and Narragansett, who were already clashing along their expanded borders, they "shall not presently Revenge it But they are to appeal to the English . . . to decide the same."

In the short term, at least, both the English and the Indian victors made important gains in the aftermath of the war. Although Massachusetts and Connecticut bickered between themselves for years over control of the old Pequot land, the Pequot defeat had lessened the immediate danger to settlers along the Connecticut River. The Bay Colony had also established itself as a prime shaper of policy well beyond its immediate borders, and the war had brutally reinforced English influence as a whole in the region. If the English could crush the Pequot, many sachems doubtless decided, they could presumably crush any Indian nation that stood against them.

The Narragansett and Mohegan leaders, anxious to avoid a similar fate while vying for control of the wampum trade, prime hunting grounds, and tributaries, worked to cement the friendship of their nearest colonial neighbors—Uncas with the river colonies in Connecticut, and Miantonomi (with Roger Williams acting as a critical liaison) with the Massachusetts Bay Colony. John Mason and Uncas formed a particularly close partnership, which was often at odds with Connecticut governor John Winthrop Jr., who never fully trusted the Mohegan leader. In fact, treading the line between the Mohegan and Narragansett became harder and harder for the English as the two Indian groups came into increasingly direct and violent conflict—and as the English realized that the sachems were pursuing their own agendas, which did not always mesh with colonial desires.

His four-year contract at an end, Lion Gardener left Connecticut in 1639. He moved his family to *manchonake,* a three-thousand-acre island between the gaping jaws of eastern Long Island—"an Island of mine owne," as he described it. Careful of both English law and Native convention, Gardener secured a royal patent for it and also paid the Montauk sachem a black dog, rum, trading cloth, a gun, and powder. The name *manchonake,* or "island where many died," was an echo of a long-ago Pequot attack, and Gardener renamed his new home Isle of Wight. When his daughter Elizabeth was born there in 1641, she became the first English child born in New Netherland. Remarkably, his descendants still retain what is now known as Gardiners Island more than a dozen generations—and some messy legal battles—later.

The influx of defeated Pequots, meanwhile, greatly bolstered the communities into which they merged. This was especially true of the Mohegan. At the height of Pequot power, Uncas had barely a few dozen followers. Now he ruled 2,500 people—although like all seventeenth-century Native communities in southern new England, his was a "tribe" born as much of amalgamation and happenstance, and bound by adoption and marriage, as of any notion of traditional collective identity.

Already married to the daughter of the murdered sachem Tatobem, Uncas married one of the dead man's widows in 1638—one of several wives he would choose from among high-caste Pequot refugees. In 1638, speaking to the Puritan elders in Boston, Uncas was questioned about the presence of Pequots in his villages. There were none, he said, only Mohegans—even though at least one of the warriors standing with him had led Pequot attacks on Fort Saybrook. From Uncas's Algonquian perspective, both facts were true and not mutually exclusive.

Along with the conquered survivors allotted by treaty to the Mohegan and Narragansett, there came others who arrived in the Indian villages far more surreptitiously. With time, it seems, almost all the Pequots enslaved after the war and not shipped abroad escaped their English masters and melted away into the diaspora. By 1644, their experiment with Indian servants a failure, the Puritans began turning to African slaves instead.

Yet as complete as the Pequot defeat appeared in 1638, when even their name was illegal, they quickly reemerged as a distinct people. Not all of the Pequots were subsumed into the Mohegan and Narragansett. Some were settled at *nameag*, "the fishing place," on the western shore of the Connecticut River, which formed the border between the Mohegan and the old Pequot homeland. Although they were tributaries to Uncas, they exercised a degree of autonomy few other Pequots enjoyed, and with time the community grew—as did their sway with the Connecticut colonists.

The Nameag leader, Robin Cassacinamon, had been a servant in the Winthrop home years before—whether voluntarily or otherwise is unknown—but because of the family ties, Governor John Winthrop Jr. trusted him enough to place an English settlement, Pequot Plantation (later New London), near Nameag in 1646. Winthrop saw Cassacinamon as a foil to Uncas's growing power, and the Pequot in turn deftly used the English against his Mohegan overlord.

With time, Cassacinamon was able to pry his people free of Uncas entirely. In 1650, Winthrop granted them a small holding on the coast at Noank, in the old Pequot homeland, where they were directly beholden to the colonial government. Eight years later, when Noank proved too cramped for the growing community—and just two decades after the edict that they "shall no more be called Pequots"—Cassacinamon and his people were given a two-thousand-acre tract at Mashantucket.

Ironically, this renaissance of a former English enemy occurred even as an English ally, the Narragansett sachem Miantonomi, found himself dangerously boxed in. At war with the Mohegan, he had to fend off allegations, fostered by Uncas, that he and his people were negotiating with surrounding tribes for a mass uprising against the English. This time, the Bay-men were the brake, reining in Connecticut's fire-breathing plan to march against the Narragansett. But when Miantonomi, leading an army of a thousand men against Uncas, was captured by his old rival in 1643 and delivered to the English, the colonists saw a way to rid themselves of a potential danger. Miantonomi was handed back to Uncas, who tomahawked the sachem to death before English witnesses.

The dread among English colonists of a coordinated Indian rebellion, planted by Opechancanough in Jamestown and fanned again before the Pequot War, never faded. No one knows if Miantonomi was really trying to raise a pan-Indian army, but in less than a generation, the threat would come true, sparking the single bloodiest war ever fought on what would become American soil, as a leader would emerge to unite the tribes in a way no sachem had ever done. His name was Metacom, better known as King Philip.

Chapter 5

Between Two Fires

DAYBREAK, JULY 7, 1677. The fishing ketch *William and Sarah* out of Marblehead, Massachusetts, rode gently at anchor off Port La Tour at Cape Sable, on Nova Scotia's southern tip. Its crew of half a dozen was still asleep, enjoying a brief respite from several weeks of hard fishing in the cold waters of the Gulf of Maine; the ship's hold was half full of cod, some of the fish longer than a man's leg. The ship's master, Joseph Bovey, planned to spend two days at La Tour, taking on water and firewood before heading back to sea.

One of the fishermen, a Marbleheader named Robert Roules, "aged thirty years or thereabouts," heard something and, in the growing light of dawn, peered over the side of the two-masted vessel. A canoe with nine or ten Indians had drawn alongside, their muskets primed and cocked. They saw him and shouldered their weapons. Roules instinctively dropped flat onto the deck as a ragged fusillade of gunfire ripped through the early-morning air, the lead balls slamming harmlessly into the ship's windlass.

"What for you kill Englishmen?" Roules shouted in the pidgin English of the frontier, as the Indians scrambled aboard.

"If Englishmen shoot we kill—if not shoot, we no kill," one of the warriors yelled back, as Bovey and the three other half-dressed fishermen staggered awake. Some of the Indians' muskets were still loaded, and the ship's guns were too far away. The crew, having no choice, surrendered.

As the tense drama aboard the *William and Sarah* played out, the frontier from the Chesapeake to Maine had already endured two years of war, after long-simmering tensions finally erupted in wide-scale violence. It was bad enough in Virginia, where tit-for-tat killings between a plantation owner and local Doeg Indians quickly escalated into a major regional confrontation with the neighboring Susquehannock, who were attacked even though they had nothing to do with the original fight.

But worse by far was the regional maelstrom that engulfed New England. The United Colonies of England (Massachusetts, Plymouth, New Haven, and Connecticut, which had confederated in 1643) faced their greatest fear—a mass uprising of tribes throughout southern New England. King Philip's War, named for the Wampanoag sachem who led it—Metacom, son of the early Pilgrim ally Massasoit and known to the colonists as Philip—was arguably the most murderous conflict ever waged on American soil, and one that saw both Europeans and Natives adopting the techniques of the other, forging the violent framework of frontier war that would endure for generations.

King Philip's War was a civil war, although no one quite agreed on what *kind* of civil war. To the English, it was a treasonous uprising by royal subjects, whose leaders had signed treaties swearing their allegiance to the Crown. In Puritan eyes, the fact that some Praying Indians—Christian converts—joined the fray added another level of perfidy.

To Philip and his followers, the fight was a just insurrection, years in coming, against outside invaders. But for many Algonquians, it was even more vividly a civil war in the purest, brother-against-brother sense, one that pitted members of the same village or the same family against one another. Loyalty to either side was sometimes bought with the head of one's own kin.

The revolt against the English started among the Wampanoag in 1675, but it quickly spread, first to other Algonquian tribes in southern New England, such as the Nipmuc, then north to many of the Abenaki, such as the Sokokis and Cowasuck. Gone was the traditional restraint of Algonquian warfare—the limited casualties, the focus on captives for adoption or torture, the sparing of women and children. The Pequot War and the massacre at Mystic Fort had rewritten the rules on both sides.

Most of the Indian attacks on frontier homes and settlements were meant to drive home a simple, terrifying message: leave the land or lose everything. This was proving especially effective in Maine (then part of the Massachusetts Bay Colony), where the widely scattered farms and villages had been all but obliterated and most of the remaining colonists forced onto outer islands, where the sea provided at least a little protection from the war canoes.

Although Roules and the others did not yet know it, the seizure of the *William and Sarah* was something altogether different—the first step in a bold Indian plan to carry the fight against the interlopers into colonial England's very heart, using the Europeans' own naval technology against them. The brainchild of a Sokokis war leader named Mugg (or Mogg) Heigon, the concept was simple but audacious. First, the Abenaki would capture dozens of the English fishing and trading vessels that still plied the waters of northern New England with impunity. With the ships, the Indians would be able to reach "all the fishing ilandes and so to drive all the contre befor them," as one English captive named Francis Card overheard Mugg say. After all, the "Eastern Indians," as the Europeans called the coastal Wapánahki, had been sailing European shallops for at least a century.

But that was only the beginning. "[H]e doth make his brag and laf at the english," Card reported, "and saith that he hath found a way to burn boston." In his fleet of captured ships, packed tight with fighting men, Mugg and the Abenaki would sail into an unsuspecting Boston Harbor without raising suspicions, then torch the city.

Before waging a ferociously effective war against the English, Mugg and another Sokokis sachem, Squando, were advocates for peace with the colonists, but their patience had been worn down by years of abuses and provocations. For Squando, who had been converted to Christianity by the Puritan missionary John Eliot, the final straw was intensely personal. Two Englishmen, rowing a boat near his riverside village, encountered Squando's wife and young son in a bark canoe. To test the widespread belief that Indian infants could, like animals, swim instinctively, they tipped the canoe into the cold water. Squando's wife managed to rescue the child and claw her way back to safety, but the boy died the following day. Squando had had enough of the English.

Although Mugg was killed before his plan for a naval assault was put into action, the Indians moved ahead with it. As the capture of the *William and Sarah* showed, the fishing boats—which, like that ketch, sailed mostly from Marblehead—were fairly easy prey, especially when the crew landed, as they had to do, to cure their catch. While their fellow Bay-men were outraged at the losses, they laid some of the responsibility on the Marbleheaders themselves—"a dull and heavy-moulded sort of People," without "either Skill or Courage to kill any thing but Fish."

In all, some twenty vessels were snapped up in short order. On the *William and Sarah,* Robert Roules found himself and his fellows tied up on the deck, stripped of most of their clothes except for "a greasy shirt and waistcoat, and drawers we used to fish in." By midafternoon, the Indians unbound the men and ordered them to sail for Pentagoet, the French outpost on Penobscot Bay, in company with four other ketches they'd taken. Otherwise, they would kill all but a handful of the twenty-six sailors onboard the tiny fleet. Chafing the blood back into their numb limbs, Roules and his companions sulkily agreed.

Two days later, the flotilla came upon a bark outbound from Boston, which, seeing nothing amiss in five Marblehead ketches sailing together, allowed them to range alongside. When Roules and the others hollered across the water demanding the boat's surrender, the captain thought it was a grand joke. Then the seventy or eighty armed Indians, who had been hiding aboard the other ships, rose to their feet, and the laughter died in his throat.

The Indians divided the crews among the various boats, bringing a man named Buswell from the bark aboard the *William and Sarah* and sending Bovey to another ship. They also reduced their own forces on the ketch, leaving just four, including two men that Roules referred to as sagamores.

Not long after taking the bark, the Natives spotted another distant sail, and the fast-growing squadron took off in pursuit. Roules was at the *William and Sarah*'s helm, and as it grew dark, one of the sagamores ordered him to bear up for the night. He refused, and as the old man began to shout, Buswell grabbed him by the throat. A melee broke out—as Buswell held down one sagamore, the crew clapped the scuttle over the cook room hatch, trapping the other. In the ensuing scuffle, the remaining two Indians were tossed over the side. Tying up the survivors, Roules had the crew set a course back to Marblehead.

When they arrived there almost two weeks later, they were met by a mob of women, frantic for news of the missing ships and clamoring for revenge. Many of the women were refugees from elsewhere in New England, their homes burned and their families dead or missing. Why were the Indians still alive? the women demanded to know. Roules and his mates said they'd lost everything, "even to our clothes, and we thought if we brought them in alive, we might get somewhat by them toward our losses." They would lead the bound men ashore, they said, and deliver them to the constable, "that they might be answerable to the court at Boston."

The women were having none of it. As Roules and the others tried to lead the Indians off the ketch, the mob began pelting the fishermen with stones, forcing them to flee. Then the crowd literally tore the two sagamores to pieces. "We found them with their heads off and gone, and their flesh in a manner pulled from their bones," Roules said. The women "suffered neither constable nor mandrake, nor any other person to come near them, until they had finished their bloody purpose."

To the English, who viewed with haughty superiority the stories of Indian women torturing captives, the Marblehead riot must have seemed like another step toward barbarism and anarchy in their own society. The wars of the late seventeenth century, of which King Philip's was the most wrenching, took place against a backdrop of two societies deeply worried about their own identities—a disturbing, in-between world in which they increasingly reflected each other.

To fundamentalists like the Puritans, life on the frontier, where colonists learned Indian woodcraft and farming techniques, thriving far from the reach of the church, posed a threat to the colonists' essential Englishness, never mind to their individual souls. "Christians in this Land, have become too like unto the *Indians* and then we need not wonder, if the Lord hath afflicted us by them," wrote the Puritan cleric Cotton Mather. One local magistrate, immigrating to the coast of Maine from England, was aghast at "the Immoralities of a People who had long lived Lawless."

If the English were worried about becoming too Indian, so were many Native leaders distraught to see how European their own people were growing. Indians were increasingly dependent on the new tools, technologies, and trade goods the colonists provided, moving them steadily farther from their traditional lifestyle. For a rising number of Indians, this change included their religion, too. Protestant missionaries such as John Eliot in Massachusetts and the Catholic orders in New France found ever-growing ranks of converts. Metacom complained through the members of his council that so-called Praying Indians "wer in everi thing more mischivous, only disemblers, and then the English made them not subject to their kings, and by ther lying to rong their kings." For literate, Christian Indians such as a Nipmuc named Wawaus—who had taken the name James Printer, had attended Harvard, and now worked as a typesetter for Cambridge's first printing shop—the growing chasm between their Indian heritage and the European culture they embraced left them mistrusted and outcast by both sides, with nowhere to turn that didn't carry the risk of imprisonment or death.

The violence that broke out with King Philip's War and continued for decades was marked by appalling atrocities on both sides, from the Indians' slaughter of entire homestead families down to the youngest infants, to a midwinter English massacre on a Narragansett town whose astonishing brutality was only amplified by the fact that the Narragansett were not at war with the English at the time. There were also occasional, oddly incongruous courtesies. When the survivors of the French and Indian attack on Deerfield, Massachusetts, stumbled away from their burning homes and barns in February 1704, they found a leather bag hanging in a tree along the main road into town. "Some of our captives in Canada, knowing the enterprize that was on foot, sent several letters unto their friends, which the enemy did carefully put into a bag, and hung it upon the limb of a tree in the high-way," wrote Judge Samuel Penhallow, noting that the letters "gave satisfaction to those that were then alive among them."

Through much of the seventeenth century, what had started as a fragile coexistence between Indians and Europeans—a wary bond aided by commerce, intermarriage, and convergent interests—was breaking down. The frontier had never been peaceful, of course. Long before the Vikings or Basques arrived, the Indian nations of eastern North America had been locked in an endless cycle of raids and reprisals among hereditary enemies. The onslaught of an alien culture, eager to trade sophisticated weaponry and household goods for skins, furs, or slaves, altered and intensified that dynamic, as well as sparking many conflicts with the intrusive, often violent, newcomers themselves. But whereas Native leaders had at one time been more likely to view Europeans as potential allies or trading partners, they now increasingly saw them as a direct and often paramount threat to their own culture's continued survival.

The whites simply kept coming. What began as a handful of coastal and riverine outposts had consolidated, by 1650, into a broad swath of seaboard under European control. In the north, the French occupied forts in Nova Scotia and Cape Breton in Mi'kmaq country, around the Bay of Fundy in Maliseet-Passamaquoddy territory, and up the St. Lawrence to Ville-Marie (Montreal), where the St. Lawrence Iroquoian villages among whom Jacques Cartier had stayed had vanished to an unknown fate.

On the New England shoreline, English colonies stretched from Pemaquid, Maine, through Boston, Plymouth, and Providence to the Connecticut coast, steadily displacing Algonquian groups such as the western Abenaki, Wampanoag, Narragansett, Nipmuc, Pequot, and Mohegan. By one estimate, there were 23,000 English settlers in the region by 1650. The Dutch held New Netherland on Long Island, Manhattan, and the Hudson Valley north to Fort Orange at present-day Albany, land that had belonged to the Munsee, Mahican, and Mohawk, among other nations. The Swedes had established themselves along Delaware Bay among the Lenape and Nanticoke.

Along Chesapeake Bay, it was Englishmen again, Virginians and Marylanders. When John Smith had explored the bay less than half a century earlier, there had been fifteen thousand to twenty thousand Indians living there, but now there were only a few thousand, their numbers in continuing decline from disease, and almost all of them had been pushed back from the fertile coastline. The farms that now stood along the tidewater grew tobacco for export. The land was tilled by English indentured servants, as well as a growing cadre of three hundred to four hundred African slaves.

From the Georgia Bight across northern Florida, and in almost forty missions and forts strung out along both Florida coasts, the Spanish remained in firm command, having put down the Guale uprisings in the late sixteenth century. The Timucua, who once numbered 200,000, could muster barely one-tenth that number by 1650. The only region where Native control of the seaboard remained relatively intact was the Carolinas, despite Spanish slaving raids and epidemics. But the diseases, like the colonists, kept coming—malaria, smallpox, influenza, and typhoid, among others—and when English immigrants finally created permanent settlements along the Carolina coast in the 1650s and 1660s, they found a land that, like Massachusetts thirty years before, had been largely emptied by pestilence. (They also imported laborers from farther south, employing Indian slavers to kidnap Timucuas from Spanish missions.)

By the time Indians woke up to the threat posed by waves of land-hungry settlers, they were often politically and militarily too weak to react decisively. Powhatan, near the peak of his power, almost succeeded in pushing the Virginians into the sea in 1610, but Opechancanough's subsequent assaults proved to be years too late for lasting success. Farther north, the callous destruction of the Pequot had cowed many sachems and sagamores into an uneasy accommodation with the New England colonists. Yet by the 1660s and 1670s, despite rapidly expanding control of the seaboard, many colonists were more frightened than ever that an Indian leader would emerge to unite the tribes against them, sweeping away what they still saw as a fragile English beachhead.

Phillip Alias Metacomet of Pokanoket
By far the most widely reproduced "portrait" of King Philip, this image is based on
an engraving by Paul Revere, who in turn lifted most of its elements from a painting of the eighteenth-century Mohawk leader Joseph Brant. (Library of Congress
LC-USZ62-96234)

It is doubly ironic, then, that Metacom—Philip—was the son of Massasoit, the Pokanoket sachem without whose help the Plymouth colony would not have survived, and who struggled for years to find a way to balance the needs of the Wampanoag confederacy with the demands of their increasingly importunate neighbors. Massasoit's death around 1660, followed quickly by that of his elder son, Wamsutta (allegedly by English poison), brought Philip to power just as the Wampanoag's situation was reaching a breaking point.

The Wampanoag had their backs increasingly against the wall. Whereas Massasoit saw the English as allies and trading partners, Philip and a growing number of like-minded sachems saw them as a plague. English birthrates were high, and shiploads of white immigrants kept arriving, pressuring the Plymouth and Bay Colony leaders, who in turn badgered the Indians to sell ever more land.

Worse for the Wampanoag, the beaver trade was flagging even as Native demand for European goods was increasing, leaving more and more Indians in debt. Selling land was frequently the only way out, and in Massachusetts, unlike in Maine, the Wampanoag had a long list of grievances involving fraudulent dealings, such as Indians made drunk so their land could be swindled from them. The Indians tried moving away from English settlements, but there was nowhere to go. The Pokanoket land around Mount Hope in what is now Rhode Island was ringed by English settlements radiating out of Plymouth, Newport, Portsmouth, and the upper reaches of Narragansett Bay.

Philip and his representatives "saied they had bine the first in doing good to the English, and the English the first in doing rong," wrote John Easton, the Quaker deputy governor of Rhode Island, who gave an evenhanded account of complaints on both sides in the run-up to war, and who tried to mediate the disputes. Easton had his own ax to grind, however. His family had been banished from Massachusetts for religious differences, and he had little love for the United Colonies. What's more, from its founding by Roger Williams, the Providence Plantations, which would become Rhode Island, had earned a reputation for relatively straight dealing with the Indians. Although Rhode Island stood apart from the United Colonies, Easton knew that any armed conflict would embroil everyone.

The Wampanoag and other Algonquians tried to seek redress through the English legal system but found the door slammed there, too; in Plymouth, the leaders banned Indians from town when court was in session, to reduce the flood of Native litigation about land cheats or livestock trampling crops. Indeed, the English couldn't even agree among themselves on issues of land; Massachusetts, Plymouth, and Rhode Island were constantly locking horns over boundaries and property sales.

If English law wouldn't respond to the Indians' pleas, there were other means of forcing the issue. The Wampanoag quietly armed and organized during the winter of 1674–75. The English, getting wind of this from another Harvard-educated Christian Indian, a minister named John Sassamon who often acted as a scribe for Philip, also prepared for war.

The fuse was lit on a frigid day in February 1675. Sassamon, his gun, and some dead birds were pulled from a frozen lake in Plymouth—ostensibly an accident, except that injuries to his neck suggested foul play. A coroner's inquest ruled it murder, and the general assumption was that he'd been killed for having told the colonists about the Wampanoag's plans. On scanty evidence, a jury of twelve Englishmen and six Praying Indians convicted three of Philip's associates for the crime, and they were hanged on June 8. The Indians massed; the English leadership began frantic negotiations with Philip, as well as with their allies among the Mohegan, Narragansett, and Nipmuc. As the expectation of violence grew, some settlers abandoned their homes.

Exactly how the war started isn't clear. Easton wrote that on June 24, several Indians were found rifling an empty house. As they fled, a boy with a gun fired at them, killing one. "The next day the lad that shot the indian and his father and fief English more were killed." Others place the outbreak of violence elsewhere, but the result was the same. As Easton said simply, "So the war begun with philop."

The conflict, which rapidly widened to encompass Connecticut, Rhode Island, and Maine, also pulled in most of the Algonquians in the region, willingly or otherwise. Many allied themselves with Philip, and the colonists saw every Native hand raised against them. "In this time of Extremitye Wee have Great Reason to beleeve that there is a universall Combination of the Indians," one settler wrote to Connecticut governor John Winthrop Jr.

How much of this was genuine fear, and how much was motivated by a desire to gain control of supposedly hostile Indian land, is an open question, but there was by no means a "universall" uprising. Some longtime English allies, such as Uncas and the Mohegan, eagerly fought alongside the United Colonies. So did the remnants of the Pequot, whose fort at Mashantucket served as a refuge for white and Indian residents alike. But loyalties were not split neatly along tribal or even family lines. The large Wampanoag population on Cape Cod, Martha's Vineyard, and Nantucket, many of whom had converted to Christianity, turned en masse against Philip. Many Pocassets and Sakonnets joined him, despite the initial misgivings of their sachems, while other Wampanoags, who had listened to English instructions to stay out of the fray, were nevertheless subsequently attacked, killed, or sold into slavery.

Many of the Nipmucs in Massachusetts needed little prompting to join Philip; they had been close allies of his father, Massasoit. But other Nipmucs, such as James Printer, had become Praying Indians, and as the war spread, they found themselves between two fires—mistrusted by both sides, shunned, murdered, or captured. James Quanapohit (also spelled Quannapaquait or Quanapaug), a Nipmuc from the Praying town of Natick, and his brother Thomas served the English for months, fighting bravely in several engagements; Thomas even killed one of Philip's chief commanders and carried the man's head to Boston to present to the governor. They were instrumental in recovering one English captive from Philip's forces and in saving the lives of two other Englishmen.

"Yet after all these services both they & their wives & children . . . were mistrusted by the English," Quanapohit later complained. By November 1675, the English were rounding up Praying Indians with only a few hours' notice. Quanapohit, his family, and the rest of the Christian Indians of Natick were herded into a wretched internment camp on windswept Deer Island in Boston Harbor, "leaving & loosing much of their substance, cattle, swine, horses & corne, & at the Iland were exposed to great sufferings haveing little wood for fuell, a bleak place & poore wigwams such as they could . . . make themselves with a few matts." Many of the internees starved or froze to death. A few, such as Tantamous (Old Jethro), managed to escape.

Their own people didn't treat them much better. At the same time as the Deer Island roundups, Nipmucs allied with Philip fell on Hassanemesit and several other Praying towns, presenting the residents with an equally forbidding choice: join Philip and fight; flee to the English, who would certainly imprison them or sell them into slavery; or be laid waste, their corn supplies burned.

"If they came to the English they knew they shold be sent to Deare Iland, as others were . . . and others feard they should bee sent away to Barbados, or other places," James Quanapohit lamented. "And to stay at Hassanamesho these indians our enemies would not permit . . . Most of them thought it best to go with [Philip] though they feared death every way."

That was the decision facing James Printer, whose brother was a prominent minister. Although James had worked for years as a translator and typesetter for the Cambridge printer Samuel Green, he had been imprisoned by the English in Boston earlier in the autumn, accused of taking part in an attack. Escaping a lynch mob, he fled to Hassanemesit, but having evaded one set of combatants, he now had little choice but to join the other side, willingly or not. The same option confronted another Christian Nipmuc, Tantamous's son Peter Jethro.

Philip's camps were populated by Algonquians of many tribal backgrounds and with deeply conflicted attitudes about the war. Many viewed it as Philip did, an overdue accounting for wrongs inflicted by invaders on a sovereign people, but others held the colonial view that Philip was a traitor and rebel against the English Crown to which he owed allegiance. Still others, regardless of their politics, focused on the consequences. One veteran of the wars against the Mohawk tried and failed to assassinate the Wampanoag leader, "alleaging that Philip had begun a warr with the English that had brought great trouble upon them."

In fact, Philip wasn't much of a leader, in the conventional sense. There is little evidence that he personally commanded more than a handful of attacks, nor that he was responsible for large-scale strategy among widely spread forces. Philip was a spark and became a symbol, but he wasn't a general. The revolt he began among the Wampanoag spread to the Nipmuc and Abenaki, but each tribe, each community, and each sachem was fighting his own war.

Still, the English were reeling, beset by Wampanoag and Nipmuc attacks from the west and Abenaki assaults to the north in Maine. They realized that if the Narragansett—the largest and strongest Indian force in the region—entered the war against them from the south, they were likely doomed. Although the Narragansett had been somewhat weakened by the execution of Miantonomi—a killing sanctioned by the English because they'd feared exactly this kind of mass uprising—the dead sachem's vigorous, charismatic son Canonchet could still field a large number of warriors.

Yet if the Narragansett were not firm English allies, like the aging Uncas and the Mohegan, they definitely did not want war with the colonists. Many Narragansett men worked as day laborers, building stone fences and doing other farm work. While some young men were certainly preying on English farms, their leaders were trying to preserve peace, and in October 1675, the Narragansett sachems signed a treaty restating their neutrality toward the United Colonies.

But the Narragansett also sheltered many Wampanoag refugees, and the English accused some Narragansett men of joining the fight. Rumors swirled of Narragansett treachery, some perhaps planted by Uncas as he had years earlier against both the Pequot and the Narragansett. The lessons of the Pequot War must have weighed heavily on Canonchet and his sachems, especially the way the Pequot had been targeted for reasons that were at least partially trumped-up. Ultimately, many of them withdrew to a massive palisaded village—an immense fort deep in the Great Swamp near modern South Kingstown, Rhode Island, where they hoped to avoid trouble.

The fort, which stood on an island of higher ground, must have been an imposing structure, enclosing four or five acres, in and around which lived an estimated three or four thousand people. It was walled with log palisades and surrounded by a thick barricade of brush, inside of which was a four- to five-foot-deep moat. It was further protected, like English forts, with "flankers," projecting structures above the walls that allowed defenders to fire down onto attackers. Inside, there was at least one fortified blockhouse that gave the defenders a clear line of fire onto the only entrances—two logs that bridged the moat on opposite sides.*

Nor were forts the only aspect of European defense the Narragansett and other tribes adopted. When Connecticut soldiers stormed Mystic Fort in 1637, their muskets were too much for the arrows of the Pequots. By the time King Philip's War broke out, most of the Indians were as well armed as the colonists—in some cases, given their insistence on flintlocks rather than the older, cheaper matchlocks, better armed—and they had the expertise, forges, and tools to make limited repairs and cast their own ammunition. It was illegal for an Englishman to fix a gun for an Indian, but some Natives now could do the work themselves. They lacked gunpowder, however, something so difficult to manufacture that even the colonists were dependent on imports from England until just before the war.

As autumn turned to an unusually hard winter, the English demanded that Wampanoag refugees be turned over to them, as the treaty required, but the Narragansett—perhaps feeling safe within the Great Swamp fort, about which the English knew nothing—refused to violate such a fundamental tenet of Algonquian hospitality. Canonchet declared that he would "not deliver up a Wampanoag, or the paring of a Wampanoag's nail."

The Bay Colony and Plymouth were ready for war. Informants had warned them that Canonchet had decided to join Philip in the spring. Connecticut wavered but finally agreed to send men. Faced with an invasion of Rhode Island territory that he did not condone but could not prevent, Roger Williams declined to contribute any soldiers, but he reluctantly agreed to allow passage of the United Colonies' force of nearly a thousand English soldiers and militiamen, as well as several hundred Mohegans and Pequots under the command of Uncas's son Owaneco.

Converging by land and sea near Wickford, the Massachusetts and

---

* Though not as large, a second, much more secret stronghold near what is now Exeter, Rhode Island, was in some respects even more impressive. Known as Queen's Fort, for the female sachem Quaiapen, it was the work of a Narragansett mason named Stonewall John or John Wall Maker, who may also have helped design the Great Swamp fort. He built Queen's Fort not from wood but from rock, elaborating on a naturally occurring hilltop system of caverns among glacial boulders, which he strengthened with walls and passageways.

Plymouth men set fire to any Narragansett village they encountered, killing and taking what captives they could. One of the Massachusetts captains seized the opportunity to buy forty-seven of the new Indian captives, "young and old, for 80£ in money." The Indians fought rearguard skirmishes to slow them down.

Bolstered by the soldiers from Connecticut and their Mohegan-Pequot allies, the combined army turned toward the Narragansett fort, about which they'd learned only a few days earlier. Clouds thickened and the temperature dropped. With no shelter, the soldiers hunkered down in the open the night of December 18 as one of the worst storms in local memory dumped up to three feet of snow. Setting out in the predawn darkness, the troops marched west fifteen miles, struggling through deep drifts, the hands and feet of the men freezing through their gloves, stiff boots, and cloth wrappings. Their commanders would likely never have found the fort had they not stumbled a few

Capture of the Indian Fortress
A typically aggrandizing nineteenth-century depiction of the Great Swamp Fight, in which colonial forces storm one of two log bridges across the frozen moat surrounding the Narragansett's elaborate fortification before going on to burn it and most of its defenders. Ostensibly English allies, the Narragansett were accused of sheltering Wampanoag refugees and taking part in Philip's uprising, which had fueled English fears of a universal Indian attack. (Library of Congress LC-USZ62-97115)

days earlier on a Narragansett turncoat named Peter, who agreed to guide them. By afternoon, they were in place, and the Narragansetts, who'd had ample time to prepare, were waiting.

Although the swamp, on whose impenetrability the tribe had depended, was frozen solid, as was the moat, the Narragansett fortification and the stiff resistance of its defenders blunted the first English attacks. Dozens of troops and a number of commanders were killed, including John Mason Jr., the son of the man who had led the bloodbath at Mystic more than forty years earlier. In fact, this new offensive was a replay of that Pequot massacre, only on a larger and still more violent scale—with the Narragansetts as victims instead of English allies.

The English, Mohegans, and Pequots torched the outlying houses, then stormed the palisades. "They within on the first onset stoutly repulsed us," an anonymous letter writer recounted nine days after the attack. "But . . . every one put forth his utmost Strength, and on the renewing of the Assault we became Masters of the place." Flames inside the fort forced the Narragansetts to flee directly into English guns. "We no sooner entered the Fort, but our Enemies began to fly, and ours now had a Carnage rather than a Fight, for every one had their fill of Blood: It did greatly rejoice our men to see their Enemies, who had formerly skulked behind Shrubs and Trees, now to be engaged in a fair Field, where they had no defense but in their Arms, or rather their Heels; But our chiefest Joy was to see they were mortal, as hoping their Death will revive our Tranquility, and once more restore us to a settled Peace."

Far from being a "fair Field" of combat, it was a sickening slaughter. According to the anonymous witness, "We have slain of the Enemy about 500 Fighting men, besides some that were burnt in their *Wigwams,* and Women and Children the number of which we took no account of." Captain James Oliver of Boston put the toll at "300 fighting men; [of] prisoners we took, say 350, and above 300 women and children. We burnt above 500 houses, left but 9, burnt all their corn." The English suffered more than two hundred casualties, including many who died in the snow on the long, bitter retreat back to the coast. But the expected counterattack, which could have been devastating, never came. By one account, the young Narragansett men were restrained by the surviving sachems because they had almost no powder for their guns.

In the days after the fight, some colonists pursuing Indians captured a white man among them, a fellow named Joshua Tift or Teffe, "a Renagado English Man." Tift, who farmed not far from the Narragansett fort, claimed he had been captured and held as a slave. But Oliver and others insisted that Tift—"a sad wretch, [who] never heard a sermon but these 14 years"—had been a willing fighter for the Narragansett and "shot 20 times at us in the swamp." Summarily tried by a "Counsel of War" and convicted of treason, he was sentenced to that most medieval of executions, drawing and quartering. On January 18, he was hanged, cut down while still alive, hacked apart, and left unburied, his head impaled on a gatepost by way of a warning.

The butchery visited on the Narragansetts in the Great Swamp Fight, the massacre of homestead families by Indians, and the shocking sentence meted out to Tift show that restraint of any sort, on every hand, had evaporated. In terms of the proportion of the population killed, King Philip's War still ranks as the bloodiest conflict ever waged on American soil, and one of the nastiest. The English death toll may have been 2,500, some "destroyed with exquisite Torments, and most inhumane Barbarities," and entire towns—Narragansett, Marlborough, Deerfield, and others—were burned to the ground.

"It is computed by the most judicious Men, That the Indians that were killed, taken, sent away . . . cannot all (Men, Women and Children) amount to fewer than Six Thousand, besides vast Quantities of their Corn, Houses, Ammunition . . . taken and destroyed," the Boston merchant Nathaniel Saltonstall wrote. By "sent away" he meant sold into slavery, for the Puritans saw the captives as instruments of both profit and policy. When soldiers killed 166 Narragansetts in 1676, mostly women and children, and took another 72 alive to be sold as slaves, a Puritan minister noted approvingly that the children were "all young Serpents of the same Brood."

John Hull, a wealthy Boston merchant, did a sporadic business in Indian slaves, buying captives hauled in by colonists and selling them on the domestic and Caribbean markets. On one occasion, his journals noted the purchase of 110 Indians for a pound or two apiece, while on another, he noted the sale of "13 Squawes & papooses wounded 1 sick" for £20—about $450 in modern currency.

Some Praying Indians also took part in the slave trade, although more than a few Christian Algonquians were themselves hauled into slavery, their religion and loyalty notwithstanding. Even the unwary household servants of Puritan masters were at risk, should they fall into an unscrupulous slaver's hands. The prospect of slavery was always on the minds of those fighting the English. When in the winter of 1675 some older men among Philip's movement counseled peace, many of the young men said it was better to go down fighting. "Why shall wee have peace to bee made slaves?" they asked. "Let us live as long as wee can & die like men, & not live to bee enslaved."

For many in Philip's camp, the time for accommodation was over. "Know by this paper, that the Indians that thou hast provoked to wrath and anger, will war this twenty one years if you will," read one extraordinary note written by a literate Algonquian attacker and left jammed in a bridge post outside the smoldering remains of Medfield, Massachusetts. "You must consider the Indians lost nothing but their life; you must lose your fair houses and cattle."

According to historian Jill Lepore, "Always brutal and everywhere fierce, King Philip's War . . . proved not only the most fatal war in all of American history, but also one of the most merciless," leaving New England "a landscape of ashes, of farms laid waste, of corpses without heads." Modern scholars estimate that 5 percent of the English population in the region, and an astounding 40 percent of the Indians, died.

In the wake of the Great Swamp Fight, the Indians settled down into winter camps of traditional wigwams, built of sapling frames covered with reed mats and slabs of bark. They had a gunsmith to keep their weapons in good repair, and although ammunition was low, they were able to trade beaver pelts and wampum to their traditional enemies, the Mohawk, for some. Food was the biggest concern. They augmented their dwindling supplies of dried corn with cattle and pigs taken on raids, and deer were easy prey in the thigh-deep snow. But it was clear to Philip and his leaders that once these supplies got too low, they would have to hit colonial outposts or starve. One of the most vulnerable towns was Lancaster, located about twenty-five miles west of Boston along the Nashaway (now Nashua) River.

English traders had been operating in the Nipmuc village of Nash-away as early as 1643, drawn by its location at the intersection of the north–south river and the east–west Nipmuc Trail, both major trans-portation routes for furs coming from the interior. When Nashaway Plantation was incorporated as the town of Lancaster, there were only nine English families there, living cheek by jowl with the Nipmuc village. By 1676, however, Lancaster had grown to fifty or sixty white families.

As the winter deepened and the Indian larders shrank, Philip planned his attack on Lancaster. But unbeknownst to him, the English had spies in his camps. James Quanapohit, the Christian Nipmuc who with his brother had fought for the English only to be imprisoned on Deer Island, and another Nipmuc internee named Job Kattananit had volunteered to seek out Philip's group and report back to colonial au-thorities. Kattananit was desperate to find his three children, who had been taken by the anti-English forces at the Praying town of Hassa-nemesit, and this was his only opportunity to look for them. Both men were taking an enormous chance, especially Quanapohit. He knew that Philip had ordered his capture or death because he'd fought with the English. Still, risking their lives as spies may have been preferable to enduring the slow, freezing misery of Deer Island.

Armed with just their knives and tomahawks, and with only a little parched cornmeal for provisions, Quanapohit and Kattananit traveled five or six days through the snow and cold of early January, infiltrat-ing a Nipmuc camp at Menameset, near modern Brookfield, Mas-sachusetts. Kattananit was reunited there with his children, who were among more than a hundred Praying Indians, including the typesetter James Printer, who had been forced to leave their homes.

The two spies learned about Philip's growing shortage of food and the planned attacks on Lancaster and other outlying towns, but as he feared, Quanapohit immediately aroused suspicion among some of the sachems. It's likely that what saved him was the public protection of a prominent warrior known as John with One Eye, alongside whom he'd fought in the past. "Thou hast been with mee in the warr with the Mauhaks and I know thou are a valiant man," John told Quanapo-hit. "None shall wrong thee nor kill thee here, but they shall first kill me."

Still, many of the Nipmucs were suspicious of his apparent change of heart and pressured him to go to Philip so that the sachem could question him directly—an interview Quanapohit knew he was unlikely to survive. Instead, he stalled for time. He said he'd be happy to go to Philip, but not until he'd had a chance to prove himself. "Hee will not beeleve that I am realy turned to his party, unles I first do some exployt & kill some English men & carry their heads to him." That seemed to satisfy the Nipmucs, but it was clear that the two spies had to escape soon with their information.

Quanapohit and Kattananit told the villagers they were going hunting, something they'd done previously. Quanapohit asked one of the Praying women for a pint of dried cornmeal, called *nokake*, before heading into the woods. Several times, as they pursued and killed four deer, they crossed human tracks in the snow; they were being watched, although they never caught a glimpse of those who were shadowing them. Nervous, the two men settled down for the night near a pond, dressing the deer and building a fire against the cold.

No one appeared; the woods seemed silent and empty. Quanapohit decided it was time to make his escape. Kattananit, however, chose to remain with his children, telling Quanapohit he would try to slip away with his family later. "If God please hee can preserve my life," he told his friend. "If not I am willing to die."

With the *nokake* and some deer meat in his pouch, Quanapohit traveled through that night and the following days, resting only briefly in order to reach Boston as soon as possible. Arriving at the end of January, he brought news of the imminent attack on Lancaster to the Massachusetts council, in Boston—which looked askance at the intelligence, brought as it was by an Indian. They sent no troops to bolster the town, and on February 10, 1676, a mixed force of roughly four hundred Nipmucs, Narragansetts, and Wampanoags attacked Lancaster at daybreak.

Although no reinforcements reached Lancaster, word of Quanapohit's warning did. Many of the residents had gathered in blockhouses, or garrison houses, a cross between a farmhouse and a fort. Usually positioned on high ground and built of thick, hand-squared beams with loopholes to shoot through, blockhouses had a second floor that overhung the first by several feet all the way around, so that water and gunfire could be poured down on attackers trying to torch the structure. Some had flankers rising from the corners that gave defenders a wide field of fire, and most were surrounded by a sturdy timber stockade with a heavily barred gate as a further perimeter of protection.

Undefended homes around Lancaster quickly went up in flames, but the garrison houses withstood the first assaults. Forty-two people had jammed inside the blockhouse of the Reverend Joseph Rowlandson, where they held off a sharp attack for more than two hours. (Rowlandson himself was in Boston, trying to enlist more soldiers for protection.) They might have held out until relief arrived, as did the other garrisons in the area, but only one of the two roof flankers was finished, and the rear of the house had been stacked high with firewood, further blocking defensive gunfire from inside. Using combustible flax and hemp they found in the barn, the Indians were eventually able to slip up and set fire to the building.

"Some in our house were fighting for their lives, others wallowing in blood, the House on fire over our heads, and the bloody Heathen ready to knock us on the head, if we stirred out," Rowlandson's thirty-eight-year-old wife, Mary, later recounted. "As soon as we came to the door and appeared, the *Indians* shot so thick that the bullets rattled against the House, as if one had taken a handfull of stones and threw them." But with the flames roaring and many of the defenders already dead, they had no choice but to stagger out. Mary Rowlandson was struck in the side by a bullet, which also hit her six-year-old daughter, Sarah, whom she was holding. Her nephew William was killed, and his mother, despairing, said, "Lord, let me dy with them," moments before she fell down dead from a musket ball.

As the fight sputtered to an end, Rowlandson and the others "[stood] amazed, with the blood running down to our heels." One victim, tomahawked in the head and stripped naked, was still trying to crawl away. Others lay dead of spear or gunshot wounds. Rowlandson had often said that she would rather die than be taken alive by Indians, "but when it came to the tryal my mind changed."

Of those inside, only one escaped. Twenty—among them Mary and her three children, including the badly injured Sarah—were taken captive; the rest died in the attack or its immediate aftermath.

For more than three days, the war party moved across the snowy landscape. Most of the English captives would have been bound with long "prisoner ropes," woven of tough dogbane fibers, their ends ornately decorated with tin bangles or dyed quillwork, or led by similar halters designed to be tied around a captive's neck, like a leash. Rowlandson and her wounded daughter, however, often rode bareback on a horse, with young Sarah crying out, "I shall dy, I shall dy," in fever and pain. They settled with the Nipmucs at Menameset, the village where Quanapohit and Kattananit had been spying. Six days later, Sarah died. Rowlandson, her own wound poulticed with dried oak leaves, survived.

King Philip's War marked several firsts in frontier history, among them the first widespread uprising of many Indian groups against a common enemy and the first time large numbers of English captives were taken by raiding parties. Captivity, slavery, and adoption of captives were, of course, traditional among the Algonquians, but this time Philip and his allies held white captives for ransom, a technique that would become common in the century to come.

Having been sold by the man who captured her, Mary Rowland-

Capture of Mary Rowlandson
This woodcut frontispiece from the 1770 edition of Mary Rowlandson's book, *The soveraignty & goodness of God*, is one of several that appeared in eighteenth-century editions showing her armed or bravely fighting off the attackers—something she never claimed to have done. (© American Antiquarian Society)

son was held for the next three months by a Narragansett sachem named Quinnapin and his wife, Wetamo (or Weetamoo). Wetamo was the widow of Philip's older brother, Wamsutta; the sister of Philip's wife; and a Pocasset sachem in her own right—a *sauncksquûaog,* or "squaw sachem," one of the hereditary female leaders among the Algonquians,* such as those who had led Wampanoag communities on Martha's Vineyard and Nantucket.

Pursued by the English, the Indians moved constantly, once burning their wigwams behind them when the army was breathing down their necks, carrying their sick and elderly on their backs through the forest. They gleaned stray grains of corn and wheat from abandoned fields and scavenged old bones "full of wormes and magots . . . scald[ing] them over the fire to make the vermine come out." In her later account, Rowlandson speaks often of her own hunger and privation, but it's clear that no one was eating well. Eventually, they joined Philip's larger forces—"I could not but be amazed at the numerous crew of Pagans"—and Philip himself asked her to sit and smoke a pipe. She primly refused, having quit tobacco in her captivity, seeing now that it was "*a bait, the devil lays to make men loose their precious time.*"

Compared with some captives, Rowlandson had a fairly easy time of it. She specifically noted that none of the men made any attempt to assault her, and she even set herself up in a minor business, parlaying her skills as a seamstress into food and the odd coin. She knitted stockings and sewed garments and caps, including a shirt for Philip's little boy. She lobbied to be taken by horse to Albany and traded for

---

* The Narragansett word *sauncksquûaog* was recorded by Roger Williams in 1643, along with other variants, including *squaw* for "woman" and *keegsquaw* for "young woman." What is essentially the same root word, variously rendered as *squa, esqua, skwe,* and *skwa,* appears in most Algonquian languages. For example, a female werowance, or subchief, in Virginia was a *werowancequa.* In its original use, *squaw* was a simple descriptive term for a woman, with no derogatory connotation.

"Squaw" has become an enormously controversial and heavily freighted word, however, thanks to centuries of racist misuse. What's more, a mistaken but persistent myth—that the word actually stems from one syllable of a crude, multisyllabic Mohawk term for female genitals—has taken root in recent years, despite rebuttals to the contrary from both professional linguists and Algonquian activists. (These include Smithsonian linguist Ives Goddard in "A True History of the Word 'Squaw'" and Margaret Bruchac in "Reclaiming the Word 'Squaw' in the Name of the Ancestors.") They point out that "squaw" was used in the early seventeenth century by English colonists recording Massachusett and other Algonquian languages—Europeans who had no contact with Iroquois speakers hundreds of miles inland. Regardless of the demeaning term it has morphed into today, "*squa* was an ancient and thoroughly decent word," Goddard wrote.

gunpowder, but she had no thoughts of trying to escape overland, later explaining, "I was utterly hopeless of getting home on foot . . . I could hardly bear to think of the many weary steps I had taken to come to this place." It was just as well, because when Quinnapin and Wetamo gave her permission to visit her son in a village just a mile away, she became so lost that she had to return to her village, find Quinnapin, and have him show her the path.

While Rowlandson was trying to convince Quinnapin and Wetamo to ransom her, the council in Boston was working toward the same goal for all the English hostages—if they could find someone to carry the offer to Philip's forces. They solicited messengers from the Praying Indians on Deer Island, who—no real surprise—showed little enthusiasm for helping the people who had incarcerated them. Eventually, however, one of the Deer Island captives, a man known as Tom Dublett or Tom Nepanet, agreed to make the trip, carrying a letter from Massachusetts governor John Leverett addressed, "For the Indian Sagamores & people that are in warre against us." Leverett asked for a written response, "if you have any among you that can write."

They had at least two—Peter Jethro and James Printer, the two Christian Nipmucs who had been among the Praying Indians captured by Philip's forces the previous autumn. First Jethro and then Printer served as scribes for the sachems, as Dublett and a second Christian Indian, known as Tatatiquinea or Peter Conway, shuttled back and forth between Boston and Philip's camp at Wachusett Mountain. After several weeks of negotiations, they settled on the rather princely sum of twenty pounds for Rowlandson. The other captives, not ranking as high socially as a minister's wife, would be ransomed later. "Though they were *Indians,* I gat them by the hand," she said of Dublett and Conway, "and burst out into tears; my heart was so full that I could not speak to them."

Mary Rowlandson was freed—"redeemed" was the usual term—in early May, just as the trees were leafing out, although none of the other captives, including her surviving children, accompanied her. Philip's luck, by contrast, was running out, as was that of his allies. The Narragansett sachem Canonchet, trying to steal corn for the spring planting, had been captured a month earlier by a group of English volunteers, Narragansetts, Pequots, and Mohegans. Like his father, Miantonomi,

Canonchet was executed by his enemy sachems. Holding wide his silver-trimmed coat in defiance, he was gunned down by Uncas's son Owaneco and the Pequot leader Robin Cassacinamon, who then carried his head to Hartford.

Stonewall John, who had built Queen's Fort and perhaps the Great Swamp fort as well, was captured by the Mohegan and tortured to death in July, his fingers and toes cut off as he stoically bore the pain, his legs broken, and his skull split. Rowlandson's former master, Quinnapin, was captured and tried by a court-martial in Newport "for these sundry crimes following, namely: for being disoyall to his said Majesty [in] sundry Ways . . . through thy wicked bloody Minde and trayterous, rebellious, roietous and routous Acts." With other "rebellious" Indians, he was executed by firing squad in August. Meanwhile, Wetamo and a small party with her were ambushed, and the *sauncksquûaog* drowned while trying to escape across a river. She, too, was decapitated, her head impaled on a pike in the town of Taunton.

A week later, Philip's wife, Wootenekanuske, and nine-year-old son were captured, and ten days after that, a Christian Pocasset named John Alderman, who had deserted Wetamo's band early in the war and was soldiering for the English, shot Philip himself. The sachem's body was quartered, and Alderman was given the head and one of Philip's hands, which bore distinctive scars. Alderman turned in the head for the standard bounty of thirty shillings, and Philip's bleached skull was displayed on a pole in Plymouth for decades thereafter. Alderman retained Philip's hand, which he kept in a bucket of rum "to show such gentlemen as would bestow gratuities on him; and accordingly he got many a penny by it."

In the wake of the war, the Algonquians of southern New England scattered like leaves. Many, like those among whom Rowlandson was held, were captured and shipped off to the Antilles as slaves, and the lands of the Nipmuc, Narragansett, and Wampanoag were seized by the English. Philip's son, and perhaps also the sachem's wife, were sold into West Indian slavery after Puritan leaders such as Increase Mather couldn't *quite* bring themselves to hang the boy for his father's sins—though they considered it, long and hard, while the lad languished in prison for months.

Rowlandson, by contrast, was soon reunited with her surviving children, Joseph and Mary. She was widowed and remarried just a few years after her release. Her story, "made Publick at the earnest Desire of some Friends," was published in 1682 under the title *The soveraignty & goodness of God, together, with the faithfulness of his promises displayed; being a narrative of the captivity and restauration of Mrs. Mary Rowlandson.*

Rowlandson's book was a landmark in several respects. It became the first American bestseller, and Rowlandson became the first American woman author published in North America within her lifetime. It was also the first of many captivity narratives, a literary form that would endure for the next two and a half centuries.

Above all else, *The soveraignty & goodness of God* was a religious testament, and in many ways the Puritan ideal. Virtually everything that befell Rowlandson during her captivity, good or ill, large or small, from her daughter's horrific death to losing two cobs of corn, is held up as a divine lesson and vital cog in God's plan. At times in the book, it seems that King Philip's War itself was staged primarily for the spiritual edification of Mary Rowlandson. When the Indians narrowly escaped the pursuing English army across a frigid river, "we were not ready for so great a mercy as victory and deliverance; if we had been, God would have found a way for the *English* to have passed this river." And when her feet remained dry when she stepped onto the raft that carried her across that river, it could not "but be acknowledged as a favour of God to my weakened body."

It is typical of the paradoxes of the First Frontier that Rowlandson was rescued from Indian captivity through the services of two Indian negotiators, with two others acting as scribes. Tom Dublett was still dunning the Massachusetts council eight years later for compensation for undertaking "that Hazardous service" on the council's behalf. He asked for twenty or thirty shillings; instead, he was finally given two coats "in sattisfaction for his paynes & travile." Peter Jethro was accepted back by the English, having first turned in his father, the sachem Tantamous, who on September 26, 1676, was marched through the streets of Boston with several other Indian leaders and hanged. Even Increase Mather had a hard time swallowing that example of familial treachery, referring to the son as "that abominable Indian."

But it is James Printer who remains the most inscrutable enigma here. When the Praying town of Hassanemesit was overrun by Philip's warriors and Printer went with the attackers, he could have been either an unwilling captive making the best of a horrible situation or a willing convert to the cause; no one knows which. Historians have viewed his role in the Rowlandson negotiations as both enlightened self-interest and selfless service. For her part, Mary Rowlandson had no use for Christian Indians, including "James the printer," considering them to be savage turncoats, despite the fact that four of them were instrumental in her release.

Two months after helping to secure Rowlandson's redemption, Printer was given a safe-conduct to return to Boston—if he was willing to make a public display of his loyalty. In most cases, this meant fighting alongside the English against Philip's diminished uprising. Instead, Printer was given specific instructions, relayed from the council by Daniel Gookin, superintendent for the Praying Indians. Printer complied, arriving in Boston with the required "two heads," which usually meant scalps, as surety of his allegiance. Whose they were and how he took them, no one knows. Still, many in Massachusetts found his professions of loyalty hollow, however many scalps he produced. "A Revolter he was," one Boston letter writer sniffed, "and a fellow that had done much mischief."

Printer would never be free of the taint of suspicion from either side. After the war, he took up preaching at Hassanemesit and resumed his work typesetting books in Cambridge. One was an Algonquian translation of the Bible. Another, in a surpassing irony, was *The soveraignty & goodness of God,* by Mrs. Mary Rowlandson.

In southern New England, Philip's death and the collapse of the Wampanoag resistance effectively marked the end of the war and of Algonquian political and military muscle there. The situation was much different in northern New England. Because King Philip's War was not a coordinated, centrally commanded offensive, other tribes continued their own feuds with the English without interruption.

Along the coast of Maine, where colonial settlements were thinly spread along rivers and bays, the English were especially vulnerable to attack. Initially, the Penobscot sachem Madockawando managed to keep most of his people out of the war, ignoring highhanded English demands that the Abenaki and other Natives unilaterally disarm and even deciding to overlook English slavers—supposedly hunting for renegades in league with Philip—who took the far easier route of simply kidnapping peaceful Wapánahkis from Machias Island.

When war did finally erupt along the Maine coast, English fears about vulnerability were realized. Scattered and far from protection, settlements were razed one after another. "Honoured Mother," wrote Thaddeus Clark of Casco Bay in June 1676, bearing tragic news:

> On Friday last in the morning your own Son with your two Sons in Law Anthony & Thomas Bracket & their whole families were killed & taken by the Indians, we know not how, 'tis certainly known by us that Thomas is slain & his wife & children carried away captive, and of Anthony & his familie we have no tidings & and therefore think they may be captivated . . . There are of men slain, 11, of women & children 23 killed & taken, and we that are alive are forced upon McAndrew's his Island to secure our own . . . We are so few in number that we are not able to bury the dead, till more strength come to us.

It was against this backdrop that Mugg Heigon's plan for an Abenaki naval sweep of the outer islands, followed by an assault on Boston, was taking shape. Whether such a strategy would have succeeded had Mugg himself not been killed in battle, we are left to guess, but after the initial successes in snatching up fishing boats—thirteen, including the *William and Sarah,* from Salem and Marblehead alone—the scheme went sour. Some of the sailors and ships managed to slip free on their own, and influential Salem merchants pressured the Massachusetts council to send north an armed ketch with forty men onboard, which succeeded in retaking the remaining vessels.

Boston was safe from seaborne attack, but by the time hostilities petered out in 1677, the "Eastern Indians" had, unlike their southern cousins, won a significant victory, all but clearing Europeans from the coast of Maine. Instead of solidifying these gains, however, the eastern Abenaki sachems (many of whom had not been active participants in the war) looked at their own decimated numbers, and at the ever-present Iroquois threat to the west, and decided they could use the newly chastened English as a bulwark against the Mohawk. They invited the colonists to return—setting by treaty what they considered inviolable conditions, including the recognition of Indian authority (with annual tribute payments of corn) and promises to remain in the lower, tidal zones of the major rivers, leaving the vast inland areas to the Natives.

How seriously the English took these promises is debatable. Within a short while, Indians were again protesting intrusions, land thefts, and expanding white settlements. War would come again soon—but this time, the reason would not be merely the sad old cycle of escalating offenses and grievances.

King Philip's War was a regional conflict fought for local reasons, as had been most of the previous wars on the First Frontier. But as the seventeenth century drew to an end, the underlying rationale for war on the eastern frontier shifted. All the earlier triggers of strife remained—indeed, in many cases grew more pronounced—but increasingly, the fundamental forces driving each side to war were global as well as regional. Monarchs and empires in Europe—forging and breaking alliances, jockeying for wealth, sending forth armies—would for the next seventy-five years ignite war after war thousands of miles away on the American frontier.

But for now, there seemed to be a chance for peace. Wanalancet, the sachem of the Pennacook in New Hampshire, had managed to steer his people clear of the worst of King Philip's War, following the course of neutrality his late father, Passaconaway, had followed for forty years. It wasn't easy. Nearly sixty, Wanalancet (who had reluctantly agreed to be baptized a few years earlier) showed Job-like patience in the face of English arrogance and double-dealing.

In the fall of 1675, for instance, he got word from his scouts that an English force led by the notorious Indian hater and privateer Captain Samuel Moseley, one of those who had slaughtered the Narragansetts in the Great Swamp Fight, was approaching. Finding the Pennacook camp along the upper Merrimack deserted, Moseley ignored his orders and burned the village, including its crucial winter supply of dried fish, before heading back to Massachusetts—never realizing his men were constantly shadowed by Pennacook warriors, who could easily have ambushed them. Instead, Wanalancet, who had moved his people out of harm's way before the English arrived, held his men in grim check, then retreated with his band still farther back into the mountains, where they scraped by on scarce game until regional hostilities simmered down in the summer of 1676.

Moseley's raiders were not the only people looking for Wanalancet. Like many neutral groups, the Pennacook accepted refugees from war-torn areas farther south, including many Nipmucs and others who had fought with Philip. In July of that year, Wanalancet signed a peace agreement with the English stating in part "that none of said Indians shall entertain at any time any of our enemies, but shall give present notice . . . when any one come among them." When Wanalancet made his mark on the treaty, he may have already been sheltering a number of former combatants, as well as many simply trying to get away from the fighting. No piece of paper, however, was going to prevent him from fulfilling what he saw as his duty to abide by common decency.

As the English leadership had already made clear in its attack on the neutral Narragansett, they did not share the Algonquian view that hospitality was an obligation. After Philip's death and the collapse of the uprising, they dispatched soldiers to hunt down fugitive fighters. Two companies were sent north to the settlement that was still known in those days by its western Abenaki name, *cochecho* or *qouchecho*, "rapid foaming water," now Dover, New Hampshire. Commanded by Captains Joseph Syll and William Hathorne, the 130 soldiers had orders to "seize all Indians." Instead of a direct confrontation, Major Richard Waldron, who had traded with the Pennacook for years at Cochecho, suggested a different approach—an elaborate trap, one for which he would eventually pay dearly.

Waldron invited the Pennacook and their guests to a parley under a flag of truce. The idea, the Indians were told, was to stage a mock fight, enroll the Algonquians into English service, and then celebrate the peace with food and drink. "Sham battles" had been a common form of public spectacle since medieval days, and everyone turned out for the festival, including roughly 350 Indians.

The Abenaki men were arrayed in their finest regalia, which showed a mix of traditional and European influences—moose-skin moccasins, wool blanket robes, cloth shirts in such magically potent colors as dark blue and red, woven sashes, breechclouts and leggings decorated with the distinctive double-curve designs of Abenaki quillwork or embroidery of dyed moose hair and beads. Some of the women would have been wearing high-peaked cloth caps, of a design based on French styles, and long skirts made of skins or trade cloth. Tin bangles, silver brooches, and even English shillings, drilled to hang as ornaments, flashed from the garments of men and women alike. On an early-autumn day, the field in front of Waldron's blockhouse trading post at Cochecho was certainly a kaleidoscope of color, movement, and sound.

The English soldiers, including two companies of local militia, formed ranks and drilled across the open land, even helpfully offering a cannon to the hundred or so Abenaki men, who grabbed the towlines and pulled it onto the field, not noticing when the English gun crew repositioned it to cover the Indians. As planned, the Abenaki "attacked" first, all the men firing their single-shot muskets into the air in one loud, harmless volley. Suddenly, their own weapons empty, they found themselves facing soldiers with loaded guns leveled at them and realized that the cannon, too, was aimed in their direction. There are reports that the cannon actually went off, killing some of the warriors, but most accounts say that Waldron pulled off his ruse without a shot being fired. The Natives were "taken like so many silly herring in a net," one smug New England historian later wrote. But many of the English living at Cochecho were uneasy about the deception, worried that a fire was now smoldering.

Only a few Pennacooks escaped, including a boy who was sheltered by a woman named Elizabeth Heard—an act of kindness that would have profound implications for her years later. The rest, including Wanalancet, were hauled off to Boston. Although he and most of the rest of the Pennacooks were eventually released, almost two hundred of the refugees living with them were sold into slavery, and eight directly implicated in the war were hanged.

Waldron—a trader and soldier much respected by the colonists, but who had an unsavory reputation among the Indians for thievery—may have truly believed that he was doing his Pennacook neighbors a favor, since a direct confrontation would have led to many deaths. The Algonquians didn't see it that way. As a result of Cochecho, and another incident the following winter in which Waldron and an English contingent killed several Wapánahki sachems under another flag of truce, the old major became a marked man. Wanalancet would go to his grave trying to live in peace with the English, but his nephew, a fiery war leader named Kancamagus, would not—and Waldron was among the first items on his long list of debts to be settled.

Chapter 6

"Our Enimies Are Exceedeing Cruell"

I T WAS THE kind of falsely hopeful dawn that winter sometimes brings: a bright start to the day, air so calm that a person's breath hangs before his face in a lingering fog, the first shafts of weak orange light peeking through the dark spires of the spruce. But a treacherous day: the clear sky quickly crowded out by clouds muscling in from the south to clot the horizon. A day that forces the hapless traveler to realize he may have made a serious mistake by straying far from shelter and a warm fire.

John Gyles sniffed the air, smelling the damp threat of snow, and pulled his worn moose-pelt jacket tighter. If he was suspicious of the weather, it was with good reason, because he was schooled beyond his years in how to stay alive when others had not. Barely a teenager, he was an English boy a long way from his home on the coast of Maine, having been captured in 1689 at age ten and taken to a Maliseet village in Canada, where he was kept as a slave. He'd watched other captives tortured; had close friends die; endured hardship, hunger, and beatings; and resigned himself more than once to his own death. But by luck and gumption, he'd survived.

In summer, Gyles and his captors lived near the palisaded village of Meductic, along the lower St. John River, growing corn and squash, gathering berries and groundnuts, fishing for salmon and sturgeon. But with the coming of winter each year, the extended family that owned Gyles left for their northern hunting grounds, moving constantly from camp to camp across thousands of square miles of what is now northern Maine, New Brunswick, and southern Quebec, rebuilding bark wigwams in which to live while the men hunted moose and bear.

A moose is a huge animal, with bulls weighing twelve hundred pounds or more. Sometimes, instead of carrying the meat to camp, the hunters would cache it at the kill site, trusting to the midwinter cold to keep it sound. Some days earlier, the hunters had killed several moose far from camp, and Gyles and a Maliseet boy had been sent to fetch some of the meat.

The morning weather had quickly soured, however, the clouds piling up thick and fast, as cold and gray as a pewter plate, and the temperature falling steadily. The two young men snowshoed all day through the deep drifts, but twilight had fallen by the time they reached the cache. It was too dark even to build a shelter; the best they could do was quickly gather wood for a small fire. The snow began lightly at first, then developed rapidly into an all-enveloping storm that roared throughout the night. The boys huddled around their sputtering fire, shivering as their bodies tried to keep them warm, their clothes growing sodden and heavy with melting snow.

In the morning, they shouldered their packs full of moose meat and began retracing their steps. In the bone-chilling wake of the storm, their wet clothing froze. "My Moose-skin Coat (which was the only Garment I had on my Back, and the Hair was in most Places worn off) was frozen stiff round my Knees like a Hoop, as likewise my Snow-shoes and Snow-clouts to my Feet," Gyles recalled years later.

The boys trudged on, the pain in Gyles's feet growing extreme, until at last the agony subsided into weary, numb relief. His Indian companion, who had better clothing and had suffered less from the elements, quickly left him behind, and Gyles moved haltingly through the silent white world alone. He was gripped with bouts of debilitating nausea, his empty stomach heaving, but there was no way to cook the frozen moose meat he was carrying. Periodically, his energy would desert him, leaving him almost unable to move, but then he would rally and feel "wonderfully reviv'd again" for a short time. Before long, he would be consumed once more by a potent desire to lie down and sleep.

Gyles knew he was freezing to death, just as his closest friend, an English captive named John Evans, had done the winter before. Weak from hunger, his legs gashed by the icy crust through which he repeatedly fell, Evans had been left behind by the rest of his master's band. Exhausted, he'd curled up in the snow with a dog in his arms. When the Maliseets found them the next morning, both he and the dog were frozen solid.

It would have been simple to lie down, Gyles recounted, "but again my Spirits reviv'd as much as if I had receiv'd the richest Cordial." Hours after dark, his erstwhile companion long since home and warmed by the fire, Gyles staggered into one of the Maliseet lodges with his snowshoes still strapped to his blue, ice-cold feet. "The Indians cry'd out, *The Captive is froze to Death!* They took off my Pack, and where that lay against my back was the only Place that was not frozen."

The Maliseets cut off Gyles's moccasins and put him by the fire. As the sensation returned to his frostbitten feet, "which were as void of feeling as any frozen Flesh could be," they began to swell and turned purplish black, erupting in blood-filled blisters. The pain was excruciating.

The Maliseets had seen it before and knew what it portended. "The Indians said one to another, *His Feet will rot, and he will die,*" Gyles recalled. Within days, the skin sloughed off below his ankles, "whole like a Shoe, and left my Toes naked without a nail, and the ends of my great Toe-bones bare, which in a little while turn'd black, so that I was obliged to cut the first joint off with my Knife."

At this point, the Maliseets knew, victims of deep frostbite can contract gangrene. But Gyles was a survivor. His captors gave him rags with which to bandage his feet and told him to treat the tender flesh with gum from the balsam fir. As a slave, he was responsible for his own care, so he scuttled about the snow on his rear end, pushing himself with sticks from tree to tree, gathering gum that he melted over the fire to make a salve.

Within a week, he could walk gingerly on his heels, supporting himself with a staff. By the time the band was ready to move a week after that, they had made two round snowshoe-like hoops for his feet, on which he could hobble behind them. "I follow'd them in their Track on my Heels from Place to Place; sometimes half [a] Leg deep in Snow & Water, which gave me the most acute Pain imaginable, but I was forced to walk or die."

Just as Gyles survived winter storms and frostbite, he survived the other dangers of a captive's life, and years later he wrote a remarkably candid account of his time with the Maliseet. More than most captive narratives, Gyles's story provides a window into a period of tremendous flux, offering harrowing glimpses of the ways in which average people on both sides were caught up in the ever-widening war along the frontier.

After King Philip's War, the northern frontier had about ten years of relative peace. The English were slow to return to the coast of Maine and New Hampshire, but by the 1680s enough had resettled there that tensions were again on the rise. The colonists were refusing to pay the annual tribute in corn that they'd agreed to, and one settler had strung so many nets across the Saco River that he'd all but blocked off the salmon run on which the Indians depended. There were "the Common Abuses in Trading *viz.* Drunkenness, Cheating, etc.," said the Cochecho minister Reverend John Pike, but once more, the biggest complaint was the English habit of stealing land, "at which [the Indians] were greatly Enraged, threatning the Surveyor to knock him on the Head if he came to lay out any Lands there."

Nor were the conflicts only with Natives. The governor-general of New France, the Comte de Frontenac, looked at the rapidly growing English presence to the south and began to wonder whether he should be at least as concerned about them as about their allies the Iroquois, who were stifling fur trade as far away as the western Great Lakes and raiding to the very outskirts of Montreal. One logical counterweight would be to further strengthen French ties with the Abenaki, especially since Frontenac had what should have been an ace in the hole: a former French officer married to the daughter of the powerful Penobscot sagamore Madockawando.

But Madockawando and the other Abenaki leaders were in a tight spot themselves. Although they had forged many bonds with the French, from intermarriage to religion (a few of the Catholic priests not only preached to them but also went to war with them), the Wapánahki depended on the resurgent English settlements and their supply lines to Boston for most of their trade goods. This was an increasingly precarious middle ground for the Indians to occupy, especially as France and England squared off more and more aggressively over control of their colonial possessions.

By 1687, shots were again being fired in anger. Tired of Iroquois depredations on western Indians, the French laid waste to Seneca villages in New York, ratcheting up tensions between New France and the English colonies. The next year, Indians from Canada struck into New England. Massachusetts buzzed with expectations of further raids, and the council in Boston ordered the capture of Indians engaged in hostilities. Unfortunately, commanders on the Maine frontier simply kidnapped about two dozen Wapánahkis, mostly women and children but also including a sagamore named Hope Hood, and shipped them to Boston. The Indians retaliated by taking sixteen English hostages, and before the royal governor, Edmund Andros, could negotiate an exchange, there was a skirmish in mid-September that resulted in the execution of several of the English prisoners.

Full war broke out not long thereafter; it must have seemed like a reprise of King Philip's War to victims on both sides. But King William's War, as it became known, was different—the first of many violent, intertwined feuds between France, England, and Spain as the three empires wrestled for dominance in the Old World and the New, even as their Indian allies maneuvered for their own benefit and defense.

In fact, what played out as King William's War in North America was simply the local manifestation of the Nine Years' War in Europe—also known as the War of the League of Augsburg or the War of the Grand Alliance, since it pitted the French king Louis XIV against the English, Dutch, and Spanish and the Holy Roman Emperor Leopold I. For the first time, imperial struggles thousands of miles away spilled directly into a new theater of war in the forests of America and Canada, and over the next three-quarters of a century, scarcely a year would pass without war or the threat of war as European armies, colonial militias, and their Indian allies contested for the First Frontier.

Wanalancet, the Pennacook sagamore who had tried for so long, and in the face of so many provocations, to keep peace with the English, had finally left the Merrimack Valley, moving to the western Abenaki village of St. Francis in Canada. As the war clouds gathered, his nephew Kancamagus assumed an important leadership role, and he was not inclined to turn the other cheek. None of the Pennacooks—and few Wapánahkis anywhere along the frontier—had forgotten the way they had been cheated and deceived, especially by Richard Waldron.

It wasn't merely the outrageous deceits, such as the sham fight thirteen years earlier or the killing of sagamores under a white flag of truce. Waldron had been the first trader on the upper Merrimack River in the 1660s, and cheating Indians seems to have been as natural to him as breathing. His thumb was always finding its way onto the scale, and he claimed his fist—pushing down heavily on the balance against stacks of beaver pelts—weighed just a pound. Abenakis paying off their accounts found that Waldron had a habit of forgetting to cross off their debts in his books.

Now in his seventies, Waldron still ran his well-fortified trading post in Cochecho. In the summer of 1689, his neighbors were doing what they could to buttress their own defenses, including reinforcing their blockhouses. During times of trouble, those living nearby in unprotected homes sheltered at night in these garrisons, emerging cautiously in the morning to work the fields. Able-bodied men were often assigned to help build, man, and defend particular blockhouses, under the command of the owner.

This was one of those troubled times. Everyone was nervous—everyone, it seemed, except Richard Waldron, who believed he knew Indians better than anyone else. That summer, when the townsfolk worried aloud to him about raids, he told them to "plant their pumpkins" and said he'd let them know when there were Indians about.

The evening of June 27 was "dull weather," chilly and damp, and as dusk came on, pairs of Indian women appeared at the door of each of the blockhouses around Cochecho, asking to be allowed to sleep by the fire. This was a common request, and except for one home, where they were bluntly turned away, they were allowed inside, even shown how to unbar the main gates so they could go outside to relieve themselves.

In addition to the women, the sachem Mesandowit appeared at Waldron's door and was invited to dinner. Over their meat, Mesandowit mentioned the presence of many "strange Indians" who had come to trade their beaver pelts and asked, "Brother Waldron, what would you do if strange Indians should come?" Waldron breezily answered that he could raise a hundred men as easily as raising his finger. Then he went to bed.

Well south of Cochecho, a runner was en route with a letter to Waldron from Boston, warning him that according to friendly Abenakis, an attack on the town was imminent, with Waldron a particular target. The messenger, delayed at the Newbury ferry, would arrive a few hours too late.

In the predawn hours, the Abenaki women in each blockhouse crept to the gates, slid back the bars, and whistled. War parties of Pennacook, Kennebec, and Maliseet men, led by Kancamagus, poured in. The Heard family—Elizabeth, the woman who had helped a Pennacook boy escape after the sham fight, and her children—were away, but their neighbor William Wentworth was watching their garrison house for them when he heard a barking dog. Racing to the gate, he found several warriors forcing their way in. He managed to shove them out, then flopped down on his back and braced the door shut with both feet. He screamed for help, holding the door closed while musket balls punched through the wood around him, until others arrived to bar it.

Generally, though, the attacks were brutally effective, with twenty-three people killed and twenty-nine taken captive. Not everyone came in for the same degree of punishment. The raiders spared the residents of one home after forcing the farmer simply to empty his moneybag on the floor, as the Indians scooped up the coins.

No such mercy was shown at Waldron's house, but the trader, though old, still had plenty of pluck. Pulling on his trousers and defending himself with his sword, Waldron forced the attackers back through several rooms. But when he turned away for a moment to grab his guns, he was hit on the head with a hatchet. Thrown into a chair, he sat stunned and bleeding as his house was ransacked and his family forced to bring the raiders food. Then the Indians, who had been cheated so often at Waldron's store, slashed their knives across his chest and stomach, each saying, "I cross out my account." They cut off his fingers, asking him how much his fist weighed now. Then they sliced off his nose and ears, stuffing them in his mouth. Finally, they let Waldron collapse on the point of his own sword to die.

Elizabeth Heard, whose home had been saved by the barking dog, was returning by boat with her children from Portsmouth just as the attacks began. They fled to Waldron's home but found Indians in control there. The family scattered, and Heard hid in the bushes through what remained of the night. At daybreak, she was found out by an Abenaki man, pistol in hand. He walked up to her and stared hard, then walked back to the house. He reappeared and stared a while longer, saying nothing. Her nerves ragged, Heard demanded to know what he was going to do. Still, the Indian said nothing, and when the war party left a short while later, they ignored her. It seems the Pennacook lad she'd saved thirteen years earlier after the mock fight was in the war party and made sure no one harmed her.

As the summer of 1689 wore on, the situation along the Maine and New Hampshire frontier only grew worse. "Our Enimies are Exceedeing Cruell & Rejoyce & say they will bee Into Boston be for Chrismass," one letter writer lamented. The danger of an Indian attack was rising just as military protection collapsed. The Catholic King James II had abdicated the English throne in favor of his Protestant daughter, Mary, and her husband, William of Orange. This delighted the Protestant New Englanders, who in turn rioted against Edmund Andros, whom James had appointed royal governor of the newly created (and largely despised) Dominion of New England.* Andros was blamed for coddling the Indians (in fact, he was one of the few colonial governors to gain their respect and a reputation for dealing fairly with them), and he was accused of colluding with the French. Andros tried to escape

---

* Unlike the United Colonies of New England, which was a voluntary confederation, the Dominion of New England was imposed by the Crown in an attempt to regain control over fractious and unruly colonies. Centralizing what had been local religious and secular authority in the person of the royal governor in chief, it was widely reviled, especially in Massachusetts, as an infringement of the traditional rights of the colonists.

Boston dressed as a woman but was caught when someone noticed his boots peeking out beneath his skirts.

Unfortunately, Andros's arrest led to mass desertions by militiamen at frontier posts. At Fort Loyall near Falmouth, Maine, for example, seven soldiers slipped away, first helping themselves to a long list of supplies, including "one watch coate Dufels one Bed case," a fair amount of powder, and "Tow hwendred musquitt & Corbine shott," not to mention swords, shoes, stockings, scarves, bottles, and "Briches" from their fellow soldiers.

At Pemaquid, Maine—a town settled by English fishermen in the 1610s and thus older than Boston (a point of fierce local pride)—Fort Charles was garrisoned by only a few dozen men, although community leaders were strenuously lobbying Boston for reinforcements. One of Pemaquid's leaders was Thomas Gyles, who had come from England to settle along Merrymeeting Bay, buying a two-mile-long tract of land at the mouth of the Muddy River in 1669. Scared off by the first round of Indian wars, he moved his growing family to Long Island. But "the air of that place not so well agreeing with his constitution," Gyles took advantage of the tenuous peace that followed King Philip's War to move his family back to the New England coast, settling at Pemaquid, in what was then part of New York.

Gyles had inherited a fair sum, enabling him to buy up several tracts of farmland, and he was appointed chief justice by the governor. He was respected by the other English families in Pemaquid—most of them, at least, since this was a region where people had long been accustomed to living as they pleased. One of his sons recalled him as a man who held strictly to the Sabbath and who had "considerable Difficulties in the Discharge of his Office, from the Immoralities of a People who had long lived Lawless."

August 2, 1689, broke with a warm morning, promising a hot day. Gyles had taken three of his sons—John, who was nine or ten, fourteen-year-old James, and nineteen-year-old Thomas—along with fourteen farm hands, to work in the family's fields, which lay several miles from the village. They toiled through the rising heat, cutting hay and barley, until their growling stomachs prompted a noontime rest for a meal. Gyles and the boys ate their dinner in a farmhouse near the fields.

As the laborers rested, armed men were slipping through the forest—a war party of several hundred Abenakis. They had paddled hard down the coast in sixty or seventy canoes, landed well away from the settlement, and then stripped down for battle. The band apparently included Algonquians from throughout the Maritime Peninsula—local Penobscots and Kennebecs, as well as Maliseets from the St. John River—led by Madockawando, the Kennebec sagamore Moxus, and two Frenchmen.

One of the Frenchmen was a Jesuit, Louis-Pierre Thury, who so frequently participated in raids against the English, whom he saw as Protestant heretics, that they called him "the Fighting Priest." In the Penobscot Bay stronghold of Pentagoet, he had left a group of Wapánahki women praying a perpetual rosary until the war party returned. The other Frenchman was Jean-Vincent d'Abbadie, Baron de Saint-Castin, a member of the French nobility, although anyone seeing him would likely have been hard-pressed to distinguish him from the Abenaki warriors. Governor Frontenac, in Quebec, often had a similarly hard time knowing which side Saint-Castin was on, since the former military officer was as happy to trade with the English as with the Indians and French.

Saint-Castin had come to Quebec in 1665 as a teenage ensign with the Carignan-Salières, the first European military regiment to fight in North America. He was one of eighty officers and twelve hundred enlisted men—all volunteers who signed on for a three-year hitch—charged with putting down the Mohawk, whose escalating war of raids and counterraids with French-allied Indians was crippling the interior fur trade.

Before coming to Canada, the Carignan-Salières had fought in the Thirty Years' War, but their initial forays against the Mohawk were a catastrophe. Even though it was winter and the men were poorly clothed, ill with flu, and without skilled carpenters, they were ordered south to Lake Champlain to build forts along the main Mohawk warpath. Rather than allow the soldiers to hole up there for the winter, the new royal governor quickly ordered them to attack deep into Mohawk territory.

The results would have been farcical had they not been so tragic. The men set out in late January without Native guides. They spent weeks wandering through the woods of New York, floundering in the deep drifts without snowshoes (the governor had turned down a merchant's offer to donate fifty pairs), dying from the cold and exposure. What's more, they encountered virtually no Mohawks, who, settled in their towns for the winter, didn't even have the courtesy to realize they'd been invaded. Eventually, the French forces found themselves in English territory near Schenectady, where they acquired from their colonial adversaries enough supplies to limp back to Canada. Without that help, doubtless far more would have died. As it was, four hundred of them perished.

If Saint-Castin was on the expedition, it must have been a sobering introduction to the New World. Yet he stayed, growing comfortable with the languages and customs of the mélange of Algonquian peoples, especially the Abenaki, who circulated through the French towns. In 1670, he was second-in-command of a French force that retook Penobscot Bay from the English. Settling into the fort community of Pentagoet, near present-day Castine, he grew to know and admire the Wapánahki who lived along the coast. This friendship paid quick dividends. When Dutch privateers in the employ of Boston attacked the fort and captured the garrison, Saint-Castin was tortured but managed to escape, hiding with the Abenaki as he made his way to Quebec.

In 1674, his elder brother died, leaving the title of baron to him. But Saint-Castin, now in his early twenties, had already cast his lot with the Wapánahki. Out of the military but acting on private orders from the governor-general of New France to encourage Indian loyalty, he set himself up as a trader at Pentagoet. His own loyalty to the French government was flexible—he often traded with English settlers and fishermen—but he cemented the only bond that increasingly mattered to him through marriage.

Like Saint-Castin, almost a third of the men who had originally come to Canada with the Carignan-Salières remained when their enlistments ended. Many of them took wives from the Filles du Roi, or "King's Daughters"—French women, frequently orphans, who were recruited to emigrate and marry settlers. Saint-Castin, however, married Pidianske (or Pidiwamiska), the daughter of Madockawando.

"The Baron of Saint Castiens," wrote Louis Armand, Baron Lahon-tan, some years later, "having liv'd among the *Abenakis* after the Savage way, above twenty years, is so much respected by the Savages, that they look upon him as their Tutelar God."

The Wapánahki did no such thing, of course, but they did respect Saint-Castin as an able leader. He rose to become a well-regarded chief among the eastern Abenaki, who, like most Indians, cared less about ancestry and skin color than about judgment and ability. Al-though his father-in-law had rebuffed French solicitation, by the late 1680s Saint-Castin's arm's-length relationship with the English had frayed to the breaking point, as had Madockawando's patience with the English. The final straw came in 1688, when English troops sacked and burned Saint-Castin's home at Pentagoet, leaving nothing un-touched except the private chapel. It's easy to imagine that in August of the following year, landing with the Abenaki force at Pemaquid, both men had revenge on their minds.

After they were finished eating, Gyles's farm hands went back to work, with his oldest son, Thomas, accompanying them. The elder Thomas and his two younger sons remained behind near the farmhouse; every-one was on edge, what with the recent attacks at Cochecho and else-where. In the distance, several cannons thundered from Fort Charles. Thomas wondered aloud to his sons if it meant good news—perhaps the council in Boston had sent soldiers to regarrison the fort. But be-fore the boys could answer, musket shots erupted from the woods be-hind the barn, as thirty or forty Abenaki warriors opened fire.

The attackers had split their forces, putting some between the fort and the village and the others between the village and the fields, where they swept in on the Gyles farm. "The Yelling of the Indians, the Whistling of their Shot, and the Voice of my father, whom I heard cry out 'What now! What now!' so terrified me," John Gyles recalled years later. His father groped for a gun as John ran in one direction and James in another. Looking over his shoulder, John saw he was being pursued by an Abenaki armed with a musket and a cutlass "glittering in his Hand, which I expected every Moment in my Brains." Instead of trying to escape, the boy simply sank to the ground and waited.

Rather than killing him, the Indian took John by the hand. He "offered me no abuse, but seized my arms, lift[ed] me up, and pointed me to the place where the People were at Work about the Hay." They passed John's father, who moved haltingly and looked very pale. John saw several dead men, while others cried out in pain, bleeding from tomahawk wounds to their heads.

James had been captured as well, although their older brother, Thomas, had managed to escape. The captives were marched a short way from the fields, then Moxus came to the elder Gyles and apologized, saying that "strange Indians" had been the ones who had shot Gyles and that he was sorry. "My Father replied, that he was a dying Man, and wanted no Favour of them, but to Pray with his Children." The boys spent a few moments with him. He commended them to God, gave them final words of advice, and told them they would all meet in a better place.

"He parted with a chearful voice, but looked very pale by reason of his great loss of blood, which boil'd out of his shoes," John recalled. "The Indians led him aside—I heard the blows of the hatchet, but neither Shriek nor Groan!" John learned later that his father had suffered as many as seven bullet wounds in the initial attack.

As the farmsteads burned and gunfire echoed around the fort, the captives were herded into a swamp, and John found that his mother and two younger sisters had also been taken. His youngest brother, though, had been playing near the fort and managed to get inside, where a small contingent held out against the attackers for two more days before surrendering the stockade and being allowed to leave unharmed, sailing to Boston in a sloop that belonged to a slain villager.

After a similar raid on the nearby town of New Harbor, the war party headed back north. The five members of the Gyles family, along with the other English hostages, were loaded into canoes, and the Abenaki paddled northeast to Penobscot Bay. John had been able to pass a few words with his mother since their capture, telling her of his father's death, and now, as they traveled, their canoes came together. "She asked me, How I did? I think I said, Pretty well, (tho' my Heart was full of Grief). Then she said, O, my child! how joyful & pleasant it would be, if we were going to *Old England,* to see your uncle *Chalker* and other Friends there? Poor Babe! we are going into the Wilderness, the Lord knows where! And she burst into Tears, and the Canoes parted!"

When they at last reached Saint-Castin's fort at Pentagoet, Father Thury showed an interest in buying John from his captor, tossing pieces of gold to the Wapánahki. The priest gave John a biscuit, but the boy was taking no chances with Catholic sorcery, of which he'd heard many stories. "I . . . dare not eat; but buried it under a log, fearing that he had put something in it to make me Love him, for I was very Young, and had heard much of the Papists torturing the Protestants."

John's mother was equally horrified when she found out. "I had rather follow you to your Grave! or never see you more in this World, than that you should be Sold to a Jesuit: for a Jesuit will ruin you Body & Soul." In that way, she got her wish. Although she and her two daughters were later redeemed, she died without ever seeing John again.

Hers was not a unique perspective. Capture and death, horrible as they might be, were not considered the worst fates that could befall a Puritan along the New England frontier. Death at the hands of Satan's agents, after all, was a sure path to martyrdom and heaven. Far worse, in English eyes, was to see a believer fall into the grip of heathenism and Catholicism, something that happened with distressing frequency and remarkable speed to even the most ardent Puritans.

For example, in the famous 1704 raid on Deerfield, Massachusetts, most of the Williams family was taken by western Abenakis and French-allied Mohawks. Those made captive included the Reverend John Williams; his wife, Eunice (a cousin of Puritan chronicler Cotton Mather), who was weak from a recent childbirth; and six of their children. Two other Williams youngsters, a six-year-old and an infant, were killed out of hand, as were many other very young children in the village, whom the attackers felt could not withstand a winter trek to Canada. Mrs. Williams was killed on the march when it became clear she could not keep up.

In the captivity that followed, Reverend Williams tried to protect his remaining family and parishioners—but less from the Indians than from the Jesuits, calling his flock "scattered sheep among so many Romish ravenous wolves." In this he had mixed success. His

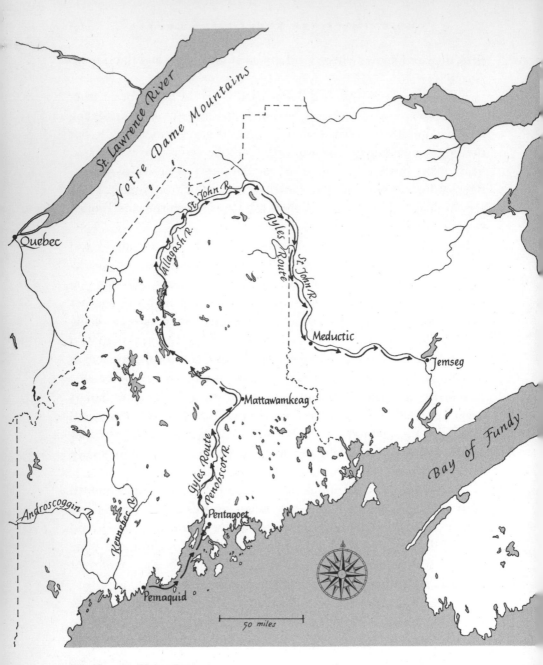

The Travels of John Gyles

Captured in August 1689, Gyles was taken first to Pentagoet, then along the well-traveled canoe paths through the interior, and finally down *wolastokuk* (the St. John River) to the Maliseet town of Meductic, where his life as a slave began in earnest. (Modern boundaries added.)

fifteen-year-old son, Samuel, ransomed by a French merchant, converted briefly to Catholicism but recanted after a series of badgering letters from his father.

Williams's daughter Eunice, however, became his greatest loss and a lifelong grief. Only seven when she was taken and cared for gently by her captors (something even Williams gratefully acknowledged), Eunice was adopted by a family of Catholic Mohawks at Kahnawake, near Montreal, and accepted as full replacement for their dead child—a degree of acceptance common among Indians but unheard-of among the English, who as King Philip's War made plain, viewed even Christian Algonquians with lasting suspicion. Within a few years, Eunice forgot how to speak English and converted to Catholicism, taking the name Marguerite, although her Mohawk name was A'ongote. Despite her father's best efforts after he and his other four children were redeemed, she remained firmly a Mohawk for the rest of her life.

John Gyles, by contrast, was having none of the papist poison, and his master soon took him from Pentagoet. The band traveled by canoe up the Penobscot, stopping at the village of Mattawamkeag (Madahamcouit), about seventy miles upriver. "At Home I had seen Strangers treated with the utmost Civility, and being a Stranger, I expected some kind Treatment here: but soon found my self deceived," Gyles later wrote. An Abenaki captive in the hands of an enemy would have known what to expect, but the boy—like most European captives—understood nothing of the highly ritualized process by which captives were either killed, sold, or accepted into the community.

Some women had formed a circle, and Gyles was pushed into the middle, "where the other Squaws seiz'd me by the Hair of my Head, and by my Hands and Feet, like so many Furies." But his captor "laid down a Pledge and released me," he recalled, explaining, "A Captive among the Indians is exposed to all manner of Abuse, and to the utmost Tortures; unless his Master, or some of his Master's relations, lay down a Ransom, such as a Bag of Corn, or a Blanket."

From Mattawamkeag, Gyles was taken by canoe another sixty miles up the east branch of the Penobscot, then along a series of lakes and portages, passing through what is now Baxter State Park and into the heart of an immense, empty plateau of low hills, bogs, and spruce woods. There were no permanent villages in this part of their territory, but because of the numerous rivers that flowed out of this region, the Wapánahki used it as a transportation hub. From here they could travel easily throughout what is now Maine, New Brunswick, and the St. Lawrence Valley of Quebec.

To a European, it would have seemed a frightening and trackless wilderness, but to the Wapánahki, it was as familiar and welcoming as the wagon roads Gyles and his family followed between home, village, and farm. And although their progress by birch-bark canoe might seem remarkable—pushing more than 160 miles upstream along the Penobscot, then another 50 miles across lakes and portages—this was workaday travel for the Wapánahki, whose canoeists often made the more than 400-mile roundtrip from the coast to Quebec City and back in as little as three weeks.

Gyles had no clear sense of where he was being taken. His captor had left him in the care of an elderly man and two or three women, and the only English the old man knew was "By and by—come to a great Town and Fort." Had Gyles known then what he would come to learn about Abenaki culture, he might have guessed his destination, for his captors were subtly different from the Penobscots his family knew near Pemaquid. They spoke a different dialect and referred to themselves as *wolastoqiyik,* "the people of *wolastokuk* (the bright or good river)"—which the English called the St. John and whose head-waters the party eventually reached. The French and English knew these Wapánahki as the Maliseet (or Malecite).

With the current now in their favor, the party raced along, fol-lowing the river east and then south another 150 miles to Meductic, the chief village of the Maliseet, which lay along the St. John about 80 miles from the coast. By following the traditional canoe trails, the party had inscribed an enormous inverted J on the region, traveling more than 350 miles by canoe to reach a village only 120 miles, as the crow flies, from Saint-Castin's fort at Pentagoet. But traveling by land would have been more dangerous, more arduous, and certainly not as fast as tracing the waterways, however circuitous the route.

By and by, come to a great town, the old man continued to assure Gyles. "I comforted myself in thinking how finely I would be re-freshed etc. when I came to this great Town," Gyles admitted. Instead, he found himself pulled into a lodge where thirty or forty Maliseets were dancing around half a dozen English captives—hostages taken during the Cochecho raid against Major Waldron.

Gyles saw one captive "seiz'd by each Hand & Foot, by four Indians, who swung him up and let his Back with Force fall on the hard Ground, 'till they had danced (as they call it) round the whole Wigwam, which was thirty or forty feet in length." Others were lifted off the ground, head down, and jolted "'till one would think his Bowels would shake out of his Mouth," or grabbed by the hair and struck on the back "'till the Blood gush out of his Mouth & Nose." A captive might be whipped, or old women would fling shovelfuls of hot embers on his chest, "and if he cry out, the other Indians will Laugh and Shout, and say, What a brave Action our Grandmother has done!"

Ritual torture among the Algonquian tribes could have multiple meanings, often centering on replacement. The loss of a respected leader in war could prompt the survivors to kill an equally high-ranking captive, and lost children, even dead spouses, might be replaced by adoption, sometimes after the captive had been tested or threatened. Not surprisingly, few Europeans understood the symbolism of what was happening to them. This may explain the story John Smith told of how the Powhatans, ready to brain him with a war club, were stopped when Pocahontas "got his head in her armes, and laid her owne upon his to save him from death." While many scholars question whether the episode occurred at all, if it did take place, the preteen girl may have been playing a well-scripted role in an adoption ritual.

Nor was youth or long captivity a guarantee of safety. If the captive didn't play by the culturally expected rules, and especially if he or she tried to escape, all bets were off. Gyles's brother James, who was captured with him, was held for three years by the Abenaki at Pentagoet. He eventually escaped with another Englishman, but they were both recaptured and hauled back to Pentagoet, where they were tortured with fire and, like Waldron, had their noses and ears cut off and fed to them before being burned to death.

As the Cochecho captives were tested (as the Maliseet saw it) or tortured (as the Englishmen viewed it), Gyles assumed his turn was next. "I look'd on one and another, but could not perceive that any Eye pitied me," he wrote. But then a woman stepped into the circle and laid down a bag of corn, while a young girl took the boy by the hand and led him away. John Gyles's life with the Maliseet had begun.

Summer passed into autumn, the sky filled with clamorous flocks of waterfowl and the maples flaming orange. The Maliseets caught fish from the river, picked pungent *alaqiminol* (wild grapes), and dug the starchy roots of *pkuwahqiyaskul* (cattails) and *skicinuwi–poceteso* (groundnuts). The family band of eight or ten people to which Gyles now belonged began traveling up the *wolastokuk* toward their winter hunting grounds, hundreds of miles inland. When the river froze, they cached their canoes and walked on the thickening ice, carrying what they needed on their backs and building rafts to cross whatever open water they encountered.

"I met with no Abuse from them," Gyles recounted, "tho' I was put to great Harships in carrying Burdens, and for want of Food: for they underwent the same Difficulty, and would often encourage me, saying, in broken English, *By-by—great deal Moose*." Gyles kept expecting the band to settle down, but instead they moved from camp to camp throughout the winter, covering an enormous swath of northern Maine and New Brunswick, eventually crossing the Notre Dame Mountains on the Gaspé Peninsula.

Although he was unaware of it at the time, the Maliseets weren't dragging Gyles around aimlessly. Among the Wapánahki, family bands had traditional hunting territories, usually delineated by watersheds and encompassing sizable, sometimes widely separated, tracts of land. Like the network of canoe paths that linked the expansive forests and mountains of *wôbanakik*, these hunting grounds, even though they covered thousands of square miles, were familiar, their routes and boundaries passed down through the oral tradition and reinforced by the annual round of seasonal travel.

In the words of two modern scholars (speaking in this case of the western Abenaki), "It is not really accurate to say that a family band actually owned its territory . . . Indeed, it might even be said that the members of the band belonged to the territory, so close was their identification with it." To an English farmer such as Thomas Gyles, the Maliseet were shiftless nomads, but to the Maliseet, the idea of voluntarily leaving one's homeland and crossing the ocean to live in a new place was incomprehensible, even pathological. Viewed from their perspective, it was the agrarian Europeans who were the nomads, each generation pushing into new lands, leaving home ever farther behind. To the Maliseet, that would have seemed to be a form of insanity.

There were two men in the band, armed with muskets, on whom the group depended for food. Winter was the best time for big-game hunting, since even with their long legs, moose bogged down in the heavy drifts, while men on snowshoes could easily cut off the fleeing animals. Sometimes the hunters found denned black bears, heavy with fat, and when one of these was brought in, the old women would stand outside the bark wigwam "shaking their Hands and Body as in a Dance: and singing *Wegage Oh Nelo Woh!* which if Englished would be, *Fat is my eating.*"

More often that winter, the band went without food for three or four days at a time, and once Gyles ascribed their salvation to divine intervention. The two hunters had chased a moose, hoping that the thin crust on the snow, which held up the men but through which the moose sank, would give them an advantage. But the bull escaped into a swamp, and they lost the trail in the dark. The next morning, they returned to find that the moose, crashing away in its haste, had broken through the ice and gotten tangled in the roots of a toppled tree, dislocating its hip. "GOD wonderfully provides for all Creatures!" Gyles concluded.

In spring, the band stitched the hides of several moose around wooden frames, calked them with charcoal and balsam gum, and floated down the St. John in the makeshift skin boats, trading them for their cached bark canoes as they made their way back to Meductic. Gyles settled into his new life as a captive—helping to plant and weed the corn that spring and summer, fishing for salmon, digging for roots, drying the mature corn in autumn, and placing it in bark-lined pits where it was safe from animals and weather, then packing up once more for a winter of travel and hunting.

Some European captives, such as Eunice Williams, were eventually adopted as full members of a community, while others were not. In Williams's case, there was a Mohawk family anxious to replace a dead daughter. Perhaps there was no similar desire among any of the

Maliseets holding Gyles. Or perhaps it was a matter of attitude: Gyles remained very pointedly English throughout his captivity. Then again, he was always treated as a slave; it's worth wondering how this adaptable, curious boy would have fared—and how firm his identity as an Englishman would have remained—if he'd been welcomed as a son into a Maliseet family.

As it was, Gyles and several other male captives in Meductic were never viewed as anything other than chattel—worked hard, poorly clothed (until it fell off his back in pieces years later, Gyles wore the same, outgrown shirt in which he was captured, along with a breechclout and a greasy blanket), always at risk of suffering the indignities and abuse considered acceptable in the treatment of captives. One old woman, for instance, took particular delight in tormenting hostages, throwing scoops of hot embers on adults or dragging children through the fire. Yet Gyles managed to avoid the worst excesses on all but one occasion.

A year and a half after his capture, Gyles's Maliseet owners went to Canada, leaving him in the care of several other Indians, who took him downriver to help plant corn. When he arrived, however, he was grabbed by Indians he didn't know and, with another English captive named James Alexander, was beaten so severely that the two could barely walk for several days.

Interestingly, it was not Maliseets from Meductic who initiated this abuse, but visiting Mi'kmaqs from southern Nova Scotia. "This was occasioned by two Families of *Cape Sable* Indians, who having lost some Friends by a number of English fishermen, came some hundreds of Miles to revenge themselves on the poor Captives," Gyles wrote. When the Mi'kmaqs and others came back for Gyles several days later, his master had him hide in a swamp until they gave up the search. Alexander wasn't as lucky and was again severely beaten.

Such heavy-handed torture aside, slaves had to accept a certain degree of day-to-day abuse. But every rope has its breaking point. While Gyles was cutting wood one day, a young Maliseet man of about twenty—"a stout, ill-natur'd young Fellow"—threw Gyles to the ground, sat on his chest, and pulled a knife, saying he'd never killed an Englishman.

Gyles tried to talk him out of it—"I told him he might go to War, and that would be more Manly"—but as the Maliseet began stabbing him, Gyles grabbed the man's hair and flipped him onto his back. Now the English boy's dander was up, the consequences be damned. "I . . . follow'd him with my Fist and Knee so, that he presently said he had enough; but when I saw the Blood run & felt the Smart, I [went] at him again and bid him get up and not lie there like a Dog,—told him of his former Abuses offered to me & other poor Captives, and that if he ever offered the like to me again, I would pay him double."

Gyles marched the bully back to the village, where the other Maliseets praised Gyles for his actions. "And I don't remember that ever he offered me the least Abuse afterward; tho' he was big enough to have dispatched two of me."

As with anyone living in the woods, Gyles had his share of narrow escapes, though none quite as racking as the blizzard that caused such profound frostbite in his feet. Like many Europeans, John never learned to swim, which the Maliseets found amusing—and no doubt a little puzzling, since most Wapánahki children learned to swim at an early age. Once, after a day of spearing salmon, his companions encouraged him to join them in a deep pool in the river. When with some reluctance he did, he dropped straight to the bottom, where he would have drowned if a young girl, following his bubbles, had not swum down, grabbed him by the hair, and pulled him to the surface.

On another occasion, he and a Maliseet were spearing giant Atlantic sturgeon when the man misstepped in their canoe, flipping it and dumping both of them into the river. Gyles floundered to the surface, grabbing the crossbar of the upside-down boat and keeping his face above water in the small pocket of air trapped beneath the canoe. His feet thrashed in the current, but he couldn't touch the bottom.

For fifteen minutes, he clung to the boat as it drifted downstream, the air quickly running out, "expecting every Minute that the Indian wou'd have tow'd me to the Bank: *But he had other Fish to Fry!*" When Gyles's feet finally found purchase on a rocky point half a mile downriver, he clambered weakly ashore and saw that the Maliseet was still hauling in the eight- to ten-foot sturgeon he'd speared. "I went to him, and asked, Why did he not tow me to the Bank, seeing he knew I could not Swim? He said he knew that I was under the Canoe, for there were no Bubbles any where to be seen, & that I should [eventually] drive on the Point: therefore he took care of his fine Sturgeon."

Another time, James Alexander, the captive tortured with Gyles, used the Maliseet's terror of the Mohawk—"a most ambitious, haughty and blood-thirsty people"—to have some fun. The weather had turned hot, and the villagers kept Gyles and Alexander running constantly to and from the spring for cool water. After dark, Alexander set a copper kettle at the top of a hill studded with tree stumps, then ran back to the palisaded village, "puffing & blowing, as in the utmost Surprize," claiming he'd seen something near the spring and thought it might be Mohawks.

Alexander's master, a respected warrior, followed him into the darkness, and as they came to the hill, the boy pointed to the stumps and nudged the kettle with his toe. "At every turn of the Kettle the bail clattered; upon which James and his master could see a *Mohawk* in every stump . . . and turn'd tail to, and he was the best man that could run fastest." The village cleared out, dozens of people fleeing for more than two weeks. By that time, "the heat of the Weather being over, our hard Service abated for this Season."

These dryly humorous episodes are only one way in which Gyles's account differs from Mary Rowlandson's. His story doesn't seem to have been intended simply, or even primarily, as a moral lesson, in which every indignity is an opportunity to see God's plan, and in which the Indians are satanic actors against whose machinations the righteous must persevere.

Gyles was hardly irreligious—he accepted as casual fact that the Maliseet "not brought over to the Romish faith" worshipped the devil, and he highlighted what he saw as divine intervention during his stay, such as the crippled moose that warded off the band's starvation. But he claimed no special providence as a Christian among heathens. In the case of that moose, for example, God was providing "for all Creatures," and Gyles explicitly lumped himself and his Maliseet captors together as deserving recipients, something few hard-line Puritans would have considered. And it wasn't until a friendly Franciscan priest, preaching against the evils of torture, pointed out that every member of the two Mi'kmaq families that so severely beat Gyles and James Alexander died before they left Meductic that Gyles, too, decided it was a case of providential justice. "Were it not for this Remark of the

Priest, I should not, perhaps, have made the Observation," he admitted.

But with age and time in captivity, Gyles seemed to grow into sensible skepticism about religion on both sides of the Protestant/Catholic divide, with little patience for zealots in either camp. The boy who once refused to eat a biscuit out of fear of Catholic magic was upbraided by a fellow English prisoner for wanting to see a ceremony in which a Jesuit was to sprinkle holy water on fields of French wheat to exorcise flocks of hungry blackbirds. "He said, that I was then as bad a Papist as they, and a d—n'd Fool. I told him that I believ'd as little of it as they did, but I inclined to [go] see the Ceremony"—and so he did.

After six years among the Maliseet, John Gyles's fortunes took a dramatic turn. When his second master died, the man's widow and Gyles's first captor argued heatedly about who should get the slave. "Malicious persons" suggested they settle the dispute by killing the captive, but the village priest proposed they sell him to the French instead—perhaps to merchants at the mouth of the river or to gentlemen visiting the colonies on a French warship.

Gyles pleaded that if he must be sold, let it be to one of the traders, since he believed that if he was taken to France, he'd never see New England again. Yet when he was informed that he had, in fact, been bought by one of the merchants, he was inconsolable. "I . . . went into the Woods alone and wept until I could scarce see or stand! The word *Sold,* and to a People of that Perswasion, which my dear Mother so much detested, and in her last words manifested so great Fears of my falling into!"

Learning that the merchant had bought and repatriated to Boston two other English slaves revived Gyles's spirits (and probably raised false hopes). Seeing his distress, the priest made a deal with Gyles: if in ten days, he still preferred living with the Maliseet rather than the French, the friar would buy him back from the merchant and restore him to Meductic.

But Gyles once more adapted quickly, this time to life at the French outpost. He referred to his new master as "Monsieur Decbouffour," but the man was actually Louis d'Amours, Sieur des Chaffours, a successful farmer and trader around forty years old, born in Quebec, recently married, and a new father. D'Amours and his brother owned large farms growing wheat, peas, and corn, as well as several dozen head of cattle and a large herd of pigs. Their establishment sat along the lower St. John River at a place the French called Jemseg and that Gyles rendered as "Hagimsack." Both names come from the Maliseet word *ah-jem-sik,* sometimes translated as "the picking-up place," a sign of the location's importance in trade long before the Europeans arrived.

Given Gyles's fluency in Maliseet and Mi'kmaq, and despite the fact that he initially spoke no French, the quick-witted sixteen-year-old rapidly assumed responsibility for the trading post. "I had not liv'd long with this Gentleman before he commited to me the Keys of his Store, &c and my whole Employment was Trading and Hunting, in which I acted faithfully for my Master, and never knowingly wrong'd him to the value of one Farthing." D'Amours and his wife, Marguerite, called Gyles "Little English," and his master trusted him enough that when d'Amours returned to France for an extended visit, he left Gyles in charge.

That trust was tested in October 1696, when word swept through the village that an English fleet with several hundred men was ascending the river to attack the French at Fort Nashwaak. To Gyles, it must have seemed that salvation was at last at hand—except that he was responsible for the store, a charge he took seriously. As the English soldiers approached, the goods from the trading post were hidden in the woods, but there was every reason to expect that they'd burn d'Amours's home, store, and farm.

"'Little English,' we have shewn you Kindness; and now it lies in your Power to serve or disserve us, as you know where our Goods are hid in the Woods,'" his mistress told him. Instead of confining him to the fort, she explained, she had decided to trust him once again. If he would protect their property from the English raiders, they would give him leave to return to Boston at the first opportunity.

It must have been an excruciating decision for Gyles. All he had to do was slip through to the English lines, and he'd be free. Instead, he agreed not to escape, nor to betray his master's family or property. Leaving a note pinned to the door, explaining that the owners had saved English captives in the past and shown him kindness, Gyles shepherded Madame d'Amours and her two young children to a remote hiding place on the far side of Grand Lake. They stayed there until the brief siege of the fort ended and the New Englanders, fearing an early freeze that would trap their boats, departed.

Gyles and the d'Amours family found their home and business untouched. Colonel William Hathorne, the English commander, had discovered the note and ordered his men not to burn the buildings or harm the livestock, save for the chickens and a couple of cattle that they took for eating. In this, Gyles and his French captors were fortunate. Some weeks earlier, Hathorne had taken command from the notoriously erratic and brutal Benjamin Church, whose men had killed King Philip and who would have been unlikely to have spared anything. Although Hathorne had been one of the commanders who had duped the Pennacook with the mock fight some years earlier, he dealt far more honorably with the French outpost.

In the spring, Louis d'Amours returned from France. He tried to persuade Gyles to remain, pledging "that he would do for me as for his own"—a not inconsiderable offer of adoption, given the family's holdings and titles. Gyles was adamant, however, and so d'Amours arranged the eighteen-year-old's passage to Boston. Gyles arrived on a dark June night in 1698, and the next morning a young man came aboard the sloop and began asking Gyles questions about his captivity. It was Samuel, John's youngest brother, who had escaped to the fort the day of the attack. Samuel told his shocked sibling that their eldest brother, Thomas, and two sisters were well but that their mother had died some years earlier.

The same familiarity with the Wapánahki language and customs that had stood Gyles in such good stead with the French proved equally valuable among the English. Within a few months, the young man was in demand as an interpreter, and in 1706, a few years after marrying, he received a captain's commission. He commanded forts at Brunswick, Maine, and along the St. George River at Thomaston, Maine, and led expeditions under a flag of truce to Port Royal, in French Acadia, to negotiate the release of captives. After his first wife died, he remarried in 1721—a year before his old Maliseet master appeared at the fort at St. George. If there had been any lingering bitterness, it was gone. "I made him very welcome," Gyles said simply.

Unlike Mary Rowlandson and the Reverend John Williams, Gyles wrote his account late in life, at the urging of his wife and after decades of acting as a liaison between the English and the Wapánahki. At first the account was meant only as a private record for his family, but as more and more people asked to read it, Gyles relented to their requests and, in 1736, published *Memoirs of Odd Adventures, Strange Deliverances, &c. In the Captivity of John Gyles, Esq; Commander of the Garrison on St. George's River. Written by Himself.*

Only a few copies of the original edition survive, but *Memoirs of Odd Adventures* has been reprinted many times, and it remains the most unusual of all the captive narratives of the eastern frontier. One of the reasons for this is the clear, uncluttered voice of Gyles himself. The reader comes away with the impression of young John as a remarkably resilient boy who, despite his travails, never lost his sense of humor or his innate curiosity.

Besides his own experiences, Gyles discusses the common animals of the Maliseet lands, exploding a few myths along the way. For instance, he calls the prevailing notion that black bears give birth to an unformed embryo and lick it into shape "a gross Mistake . . . I have seen their Fœtus of all Sizes, taken out of the Matrix, by the Indians, and they are as much, and as well Shap'd as the Young of any Animal." He also gives a brief but fairly evenhanded overview of Maliseet life and culture. Two modern historians have called the short book "part horror story, part ethnography, part natural history, and part sermon."

There is a fascinating postscript to Gyles's story, emblematic of the way cultures and personal histories became intertwined on the First Frontier. When the English raiders were advancing in 1696, Gyles rowed Madame d'Amours and her children across a lake to evade capture. One of those children was a girl, Marie-Charlotte, who in 1707 married Bernard-Anselme d'Abbadie—a French officer, Abenaki chief, and privateer, and the son of the man who led the raid in which Gyles had been captured. With his father's death that year, Bernard had also assumed the title, and there was a new Baron de Saint-Castin in Pentagoet.

If John Gyles, looking back on his youthful captivity from the safety of old age, tells an Englishman's tale with a fair bit of sympathy for the Indians, then the story of Hannah Duston provides a jarring contrast—one that reinforces not only the differences between white settlers and Natives but also those between eighteenth-century and modern sensibilities. More than three hundred years after the events, its emotionally charged echoes continue to reverberate to this day.

In the winter of 1697, Thomas and Hannah Duston lived on a small farm along the lower Merrimack River in Massachusetts, two miles from the town of Haverhill, which in turn lay about thirty miles north of Boston. With a good supply of clay on the farm, Thomas was making bricks for sale—and with seven children and an eighth on the way, he was also working on a larger, brick house for his family a short distance from their existing home.

Hannah came from what might today be called a troubled home. When her father, Michael Emerson, first moved the family into Haverhill in 1656, at least one neighbor was alarmed enough to protest to the owners of the plantation. Why the Emersons were considered bad neighbors doesn't appear in the records, but Michael allowed himself to be bought off, accepting a gift of land if he would "go back into the woods," which he did. In 1667, he was fined five shillings for "his cruel and excessive beating of his daughter [Elizabeth] with a flayle swingle, and kicking of her." (A flail, used to pound flax to remove its seeds, has a long handle to which is bound a heavy wooden truncheon—a vicious weapon to use against anyone, much less an eleven-year-old child.)

Still another daughter, Mary, was sentenced to be fined or whipped for fornication before marrying—a common enough charge in those days, and one that didn't carry a heavy stigma if the accused couple married, as did Mary and her beau. Elizabeth, however, faced fornication charges of her own in 1686, and no one stepped forward to claim the child. In the messy paternity fight that followed, during which Elizabeth claimed she was raped by a neighbor, the Emerson home was publicly denounced as "that wicked house." Hannah, the oldest child of the family, was deposed as a witness against her younger sister.

Elizabeth, still unmarried, became pregnant again in 1690 and this time managed to keep the pregnancy a secret—not that difficult in colonial days, when heavy winter clothes concealed a great deal and with Elizabeth's excessive weight aiding in the deception. More remarkable, even though she slept in a trundle at the foot of her parents' bed, Elizabeth claimed to have somehow given birth to twins in May 1691 without waking anyone. Whether at least one of the babies was stillborn, as Elizabeth claimed, or she killed them both, she hid the bodies in a chest, then sewed them into a cloth bag and buried them outside. Several women of the community, however, were suspicious of her condition and confronted her with a legal order permitting a physical examination. This was executed while her parents were in church, which suggests that the Emersons may have known of their daughter's actions. While the women examined Elizabeth, their husbands discovered the makeshift grave in the garden.

On the morning of June 8, 1693, two years after the twins' death, Elizabeth was hanged in Boston. Although she was convicted of murder, it was a capital crime merely to have concealed the death of a bastard child, even a stillbirth. A few hours before her execution, with Elizabeth apparently in attendance, the Puritan minister Cotton Mather (fresh off the Salem witch trials) preached a stem-winder of a sermon in which he read what he said was Elizabeth's confession and contrition. "I was always of an haughty and stubborn spirit," she purportedly told him, blaming her troubles on "Disobedience to my Parents."

Whether Elizabeth Emerson actually said any of this is guesswork; the reputed confession reads more like the sermon it begat than a young woman's sincere unburdening. Mather, in almost gleeful Puritanical form, saw the fact that Elizabeth's execution fell on a Lecture Day on which he was to preach as a sign of God's favor. He later preened in his diary that the sermon was heard by "one of the greatest Assemblies, ever known in these parts of the World" and that printed copies (titled *Warnings from the Dead, or Solemn Admonitions unto All People*) were "greedily bought up."

Hannah Emerson Duston seems to have left the troubles of her childhood home behind. By 1697, she and Thomas had been married for almost twenty years, and although they'd lost a number of children—their newest daughter, Martha, was their eighth living, out of thirteen she'd borne—the family was well established. Unlike her father, her husband was respected in Haverhill, having been elected constable for the town's west end.

On March 15, the weather was profoundly cold, a sharp return of winter that kept the fires stoked throughout Haverhill. Hannah was bedridden, still drained from giving birth to Martha a week earlier. A widowed neighbor, Mary Neff—one of the women who had confronted Elizabeth six years earlier—was nursing them both while Thomas worked on the new house, his horse tethered nearby and a gun at hand. Worries about Indian attacks ran high. The previous August, a neighbor named Jonathan Haynes and his four children had been captured while working in the fields, and rumors of further raids were rife.

Chatter about Indians was incessant on the frontier, both produced by and reinforcing the cycle of tensions born of life at the edge of seemingly endless war. Although King William's War had begun with grand assaults such as the English siege of Quebec in 1690, it had devolved into a series of guerrilla raids over the years that followed. The Iroquois struck into Canada, though with diminishing strength as the war dragged on, and the English refused to back them up. In turn, the French and their Indian allies repeatedly hit settlements in New York and New England—Schenectady in 1690 and York, Maine, in 1692 being among the bloodiest. In 1693, the French destroyed winter villages and food supplies throughout Mohawk country and took hundreds of captives, while bringing a vast, heavily guarded convoy of more than four hundred canoes laden with a million pounds of furs to Montreal, a prize the Iroquois would have snapped up in earlier years, when they were stronger.

In New England, the combination of Puritan fundamentalism and constant anxiety over Indian attacks was a powder keg. No one felt safe from the drumbeat of violence and the notion that evil could be anywhere—personified by Indians sweeping out of the darkness or hidden in the shadows of even the most seemingly righteous home. As refugees from the Indian war poured into towns such as Salem, the paranoia and hysteria grew. The ravings of one redeemed captive, Mercy Short—who was held for eight months by the Abenaki and who, after resettling in Salem, began ranting about "Tawney" devils in league with the French and with "Indian Sagamores"—added fuel to the witch trials of 1692. Cotton Mather noted that almost all the confessing "witches" said that Satan looked like an Indian, and many scholars have pointed out that most of the so-called afflicted—

Dustan Covering the Retreat of His Seven Children (1851)
Thomas Duston (the spelling of his last name varies) was faced with a stark choice in March 1697—protect his bedridden wife, Hannah, and their newborn daughter from attacking Abenakis or shepherd their other children to safety. Bluffing with his single-shot musket as they retreated, he managed to save the children, but Hannah, the baby, and a neighbor were taken captive. (Library of Congress LC-USZ62-38192)

those women whose fits and hallucinations stoked the witch-hunting flames—had, like Mercy Short, suffered losses and trauma on the northern frontier.

So as Hannah Duston nursed her newest infant and Thomas made bricks, they were both aware of fresh speculation about skulking Indians. This time, the rumors were accurate. A band of Cowasuck Abenaki had slipped into the area overnight, fanning out through Haverhill. In many cases, the surprise was complete, but just as the Indians opened their attack on multiple fronts, Duston stumbled upon one of the bands. Leaping to his horse, he made a run for the house, with the warriors moving quickly behind him.

Yelling to the children, Thomas sent them scrambling up the road toward the village blockhouse. But the youngest child was only two, and it was clear that they'd never make it in time; they were only a few hundred yards away from the house when the war party arrived. Inside, Hannah was in bed with the baby, whom she refused to leave. She sent Thomas after the children, knowing there was no hope for her. After a moment's hesitation—one can only imagine how Thomas must have been torn—he ran out of the cabin.

Thomas lashed his horse, thinking he'd snatch up one child at least and make an escape. Faced with yet another agonizing choice, however, he decided he couldn't take one and simply abandon the others. Instead, he would make a last stand with his children. While most of the raiders swarmed into the farm buildings, a few advanced toward them. Thomas dismounted, using his horse as cover while the Abenakis opened fire.

It was a strange and hapless procession, creeping up the road only as fast as the older children could drag or carry their younger siblings, with their father presenting a near-hopeless rear-guard defense. Although he was armed, he couldn't risk actually shooting his musket, because in the thirty or forty seconds it would take him to pour powder down the muzzle, ram a ball home, and reprime the firing pan, the Abenakis would overwhelm them. So he had to bluff, raising the gun as if to fire, forcing their pursuers to take cover again and again as his covey moved up the road at what must have seemed like an agonizing crawl.

Incredibly, he succeeded, reaching a garrison churning with frantic activity. Indians were striking across the town, eventually killing twenty-seven and burning nine homes.

As Thomas Duston shepherded most of his family to safety, Mary Neff tried to escape with the Dustons' baby. She was quickly captured and taken back to the Duston home, where Hannah stood mutely, one foot bare on the ice-cold dirt, while it was sacked and torched. The captives were marched out of the village, with Mary cradling the baby. When the raiders saw that the child was slowing her down, they smashed Martha's head against an apple tree and left the small body behind.

They set out with eleven other captives, although in typical fashion several were killed when they, too, began to flag and tire. The war party covered twelve miles the first day to put distance between themselves and their pursuers. Over the next several days, the survivors were marched up the Merrimack Valley, covering about forty-five miles as the raven flies, though the paths wound through deep forests, mud, and the remains of the winter snow. It must have seemed an endless nightmare for the English captives, not least for Duston, with only one shoe and just one week after giving birth.

The Cowasuck, whose name means "the people of the white pines," were on familiar ground, however, moving deeper into the heart of *n'dakina,* their homeland. It was not the end of March to them, but the beginning of *sogalikas,* the Sugar-Making Moon, with *mozokas,* the Moose-Hunting Moon, just past. The river along which they traveled was *mol8demak,* "deep river," and the smaller tributary at whose mouth they eventually camped, on an oxbow near the present town of Penacook, New Hampshire, was *nikn tekw ok,* which the English pronounced Contoocook.

Here the war party began to split up. The two *iglizmôniskwak,* as the English women were known, were taken by an extended Abenaki family—two men, three women, and seven children, who also held a fourteen-year-old English boy named Samuel Lennardson (or Leonardson), captured a year and a half earlier in a raid on Worcester. The plan, as the captives understood it, was that their party would rest for a time with their captors' relatives, then rejoin the rest of the raiders at a large Abenaki village in Maine.

When they got there, the three colonists were told, they'd be stripped and forced to run the gauntlet—the typical test of bravery and strength for male captives, racing through a rain of blows, jeers, and abuse. To the horrified women and teenage boy, however, it was simply torture. And while such treatment was rarely, if ever, extended to women, they may not have known this. Duston, in fact, was no doubt aware of hair-raising accounts such as those of Mercy Short, who saw a five-year-old boy killed with a tomahawk and dismembered and a teenage girl beheaded and scalped as an explicit warning to the other captives to behave. A pregnant woman taken captive with Mary Rowlandson, who begged constantly to be sent home, was likewise killed out of hand as an example to the others. And as Duston had already seen, weak or sick captives, regardless of age or sex, were often executed on the march.

European captives knew that death was always possible, but interestingly, sexual assault seems to have been all but nonexistent on the eastern frontier, perhaps because of deeply ingrained cultural taboos against it among many Woodland Indian cultures. Rowlandson could have spoken for the vast majority of female captives when she noted, "Not one of them ever offered the least abuse or unchastity to me in word or action." Such constraints were decidedly absent on the other side, however, and the rape of Indian women by Europeans was depressingly common.

Like Rowlandson, Duston and Mary Neff could expect rough treatment for a time as slaves, with the likelihood of eventual ransoming and a return to their homes. Full adoption into Abenaki society was possible, but this was less likely than among tribes such as the Mohawk, who made a regular habit of accepting white captives into society. More likely, they would be transferred to French control in New France, where a surprising number of English girls found an agreeable new life. Some, like Eunice Williams, embraced Indian culture, but far more assimilated into New France society, converting to Catholicism and in some cases becoming influential nuns.

That so many women captives welcomed this change, while very few men did, may reflect the relative freedom and authority that the women enjoyed in both Indian culture and New France, compared with the strictures facing them in New England. In at least one case, captivity may have brought a strange kind of relief. Abigail Willey was captured with her two daughters in 1689 at Oyster River, New Hampshire. She was thirty-two and married, but six years earlier she'd unsuccessfully sought legal protection from her abusive husband, having suffered "sore and heavy blows . . . too much for any weak woman to bear." Willey never returned to New England, although her children did.

To Duston and Neff, the immediate prospects were frightening. No one knows whether their captors were simply tormenting them with the prospect of the gauntlet, whether the *iglizmôniskwak* misunderstood the Abenaki plans, or whether they leapt to conclusions based on stories they'd heard from other captives. They tried to pray but had to be surreptitious, for the Abenaki would stop them—though not from a lack of Christian faith themselves. The Indians had been baptized Catholic by the French missionaries; they prayed three times a day and wouldn't let their children eat without saying grace. Instead, it was the English *style* of prayer that bothered them, just as their Catholicism was anathema to Protestants such as Duston. One of the Abenaki men had lived for a time with Mary Rowlandson's family in Lancaster, and he told Duston that he had "pray'd the English way [and] thought that was good: but now he found the French way better." He also told the women not to worry: "If your God shall have you delivered, you shall be so."

The same man also explained to Samuel—apparently at the boy's request and possibly at Duston's secret bidding—how he killed Englishmen with a tomahawk. "Strike 'em dere," he supposedly said, pointing to his temple, then answered further questions by explaining how to take a scalp.

What happened next has been the center of controversy for more than three hundred years. Some see Duston as a plucky symbol of frontier self-reliance. Others, like Cotton Mather, who wrote the first account of her captivity, see her as the weapon of divine justice, like the biblical Jael killing the Canaanite king Sisera by driving a tent peg through his head. Still others, notably Abenaki descendants, dismiss her as a cold-blooded murderer.

That night, as the camp slept, Duston "heartened the nurse and the youth to assist her in this enterprize," Mather wrote, "and all furnishing themselves with hatchets for the purpose, they struck home such blows upon the heads of the sleeping oppressors" as to kill ten of the twelve Abenakis. Samuel killed the man who had helpfully explained how to wield a tomahawk. One woman, left for dead, escaped despite seven deep hatchet wounds in her head, wandering in the woods for several days before finding other Abenakis. A boy of unknown age was spared or escaped, the only one of the seven Indian children to live through the attack.

The three English captives grabbed a gun and the hatchets, gathered some supplies, and punched holes in all the canoes except one before heading downstream. But then Duston did something that makes her story even more of a moral quagmire. Having made their successful escape against immense odds, she turned her party around and returned to camp. There she straddled each body and, with a knife, peeled the scalps off the ten dead Abenakis—including the children—wrapping the bloody skins in a piece of fabric.

One can imagine a kidnapped woman doing almost anything to escape, and it's equally easy to imagine a mother who saw her newborn's head smashed open, and who had every reason to believe the rest of her family dead, wanting to exact revenge. Two weeks of fear, grief, and exhaustion could unhinge almost anyone and make her capable of great violence. But the deliberate slaughter of sleeping children, compounded by the methodical scalping of all the bodies, has made Duston's story a hard one to stomach.

Should it matter? Casual, often incomprehensible cruelty was a hallmark of the frontier wars on both sides, driven by deep-seated hatred and the desire for revenge. In March 1690, the Abenaki sagamore Hope Hood, whose kidnapping had precipitated the war, was among those attacking Salmon Falls, across the river from Cochecho, where he kidnapped a five-year-old boy named James Key. For days, the lad cried for his parents. So first Hope Hood beat him bloody, and then when the boy complained about a sore eye, Hope Hood "[lay] Hold on the Head of the Child with his Left Hand, and with the Thumb of his Right he forced the Ball of [the boy's] Eye quite out," warning the boy that if "he heard him Cry again he would Serve t'other so too, and leave him never an Eye to Weep withal."

Certainly, no one in the English settlements was the least disturbed by what Duston and her comrades did—quite the contrary. With the fresh scalps wrapped in Duston's apron, the trio traveled by night down the river, making their way safely back to Haverhill. Less than two weeks later, Thomas Duston took all three to Boston, where he petitioned for a payment. Even though the bounty for Indian scalps had been revoked the year before, the General Court granted Hannah fifty pounds (some accounts say twenty-five, with a similar amount later to reward Thomas). Mary and Samuel each received twelve pounds and ten shillings. With the money, Thomas Duston purchased an additional twenty acres of land five months after his wife's escape.

Hannah Duston was a celebrity. The governor of Maryland sent his compliments and an unspecified gift, perhaps the silver tankard that became a family heirloom (as did a silver earring Duston may have taken, along with a scalp, from one of the dead women). Cotton Mather, four years after making a stern sermonly example of her condemned sister Elizabeth, now ringed Hannah with praise. He later included her tale in his 1702 book *Magnalia Christi Americana*. Less publicly, her account was included in the diaries of several Puritan notables, such as Samuel Sewall, who spoke with her shortly after her escape.

Over the next two centuries, Duston became a New England folk legend, the gruesome story abridged or embellished as necessary, used as a morality tale for schoolchildren, reenacted in public pageants, translated into heroic artwork, and referenced by serious writers like Hawthorne and Thoreau. A mountain in northern New Hampshire was named for her. Thomas Duston, having shielded his children from the approaching Indians, also was lionized in art and poetry, but not to nearly the same degree. The fact that a *woman* had committed such acts was what made Hannah Duston stand out through history.

In 1874, a statue of Duston—said to be the first monument to a woman erected anywhere in the United States—was placed on a small island at the confluence of the Contoocook and Merrimack rivers near Boscawen, New Hampshire, roughly where she, Mary Neff, and Samuel Lennardson killed the Abenaki family. It depicts her gripping a hatchet in one hand and scalps in the other. In 1879, Haverhill raised a statue to Duston, requisite ax again at the ready.

Unlike Mary Rowlandson and many other captives, Duston never committed her own story to paper. We have only secondhand accounts of what happened in the cold predawn hours along the Contoocook. It is the uncertainty about her state of mind and motives that leaves us with two central questions: why kill the children, and why return for the scalps? And it's the latter act of apparent premeditation that turns Duston from a merely vengeful captive into something darker and far more disturbing.

Americans of all backgrounds have been trying to make moral sense of Duston's story for a long time. Starting in the early 1800s, local historians suggested that she needed the scalps as proof that two English women and a boy could overpower an entire band of Indians. Certainly, that is the most sympathetic rationale. Strikingly, however, neither Mather nor Sewall—the only two writers who heard the story from Duston herself—give any hint of that justification. If anything, Mather's comment about how the trio "[cut] off the scalps of the ten wretches, they came off, and received *fifty pounds* from the General Assembly . . . as a recompense of their action" suggests that the intent was for a bounty of some sort.

Nathaniel Hawthorne found her an impossibly conflicted figure, calling her—all within a single 1836 essay—a good wife, a "raging tigress," and "a bloody old hag" who would have been better drowned or lost forever in the forest, "nothing ever seen of her again, save her skeleton, with the ten scalps twisted around it for a girdle!"

Henry David Thoreau, floating down the Merrimack River in 1839, imagined the scene 142 years earlier: "These tired women and this boy, their clothes stained with blood, and their minds racked with alternate resolution and fear, are making a hasty meal of parched corn and moose meat, while their canoe glides under these pines . . . They are thinking of the dead whom they have left behind on that solitary isle far up the stream, and of the relentless living warriors who are in pursuit." Perhaps his more empathetic turn came from thinking about the apple tree against which the infant Martha was killed; in Thoreau's day, many an old-timer claimed to have eaten fruit from that tree.

Not surprisingly, the Indians had a starkly different view of what transpired. Apparently, both the escaping boy and the badly wounded Abenaki woman made it to safety with their story, and in the generations since, Duston has been a symbol of treachery and murder. Abenaki historian Margaret Bruchac has suggested that in the Native view, Duston "shifted into a dangerous non-human form" in order to commit the acts she did—although in the same breath, Bruchac dismisses the murder of six-day-old Martha Duston by the Abenaki raiders, saying the infant and other captives were "mercy-killed," in keeping with tradition, to spare them the rigors of a winter trek.

Historians caution that it is unfair to judge anyone in the past except by his or her own standards and the standards of the time. In this sense, what any modern bystander would view as atrocities—smashing a baby's head against a tree or tomahawking a sleeping child and ripping off its scalp—were acceptable reactions to frontier war. What is self-evidently wrong to us was not so to them, and although we can take comfort from humanity's albeit slow progress away from the acceptance of offhand savagery, we need to be careful to view actions on both sides through the prism of the times.

Except Hannah Duston and the others bound up in her story won't remain quietly part of the past. Each generation of New Englanders, white and Native, has had to wrestle with her. Whereas nineteenth-century writers seemed unsure what to make of this episode of violence by a woman, some twentieth-century feminists contended that Duston's actions were those of an oppressed woman substituting Indian victimizers and victims for the white patriarchy.

Not all the attention has been academic. In the 1970s, a rather busty version of Duston was turned into a commemorative whiskey decanter by the Jim Beam distillery, complete with her fistful of scalps. Nor, hard as it may be to believe, was that the nadir of Duston's commercialization; that dubious honor goes to a recent Hannah Duston bobblehead doll.

Duston again became a lightning rod in 2006, when the Haverhill Rocks music festival used a doctored image of her statue showing her holding a red electric guitar instead of her trademark hatchet. Some people were bothered by the choice. "What, was Lizzie Borden busy?" one resident quipped to a reporter. Others defended it. A consultant helping Haverhill's revitalization efforts called Duston part of the town's unique identity and suggested that her statue should be no more controversial than one on Boston Common honoring William Tecumseh Sherman, a man seen as both a hero and "a psychotic murderer and arsonist," depending on one's background. Hollywood, sensing a good story, began talking about a film version of Duston's capture and escape, and soon Haverhill was discussing an annual Hannah Duston Day to breathe life back into the aging town.

As one would expect, Indians in general and Abenakis in particular weren't happy with the renewed attention given a woman they feel has already garnered more fame than she deserves. "More than being an Indian, being a mother I find it absolutely appalling that a community would promote violence and a violent act in a racist manner to young people today," said Chief Nancy Lyons of the Koasek Traditional Band of the Koas Abenaki Nation, descendants of Duston's captors and victims.

More than three hundred years after Hannah Duston canoed back to Haverhill, scalps wrapped in her apron, tensions of the war years on the First Frontier still simmer, and Duston's bloody legacy continues. As Margaret Bruchac has asked, "Must we be haunted by Hannah for eternity?"

Chapter 7

"Oppressions, Grievances & Provocacons"

S HORTLY AFTER SUNRISE on a January morning in 1704, an army of about sixty Englishmen and Indians converged on the center of La Concepción de Ayubale, one of the largest of the dozens of mission towns strung across the neck of Spanish Florida. These towns were connected by deeply worn trails that stretched from the cypress swamps and pine woods of the Apalachicola River near the Gulf of Mexico east to the huge presidio at San Agustín de la Florida (St. Augustine), whose cannons guarded the Atlantic Ocean.

Normally, hundreds of Apalachee Indians tilled the fields of maize and squash that surrounded Ayubale, and they attended Mass in the large mud-walled mission church where the Franciscan priest, Fray (Friar) Angel de Miranda, presided. Traders brought tanned deerskins to the village to be painted and decorated, for the artisans of Ayubale were famed for this craft. This morning, however, the town seemed deserted, except for the crowing of roosters, the scuffling of a few pigs, and the worried barking of dogs. The dry, wheat-colored palmetto fronds, with which all the mission buildings were thatched, rustled softly in the breeze.

As the invaders reached the edge of the town proper—the church and convent, the storehouses and other buildings, all shielded by a palisade—arrows began to flash through the air, striking the packed dirt streets and sinking into the unwary attackers' flesh. The advancing men scurried for cover behind a large wattle-and-daub house while their commander, Colonel James Moore, considered his next step.

Moore—born in Barbados, the former governor of English Carolina, and the leader of a faction of wealthy, staunchly Anglican plantation owners from Charles Town (later Charleston) known as the Goose Creek Men—was officially on a diplomatic mission to "Apalatchia." It must have seemed a strange kind of diplomacy to the defenders of Ayubale. His small band of soldiers was accompanied by almost a thousand Creek warriors, nearly all of whom were, at this moment, attacking the surrounding villages and farmsteads—a frequent occurrence for the local Apalachee, who were the victims of regular Creek slaving raids.

Talk of diplomacy was political window-dressing. Moore's mission was widely understood to strike a powerful blow against Spain, whose missions and military had dominated the colonial Southeast for more than a century and a half. But Moore, who had offered to finance the months-long expedition out of his own pocket, also knew it was a golden opportunity to reap a windfall in plunder and slaves.

Swinging axes, Moore's men raced to the church door and hacked at the wood, while the Christian Apalachees inside, trapped with Fray Miranda, rained arrows down on them, killing several of the Carolinians. The battle seesawed through the morning and into the early afternoon, until the Apalachees ran out of arrows and the attackers finally set fire to the church roof. When Miranda emerged to beg for mercy, more than two dozen Apalachees lay dead behind him, and the eighty survivors faced a grim future in which slavery was probably their most hopeful fate.

Ayubale was the first stop in Moore's sweep through the Apalachee missions. In the weeks ahead, he and his army, brushing aside Spanish counterattacks, stripped and plundered the Apalachee towns that resisted them. Those willing to pay a ransom, such as San Lorenzo de Ivitachuco, about three miles from Ayubale, bought their survival—in Ivitachuco's case, with ten horseloads of food and the silver censers, patens, chalices, and other valuables from its church.

By April, Moore was shepherding north thousands of Apalachees and other mission Indians. "I now have in my Company all the whole People of three Towns, and the greatest part of four more: we have totally destroyed all the people of two Towns," he reported to the governor of Carolina. "The waiting for these People make my Marches slow; for I'm willing to bring away with me free, as many of the Indians as I can."

What Moore meant by "free" is a window into the tortured conventions of the day. The "free" Indians were those Apalachees who had surrendered, and who would be resettled hundreds of miles to the northeast, along the lower Savannah River, joining a growing community of displaced Natives protecting the colony's southern flank and providing customers for its traders. Not all the Apalachee needed convincing. Tired of being the victims of English-sponsored attacks, some felt it was safer to live under Carolina's protection instead of its sword. Even the royal governor of La Florida, José de Zúñiga y la Cerda, admitted that "the entire population of two [towns] accompanied them voluntarily."

But outnumbering the thirteen hundred or so "free" Indians were more than four thousand others, mostly women and children, who came from towns that had fought back and lost. Under the rules by which the slave trade was conducted, they were destined for the plantations of the Carolina Low Country or the slave markets of the Caribbean and New England. Besides producing a handsome profit for Moore and his Creek allies (who made additional raids in the months ahead), this campaign devastated the breadbasket of the Spanish mission system; so terrorized the remaining Apalachees that many fled west to French Mobile or Spanish Pensacola, or east to St. Augustine, reducing the province's prewar population of more than eight thousand to barely a few hundred; and gave notice that Carolina was now a power to be reckoned with in the Southeast.

The odds against Carolina rising so high, so quickly, were steep. History was against it. Spain and France had both tried and failed to settle the region, and the first English colony at Roanoke had famously perished. There were other desultory attempts by the English, who tried to send a small band of French Huguenot exiles to Carolina in the 1620s. That effort failed, as did another in the 1630s. Not until the 1650s did settlers from Virginia filter south into Albemarle Sound, followed by a formal colony beginning in 1670.

The tiny English beachhead was wedged amid Spain's extensive mission system to the south, French Louisiana to the southwest, and, most especially, powerful Native societies on all sides, whose strength, though diminished by disease, was still tremendous. Yet within just three decades, Carolina would leapfrog past Spain as the major colonial force in the region, dismantling the mission network, and through the colony's booming traffic in Indian slaves, set off a chain reaction that would reshape virtually every aspect of the southern frontier.

Slavery was not a new concept introduced by morally corrupt Europeans to innocent Indians. It had a long pre-contact history among southern chiefdoms, although the original scale was relatively small, and the understanding of what constituted slavery—among Mississippian Indians, a condition that could be transitory and might end in adoption, versus the usually permanent and hereditary status of slaves among Europeans—was markedly different.

What was a radical change was the idea of a global slave *trade*, linking Native slave catchers deep in the interior of the American Southeast with middlemen in Carolina, sugar plantations in the West Indies, farmers and storekeepers in Boston, and manufacturers in England, who made the highest-quality trade goods available in the New World. (In this respect, it was similar to the African slave trade, which similarly bound captives, native middlemen, and traders in an international economy.) When it became clear that the English traders would give the best exchange not for beaver pelts or deerskins, but for human beings, virtually every tribe enthusiastically took to slave hunting. That the founders of Carolina had expressly forbade the trade made not the slightest difference to Englishmen such as Moore, and still less to their Indian partners, who saw a lucrative path to European goods. By 1715, it is estimated that the Carolinians and their Native allies had enslaved up to 51,000 Indians, perhaps a quarter of the entire population of the Southeast.

The result was cataclysmic. From Cape Fear south to the Florida Keys and west to the Mississippi, almost every society—French, Spanish, English, and dozens of Indian nations encompassing hundreds of thousands of people—was shaken by war, disease, waves of refugees, and cultural chaos. In the space of barely half a century, the Indian slave trade radically altered the face of the South, emptying enormous and once heavily populated regions; smashing established Indian societies and scattering the survivors, who blended into new tribes and confederacies; shipping tens of thousands of people into bondage in distant lands; and crushing the Spanish mission system

It also lifted the English Carolinians on a rising tide of wealth, power, and political success they could scarcely have imagined, permitting them to handily crush one Indian revolt. But then their Native allies, fearful of becoming slaves themselves, turned against them and, in the single greatest Indian uprising in colonial history, came within a whisker of wiping the Carolinians off the map.

For James Moore, the expedition against the Apalachee promised several benefits. Not only would it burnish his political reputation, tarnished when he launched a futile assault on St. Augustine in 1702, but it was also likely to pay off handsomely in terms of slaves, although he was officially instructed to "gain [by] all peaceable means if possible the Appalaches to our interest." But those orders were a fig leaf. There was no public money to pay for the undertaking, meaning that everyone—Moore, the members of the Commons House of Assembly, and probably many of the Creeks who joined him—knew that the only way he could recoup his expenses was through booty and slaves. Peace would be very much a secondary consideration.

Although the Spanish had initially established military outposts as far inland as what is now eastern Tennessee, these had largely been abandoned by 1570, and Spain's focus turned instead to creating a network of Franciscan missions. By the 1650s, these missions stretched from the province of Guale, along the Georgia coast, to Timucua in northern Florida, Apalachee in what is the modern Florida Panhandle, and Mayaca-Jororo in central Florida. The well-fortified *castillo* at St. Augustine anchored Spanish interests in the region, while the missions, especially the food-producing region of Apalachee, supported colonial activities throughout La Florida.

The Apalachee had been among the first southeastern Indians to encounter the Spanish, having mauled (and scalped) Hernando de Soto's men in 1539. At one point, the Apalachee had fired their own town of Ivitachuco rather than have it fall into Spanish hands. By the turn of the eighteenth century, however, most Apalachees had become devout Catholics—although at the same time, they lived in a Native society whose structure was little changed from the prehistoric Mississippian culture, built around chiefdoms ruled by hereditary leaders known as *holahtas*. The defiant town of Ivitachuco that de Soto had encountered had become the quiet village of San Lorenzo de Ivitachuco, home to a *doctrina,* or central church, whose leader, Don Patricio Hinachuba, had bribed Moore with silver treasures and ten loads of food. Despite his Hispanic name and fluency in spoken and written Spanish, Hinachuba was able to trace his lineage back, *holahta* by *holahta,* into the dim pre-Spanish past, even as he begged Governor Zúñiga for arms and a dozen soldiers to defend God, "the crown of my king, and the many children of this village."

The Spanish had, in a way, simply grafted themselves onto a society already well suited to the colonial demands of food production and institutionalized labor. Unlike the English, who populated their colonies with farmers and tradesmen, the Spanish immigrants consisted of soldiers, religious leaders, and bureaucrats, supported by a royal subsidy, or *situado,* and fed by the maize, squash, and bean fields of the mission Indians, without which they would have starved.

Apalachee Province was one of the richest farming areas in Spanish Florida, and whether they liked it or not, the Apalachee were important cogs in the Spanish *repartimiento* system. Leaders such as Hinachuba were responsible for filling annual labor quotas, providing human beasts of burden to cart food and other commodities along the footpaths that wound hundreds of miles through forests, across rivers such as the Río San Juan de Guacara (today's Suwannee), and around cypress swamps. The work was not only backbreaking; it was also dangerous. In one baggage train comprising two hundred Apalachee men hauling tons of corn to St. Augustine, all but ten eventually died of hunger and fatigue.

Because the Spaniards would not supply the Apalachee with muskets, they were ripe for the picking by English-armed Indians from the interior, such as the Creek. Not a single tribe, the Creek were an alliance of many different people drawn together by the depopulation, war, and political upheaval of the late seventeenth century. They eventually consolidated into the Lower Creek in the east, across what is now Georgia, and the Upper Creek along the Coosa and Tallapoosa rivers, in modern Alabama. The southern and eastern fringe of their land bordered the Apalachee and Timucua mission provinces—fertile slave-hunting grounds.

The slave trade had long troubled the lords proprietors, eight powerful men to whom King Charles II had granted control of Carolina in 1663. The issue of slavery itself was of little concern to them. Their gripe was that the trade, which rarely differentiated between local allies and distant enemies, lined the pockets of colonials such as Moore while robbing the proprietors of customers—and thus profits that should accrue to them through their monopoly of the trade in deerskins and pelts.

From England, Daniel Defoe, author of *Robinson Crusoe* and a prolific pamphleteer, attacked Moore for "having already almost utterly ruin'd the Trade for Skins and Furs . . . and turn'd it into a Trade of Indian or Slave-making . . . a Trade so odious and abominable, that every other Colony in America (altho' they have equal temptation) abhor to follow." Defoe's ire had more to do with Moore's High-Church persecution of non-Anglicans, including Presbyterians, Quakers, and Baptists—known in Carolina as Dissenters—than with his enthusiasm for human trafficking. Many other colonies had embraced this "odious and abominable" trade, but none had done so with such complete and utter abandon. Nor had any perfected the use of almost every instrument of public policy to maximize profit as had the Goose Creek Men, who had settled along their namesake waterway near Charles Town.

Even compared with some of the spectacularly venal behavior in colonies such as Virginia, the actions of the Goose Creek Men took the prize. They were, in a very real sense, a brutal trading cabal masquerading as a government. Early Carolina, concluded one historian, was "a dark kingdom ruled by cruelty, partiality, pride, revenge, and gain."

It wasn't supposed to be that way. On paper, at least, Carolina was going to be an enlightened Elysium. The "Fundamental Constitutions" proposed in 1669 were an odd mix of democratic, monarchic, and feudal ideas, blending religious tolerance with state-sponsored Anglicanism, creating two categories of landed nobility, and codifying systems of serfdom and slavery—yet guaranteeing even small landholders a voice in the government and the right to vote by secret ballot.

In practice, however, Carolina's utopian dream was hijacked right out of the blocks by avarice and political factionalism, and the Fundamental Constitutions were never ratified. From the very start, divisions arose between northern Carolina, with settlements around Albemarle and Cape Fear, and the southern power center at Charles Town, laid out in 1670 at the confluence of the Ashley and Cooper rivers. Most of the influential southern Carolinians came from Barbados, where slave-based plantations were already well established.

Barbadian planters weren't the only colonists. Carolina attracted a bewildering array of people, in part because of its early religious tolerance. English Dissenters joined Protestant French Huguenots, who were early players in the Indian trade, while Sephardim from Spain and Portugal established what would eventually become the largest Jewish community in the New World. A boatload of Scots founded Stuart's Town on Port Royal Sound, where earlier French and Spanish settlements had failed. Up and down the Carolina coast, pirates (both English and Huguenot) preying on Spanish ships and missions operated out of snug and secret harbors.

Taken as a whole, the Carolinian leaders were, in the words of the Anglican envoy Gideon Johnson, "the vilest race of men upon the earth. They have neither honor, nor honesty, nor religion enough to entitle them to any tolerable character, being a perfect medley or hotch-potch, made up of bankrupt pirates, decayed libertines, sectaries, and enthusiasts of all sorts . . . the most factious and seditious people in the whole world."

Not only did the Carolinians ignore the utopian constitution, they ignored the lords proprietors' pointed instructions to keep their hands off the local Indians, whose goodwill was considered essential for security and trade. Instead, they immediately began buying war captives offered by neighboring tribes, either using them as plantation labor or selling them to slaveholders in the West Indies.

Although the Goose Creek Men and their like were raising rice and other crops on their Low Country plantations, the real riches of Carolina—slaves and deerskins—were supplied by Native middlemen. Beavers, the mainstay of the northern fur trade, weren't as common in the South, but white-tailed deer were everywhere, exciting wonder from the earliest years of exploration. In the 1580s, Thomas Hariot at Roanoke wrote, "*Deareskinnes* dressed after the manner of *Chamoes* or undressed are to be had of the naturall inhabitants thousands yeerely by way of trafficke for trifles: and no more wast or spoyle of Deare then is and hath beene ordinarily in time before."

A century later, Thomas Ashe said that white-tailed deer were still in "such infinite Herds, that the whole Country seems but one continued Park." A single Indian, Ashe said, might bring in as many as two hundred skins a years, and by the turn of the eighteenth century, Carolina was exporting more than sixty thousand skins a year through Charles Town. By 1706, that number had almost doubled. A single hide would bring a pair of scissors or thirty round musket balls; five, a broad-hoe blade; sixteen, a woolen blanket; and thirty, a medium-weight "half-thick" coat. A musket required thirty-five skins. For the traders, the markup was immense; a hoe blade that cost six shillings to make and import turned a profit of more than 400 percent. Eventually, the term "buckskin"—shortened to "buck"—became frontier slang for American currency.

Valuable as deerskins were, Indian slaves were far more so. A single slave might bring as much in English goods as a winter's worth of deerskins, which explains the lengths to which Carolina traders went to get them and the eagerness with which Natives set about capturing their enemies. This last was a key point; although proprietary policy forbade slaving, it made an exception for those taken in war. Slavery was already a fact of daily life in Indian society, but with the opening of English trade, the scale on which it was practiced ballooned.

The first Indian group to take up commercial slaving were the Westo. Originally from the southern shore of Lake Erie, they became embroiled in the trade-driven conflicts over access to furs and Eu-

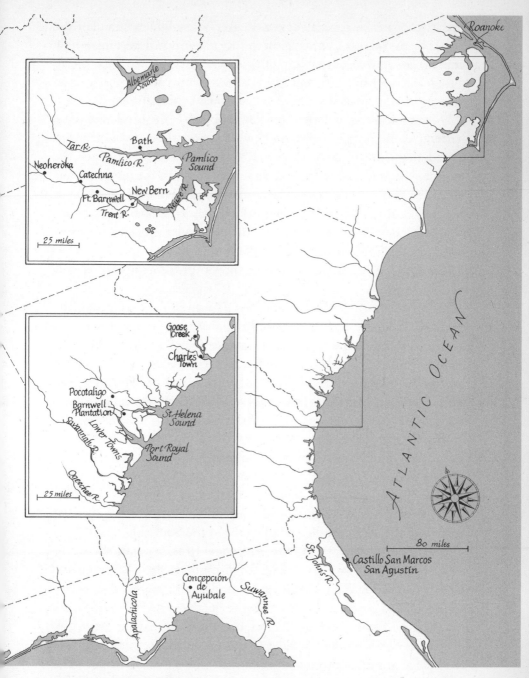

Albemarle Sound
Roanoke
Tar R.
Bath
Pamlico R.
Pamlico Sound
Neoheroka
Catechna
New Bern
Ft. Barnwell
Trent R.
Neuse R.
25 miles

Goose Creek
Charles Town
Pocotaligo
Barnwell Plantation
St. Helena Sound
Lower Towns
Savannah R.
Port Royal Sound
Ogeechee R.
25 miles

ATLANTIC OCEAN

80 miles

St. Johns R.
Castillo San Marcos
San Agustín

Apalachicola R.
Concepción de Ayubale
Suwannee R.

Landscape of the Tuscarora and Yamasee Wars, 1711–1715
(Modern boundaries added.)

ropean goods known today as the Beaver Wars, which engulfed the Northeast and Great Lakes region in the mid-1600s. These incredibly violent wars played out largely outside the direct view of European chroniclers, so what little we know about them comes mostly from haunting, blood-soaked stories handed down by the survivors.

Driven south to the upper James River by 1656, the breakaway Erie befriended the Virginia colonists, who armed them with muskets, named them the Rickohockan, and facilitated their entry into the nascent slave trade. Moving once again, this time to the upper Savannah River in the 1660s, they chased many Natives to the protection of the French in the lower Mississippi Valley and the Spanish missions along the coast. Such security was mostly academic, however. Now known as the Westo, they destroyed the mission at Santo Domingo de Talaje, along the Altamaha River, and swept through what is now inland Georgia.

Refugees from the Westo raids were washing up throughout the South. In 1663, the Spaniards noted the arrival in Guale Province of a group they called the Yamasis, or Yamasee—not a tribe in the usual sense, but a composite people thrown together by circumstance, most coming from the defunct chiefdoms of central Georgia. The Yamasee's ranks continued to swell as fresh Westo attacks drove them to the missions, where even the annual labor drafts of the *repartimiento* system seemed at first preferable to endless raids and the risk of death or enslavement. Similar fracturing and coalescing was occurring throughout the region—among the Catawba confederacy in the Carolina-Virginia borderlands, the Cherokee in the mountains, the Chickasaw and Choctaw to the west, and especially the Creek, whose newly cemented coalitions controlled a large area of the South.

After a rocky start, the Westo had come to terms with the Carolinians—a nearer, surer source of English goods than Virginia, and a ready outlet for their slaves. Again and again, they and English-backed pirates struck at Guale, now the northern rim of the Spanish system, until the last outpost there fell in 1684. But having the Westo as neighbors was a double-edged sword for Carolina. Although they provided a convenient source of slaves and skins, their ferocious presence blocked access to trade with any of the other tribes farther inland.

The equation shifted with the appearance of another refugee group from the north, probably (like the Westo) driven south by the Beaver Wars between the Iroquois and northern Algonquians, although no one is sure of their origins. Known to the Carolinians as the Savannah, and to later generations as the Shawnee, these new people provided an excuse for James Moore and others to turn on their erstwhile allies, the Westo. With Savannah help, they killed or enslaved all but fifty of the Westo and made the Savannah their new middlemen. The surviving Westos sent an envoy to plead their case, but he was shipped off into slavery, too. His fate should have been a warning to the Savannah and any other Indian allies of the English: Carolinian goodwill was a limited commodity.

Spain officially outlawed Indian slavery, but in practice the mission-based *repartimiento* system was little different, especially when the Spaniards tried to ship some of the Yamasee to Caribbean plantations to work off their labor debt. Between that and the repeated Anglo-Indian raids, the Yamasee leaders eventually decided it would make sense to switch sides. Hundreds of Yamasees, joined by Guales and scraps of other decimated tribes, moved north to Port Royal Sound. Eventually, ten Yamasee towns, with more than thirteen hundred inhabitants, lived under the English umbrella. There the Yamasee turned wholeheartedly to slave raiding themselves, attacking Timucua missions under Spanish control in the late 1680s and early 1690s.

In the Yamasee, the Carolinians had finally found a partner whose zeal and expertise at catching slaves matched their own rapacity for selling them. Like the Westo before them, the Yamasee had English flintlocks, which meant that no group of mission Indians, armed only with bows and arrows, could hope to withstand their assaults. And as was increasingly the case across the frontier, political developments thousands of miles away in Europe drove events—in this case, in ways that made the Yamasee-English partnership especially successful for both parties.

During the Nine Years' War of the 1690s (known as King William's War in the northern colonies, the conflict during which young John Gyles was kidnapped by the Abenaki), Spain was an English ally against the French. Slave raids against the Florida missions were, at least on paper, an affront against an imperial partner. But when war erupted in 1702 over the disputed Spanish throne—the War of the

Spanish Succession, which in the colonies was called Queen Anne's War—Spain was aligned with France against England. The Goose Creek Men had the perfect geopolitical excuse to launch an all-out assault on Spanish and French holdings in the South.

With the Yamasee and Creek at their backs, the Carolinians went after what was left of the vulnerable Spanish mission network. Starting with Moore's attack on Apalachee in 1704, the mission system all but crumbled over the following four years, leaving the Spanish essentially marooned in the fortress of St. Augustine, the rest of northern Florida all but deserted.

English successes against the missions alarmed France, which saw its own small presence in Louisiana at growing risk. They reached out to potential allies in the interior—the Choctaw, the Yazoo, and other tribes—and while they couldn't match the quality or quantity of English goods, the French offered a more sophisticated understanding of Native customs and social ties, and a firm promise never to enslave their allies, all of which they played adroitly to maintain their Indian buffer.

The French also drew up plans for a massive Franco-Spanish thrust designed to drive the meddlesome English out of the Atlantic seaboard entirely—beginning in the Caribbean, advancing rapidly to Charles Town, and destroying coastal settlements all the way to New York. Instead, the coup de main became a farce. The joint fleet floundered through the Caribbean, arriving at Charles Town to find the Carolinians forewarned, armed, and contemptuous of demands for ransom—and able handily to repel the few sorties for slaves and booty.

The grand armada caused barely a hiccup in Carolina's success. The Creek were raiding throughout the Gulf of Mexico drainages, and the Yamasee continued to range at will. "We have been intirely kniving all the Indian towns in Florida wch. were subject to the Spanish," wrote Thomas Nairne, the Scottish-born Indian agent for the colony. A few years later, in 1708, he noted that the Yamasee and other Carolina Indians "are now obliged to goe down as farr on the point of Florida as the firm land will permitt," having "drove the Floridians to the Islands of the Cape." No longer safe even among the mangrove islands of the Keys, some of the Florida Indians gambled on the only avenue still open to them—the ninety-mile voyage to Cuba, if they could gain Spanish permission. There, at least, the Yamasee could not reach them.

One of the few things that can be said with any certainty of John Lawson, before his sudden appearance in Carolina, is this: the man was not shy about taking a chance.

In the spring of 1700, about to embark from London on a trip to Rome, Lawson impetuously changed his mind—and his life—after a casual conversation with an unnamed gentleman, who "assur'd me that Carolina was the best country I could go to." Without an apparent second thought, Lawson booked immediate passage for New York and thence to Charles Town. He seems to have been well schooled—he later disparaged most of the travelers to America as "the meaner sort, and generally of a very slender education"—and he proved to be a skilled naturalist. But when he was born and where he was raised and educated—perhaps the north of England, perhaps in London itself—no one really knows; he appears on the historical stage in Carolina as a blank slate.

Although he apparently lacked any backwoods experience, within a few months he managed to talk himself into an appointment to explore the hinterlands, a rugged trip with five other Englishmen and four Indian guides that inscribed an immense, five-hundred- to six-hundred-mile arc from the Santee River in the south, far up into the mountains where few English explorers had penetrated, and down the Tar River to the coast. It took almost two months, beginning in the dead of winter and proceeding across rivers in dangerous flood, through swamps in which they had to swim naked, and to dozens of Indians towns on which Lawson cast a sharp and observant eye. They shot passenger pigeons, "which are so numerous in these Parts, that you might see many Millions in a Flock," and encountered bears, wolves, bison, and panthers. When their canoes finally reached Pamlico Sound, John Lawson knew he had found his place in the world.

For the next eight years, Lawson traveled through the Carolina backcountry, observing the wildlife, and especially the Indians, and learning the art of surveying from Edward Moseley, the surveyor general of the colony. Lawson built a house along the lower Neuse River, not far from the Neusiok town of Chattooka, and he speculated in land. In 1705, he cofounded Bath, northern Carolina's first town, and laid out its 70 one-acre plots. When he returned to England in 1708, it was to publish the first version of *A New Voyage to Carolina,* his brisk and (except for a few chamber-of-commerce flourishes) reasonably accurate assessment of the land, people, and natural bounty of the region. Named by the lords proprietors to be the new surveyor general, Lawson returned to Carolina to map out the northern border with Virginia.

This new job brought him into even closer contact with the Indians of Carolina, especially the "Tuskeraroo." Like Lawson, the Tuscarora are of mysterious origins. A people speaking an Iroquoian tongue, they probably moved south into Carolina around A.D. 600, based on current archaeological evidence. It's ironic that we have Lawson to thank for the clearest picture of Tuscarora life—just before it was irreparably ripped apart.

Lawson described a network of about fifteen Tuscarora villages, with a total of about twelve hundred men of fighting age, suggesting a population of about five thousand to six thousand people. There were two divisions, the southern Tuscarora along the Neuse River and Contentnea Creek, and the northern Tuscarora, along the Tar and Roanoke rivers. Each village, Lawson said, had its own *teethha,* or chief, and took care of its own affairs. Although Tuscarora communities were described as towns, they didn't always appear so to English eyes, especially the widely separated farmsteads of the southern Tuscarora. "Tho' this be called a town, it is only a plantation here and there scattered about the Country," one observer wrote, "no where 5 houses together, and then ¼ a mile such another and so on for several miles." Northern villages, on the other hand, were more centralized, and some were palisaded, given their proximity to the once powerful Tsenacommacah confederacy in Virginia.

Lawson's long, detailed account describes Tuscarora life, customs, and social structures in the early years of contact. Few explorers wrote with such evident admiration, although he found their treatment of captives "a natural Failing," especially one horrific method of torture, in which they "split the Pitch-Pine into Splinters, and stick them into the Prisoner's Body yet alive. Thus they light them, which burn like so many torches; and in this manner they make him dance around a great fire, every one buffeting and deriding him, till he expires."

The kind of natural delights and fertile landscapes Lawson described in his book attracted more and more settlers to Carolina, and by the first years of the eighteenth century, the immigrants were coming from new areas of Europe. One group that would make an enormous mark on the frontier—and whose fortunes would collide with those of the Tuscarora—were the Palatines, named for the fertile but war-scarred land along the Rhine in what is now southwestern Germany.

Lying on the French border, the Palatinate had been torn by competing armies for generations and laid particular waste in the Thirty Years' War of the early seventeenth century, one of the most destructive periods in European history. Even with the end of that conflict, warfare continued to roll across the region, and even in neutral countries such as Switzerland, floods of refugees and growing religious persecution against controversial pietist sects such as the Mennonites and Amish prompted an exodus. An estimated thirteen thousand Palatines fled to London, so straining the city's ability to house them that Queen Anne and her advisers offered to pay for their resettlement in the colonies.

Most of these Palatine immigrants wound up in New York and Pennsylvania, but one group sailed for Carolina under the hand of the Swiss baron Christoph von Graffenried, who'd been planning a New World settlement for years and who found in John Lawson the man who could make it happen. With title to more than seventeen thousand acres near Lawson's home on the Neuse and Trent rivers, Graffenried dispatched the English surveyor and more than 650 Palatines in January 1710, following himself with another hundred Swiss settlers some months later.

Initially, the undertaking was a disaster. Crammed belowdecks with poor food and water, and riddled with disease, almost half the Palatines died on the sea voyage or shortly after their arrival in North Carolina. (The political divisions between northern and southern Carolina had become so great that the colony was by then splitting in two, and the lords proprietors had named one of their own, Edward Hyde, as governor of the newly independent north.) Graffenried later accused Lawson of settling the Palatines on hot, unhealthy land on the Trent River that Lawson himself owned and wanted cleared, instead of in a better location on the Neuse. But whatever the initial hurdles, the industrious Germans and Swiss soon put their affairs in order, and the town of New Bern began to thrive. "Inside of 18 months these people were so well settled . . . that in this short time they had made more advancement than the English inhabitants in four years," Graffenried bragged.

Those who were working the ground had, perhaps, a more level-headed view of the colony. "The land is good, but the beginning is hard, the journey dangerous," one Swiss settler wrote to his relatives back home. "This country is praised too highly in Europe and condemned too much . . . Of vermin, snakes, and such like, there is not so much as they tell in Europe. I have seen crocodiles by the water, but they soon fled. One should not trust to supporting himself with game, for there are no wild oxen or swine. Stags and deer, ducks and geese and turkeys are numerous."

Another, more serious complaint Graffenried leveled against Lawson was that, the surveyor's assurances to the contrary, the land around New Bern was already inhabited by several tribes, including the Neusiok and Coree, whom the Swiss had to pay to relinquish their land. Graffenried sensed the Indians' increasing frustration with colonial encroachments, of which the Germans and Swiss were only the latest and most egregious example. More and more of the tidewater villages were being forced inland, gathering near the Tuscarora, higher up the Neuse and Trent rivers.

Things were getting so bad that some Tuscaroras wanted out of Carolina completely. Slaving raids snatched Tuscarora women and children at will, and plantations gobbled up much of the best lowland hunting ground. The year before, the Tuscarora had sent an extraordinary mission north to Pennsylvania, which had a few years earlier become the first English colony to ban the importation and sale of Indian slaves. On June 8, 1710, they met with two commissioners from Pennsylvania, asking permission to relocate to the Susquehanna Valley, which was becoming a magnet for Native refugees. In fact, the Tuscaroras were hosted by leaders of the Conestoga and Shawnee, the latter the former Savannah, who were themselves recent migrants to Pennsylvania.

As was the custom, the Tuscaroras presented their request in the form of eight long wampum belts. The designs, speaking in the ancient, beaded language of Indian diplomacy, echoed what the orators said. One belt represented the wishes of "elder women and mothers . . . so they might fetch wood and water without risk. By the second, the children born and those about to be born, implored for room to sport and play without the fear of death and slavery." Old men asked for forest paths as safe as forts; the chiefs sought a stable government; and all the Tuscarora asked for an end to murder and slavery. With each petition, the Tuscarora speakers held up the corresponding wampum, elaborating on their meaning, until they reached the eighth and final belt, perhaps the most important—one that asked the Pennsylvanians to show them the way to bring this vision of peace to pass.

The colonial officials were willing to consider the Tuscarora request, but they asked for an impossibility—a letter from the North Carolina government attesting to the Tuscarora's "past carriage toward the English" and good behavior. The Carolinians, unwilling to lose a source of slaves, refused to provide such a letter, and the Tuscarora felt the screws tighten further.

In the reckless approach North Carolina took toward its Indian neighbors, Alexander Spotswood, the governor of Virginia, saw trouble ahead for all English colonies. "I am credibly informed the Indians have more reason to complain of injustice from the people of Carolina, who are daily trespassing upon them," he warned Hyde in 1711, "and if they do sometimes retaliate it is more excusable because your people have been the first aggressors."

Spotswood went on to say he'd learned that some Carolinians intended "to fall upon the Indians, and to compell them by force to yield to their unreasonable pretensions." Such a move, he warned, would be followed by "a Train of ill consequences, by involving both Governments in a War with the Indians, for tho' [the Carolinians] may perhaps surprize one Nation, they ought to consider that there are a great many others that will take the alarm when they find the English have broke faith with them." Such an uprising, Spotswood said, would be "a just reproach on us."

In September 1711, a year after Graffenried arrived in North Carolina, Lawson came to New Bern and asked the baron to join him in a trip up the Neuse River, "saying there was a quantity of good wild grapes, that we could enjoy ourselves a little with them." The weather was sweet, and Lawson, who relished his time in the boondocks, was obviously looking for an excuse to get into the woods.

Graffenried scoffed at the idea; he had better things to do. But Lawson kept after him. What if, instead of just chasing wild grapes, they also investigated how far upriver the Neuse was navigable? What if they determined how far a journey it was to the Appalachians and whether there might be a shorter, easier route to Virginia? Here he hit the mark, because Graffenried had been curious about those questions himself. Maybe he didn't need much convincing; the weather was "fine and apparently settled," so they quickly packed two weeks' supplies, gathered two black slaves to paddle their canoe and two local Indians to act as guides, and set off. When Graffenried asked if they should worry about Indians, Lawson replied that "this was of no consequence . . . that he knew of no wild Indians on this arm of the river."

Less than three days into the trip, however, they ran into trouble—not from "wild Indians," but from the Tuscarora Lawson knew so well. Already alarmed by the rumors of English treachery, and angry at the growing intrusions on their land, they were upset to find the surveyor general sniffing around. Charged with trespassing, Lawson, Graffenried, and the two slaves were seized and taken to the southern Tuscarora village of Catechna, where, to great excitement, the Swiss baron was initially mistaken for Governor Hyde.

There followed an extraordinary series of what Graffenried characterized as "trials," argued before elders, including the Tuscarora *teethha* known as Hancock. "There came . . . a general complaint, that they, the Indians, had been very badly treated," Graffenried said, and Lawson was singled out for his actions. Hancock was initially satisfied by their answers, though, and agreed to free them the next morning. Their departure was delayed just long enough for the chiefs from some neighboring villages to arrive, including those from the Neusiok village of Chattooka, still smarting from their eviction by the Palatine colonists. Lawson, who seemed to have trouble keeping his mouth shut even when his life depended on it, got into a roaring argument with Coree Tom, the chief of Chattooka, in which Lawson threatened to exact a horrible revenge on them all. "This spoiled everything for us," Graffenried said, with considerable understatement.

Tempers rose on all sides. The furious chiefs grabbed the Europeans, hauled them back to the council grounds, and bound them, throwing their hats and wigs into the fire, while some of the younger men rifled their pockets and the elders debated. To their shock, the captives were this time sentenced to death.

In the morning, they faced their third and final tribunal—a sign that the Indians were still undecided about their fate. They did not meet the challenge as a unified front, however. "I turned toward Mr. Lawson bitterly upbraiding him, saying his lack of foresight was the

Trial of Lawson and Graffenried
This sketch, made by a colleague of Christoph von Graffenried based on the Swiss baron's description, shows him, the English surveyor John Lawson, and Graffenried's African slave under guard while a shaman "conjures," dancers wheel around a fire, and a council of Tuscarora elders decides their fate. (Burgerbibliothek Bern Mss. Muel. 466[1]p1)

cause of our ruin," Graffenried said. Spying an English-speaking Indian "dressed like a Christian," Graffenried pulled him aside for some frantic, backdoor diplomacy. He was sorry, he said; obviously, this wasn't *his* fault; *he* had threatened no one. The protestations seemed to fall on deaf ears.

The two prisoners, plus one of Graffenried's slaves, were stripped of their coats, tied hand and foot, and guarded by a hatchet-wielding Indian "in the most dignified and terrible posture that can be imagined," as well as by four warriors armed with flintlocks. While women and children danced and a shaman conjured, the council met through the day and into the night. Graffenried, twisting around in his bonds, "made a short speech, telling of my innocence, and how if they did not spare me the mighty Queen of England would avenge my blood."

In the dark, Graffenried made his peace with God, tried to instruct his slave in proper contrition, and ignored the man who'd gotten them into such a mess. "Surveyor General Lawson, being a man of understanding though not of good life, I allowed to do his own devotions," he said.

The Tuscarora weighed more than just the sentences for the two whites and their slaves; it's likely they were deciding whether the time had come to send a much more direct and pointed message to the English. They freed Graffenried and the slave, but Lawson was executed—just how, the baron did not know. "The threat had been made that he was to have his throat cut with a razor that was found in his sack," Graffenried said, but later reports contended that Lawson was hanged, or—a morbid irony if true—slowly burned to death with pitch-pine slivers, exactly the way he had described in his book.

That was just the beginning. More than five hundred fighters drawn from Hancock's Tuscarora and a coalition of smaller tidewater tribes—the Neusiok, Coree, Mattamuskeet, Pamlico, and Machapungo, among others—swept the Neuse, Trent, and Roanoke River settlements on September 22, 1711, killing between 120 and 140 English, Swiss, and Palatine settlers and taking dozens captive. The Carolinians were shocked and badly unprepared. For most of the year, their attention had been fixed on an armed struggle between Hyde and his opponents, a dustup known as Cary's Rebellion, which had been put down in July by British marines. Powder was short, there were no fortified refuges to which they could retreat, and by winter the residents of New Bern (which had been spared the initial attack because of Graffenried's efforts) had burned their homes and fled rather than remain on the frontlines. Food, already in short supply because of a drought and the chaos of the rebellion, was so scarce that famine was a ready danger, and illness struck down so many that Governor Hyde "cannot find a member to advise with . . . nor Assembly that will meet to do business." An attempted counterattack that winter resulted in a rout of the colonial militia and still more deaths.

Not all the Tuscarora took up the hatchet, however. The northern Tuscarora towns between the Roanoke River and Albemarle Sound, generally united behind a *teethha* known as Tom Blount, stayed out of the fight. Blount had much closer relations with the colonists and had been instrumental in convincing the council to spare Graffenried. Now he sent emissaries to Virginia, testifying to Tuscarora friendship and even offering to help against their own people. Graffenried backed them up, and Virginia's Indian traders spoke of their sincerity as well. Regardless—and to Governor Spotswood's enormous frustration—the House of Burgesses voted to raise £20,000 "for carrying on a war against the Tuscaruro Nation in general, upon a bare Surmise that the whole Nation was concerned in the Massacre, tho' it plainly appeared otherwise." Only when it became clear that they didn't have the money did the burgesses relent.

Virginia and the northern Tuscarora stayed out of the fight, but South Carolina finally came to the aid of its beleaguered neighbor. Some historians have argued that Hancock and the southern Tuscarora, having made their point with the September attacks, expected a negotiated settlement. If so, they were badly mistaken. South Carolina dispatched Colonel John Barnwell, a forty-year-old Irishman who'd come to the colony only a decade earlier, "out of a humor to goe to travel," and parlayed his success in raiding the Spanish missions into ever greater roles in the government.

Backing his force of thirty colonials were almost five hundred Natives from a variety of tribes. Most eventually slipped off as soon as they'd taken some slaves, but not "my brave Yamassees," as he called the Indians that formed the heart of his auxiliaries and whose towns lay close to his plantation on Port Royal Island. Marching through winter floods, on roads so swampy they had to repeatedly rig ropes to haul out deeply mired horses, Barnwell's men arrived in Tuscarora territory in late January, to find that their enemies had not been idle.

Barnwell confronted a string of "small forts at about a miles distance from one another . . . I have seen 9 of these Forts and none of them a month old." Egged on by the Yamasees, Barnwell's men attacked the first one they encountered. Breaking in despite heavy fire, they were stunned to find that "within the Fort were two Houses stronger than the fort which did puzzle us and do the most damage." The defenders, including desperate Tuscarora women who took up bows, "did not yield untill most of them were put to the sword." Busy with executing the men, Barnwell complained that the Yamasees "got all the slaves & plunder, only one girl we gott." A few days later, having swept through a large Coree village, they cooked and ate one of their Indian victims "to stimulate themselves still more."

Barnwell was the first to face the Tuscarora genius for improvised defense. If the small forts were a surprise, he got a genuine shock when he reached Hancock's town of Catechna. The village was deserted; the men had withdrawn into a newly built fortification nearby, while the women and children were hiding in the swamps. This wasn't the simple ring palisade of the Pequot's Mystic Fort, and it was considerably more sophisticated than even the Narragansett's Great Swamp fort, with its blockhouse and moat. At Catechna, the Tuscarora had built one of the most ambitious defensive works any Indian nation raised in the entire colonial era, and they'd done it in very little time.

Perched along the banks of the Neuse, the fort was "strong as well by situation . . . as Workmanship," Barnwell admitted. There were several layers to the defense. A jungle of large tree limbs surrounded the fort at some distance to slow assaulters. Inside this, the ground was spiked with "large reeds & canes to run into people's legs." A deep trench lay immediately outside the heavy palisade walls, which were peppered with two tiers of loopholes through which defenders could fire, and which they could then plug up for safety. Four round flankers built into the walls gave unobstructed lines of fire.

Barnwell's initial attacks—in which his Carolinian and Indian troops rushed the ditch in the pouring rain, hoping to bridge it with bundles of sticks known as fascines—withered under Tuscarora arrows and musket shot. So Barnwell settled into a siege, the fort's one evident weakness being its lack of a ready water supply. After several days, the Tuscarora responded by torturing their English and German captives. After they killed an eight-year-old girl within earshot of South Carolina forces, Barnwell finally offered a truce.

Although the Tuscarora didn't know it, his men were low on food and almost out of ammunition. Barnwell agreed to withdraw, and the Tuscarora promised to relinquish their hostages. Marching twenty miles downriver, he waited for resupply from the North Carolina government he was supposed to be rescuing and for the Tuscarora to fulfill their end of the bargain. He was disappointed on both counts. After two impatient weeks, Barnwell marched his men back to Catechna and found that the Tuscarora had used the time to greatly expand the fort, encompassing the land from which he'd launched his first assault.

Even worse, his meager North Carolina reinforcements showed up with no supplies, "without a dram, a bit of meese bisket or any kind of meat but hungry stomachs to devour my parcht corn flower," Barnwell fumed. When the newcomers had the temerity to grouse about the food, Barnwell flew into such a rage "that I ordered one of their majors to be tyed neck and heels & kept him so, and whenever I heard a saucy word from any of them I immediately cutt him, for without this they are the most impertinent, imperious, cowardly Blockheads that ever God created & must be used like negros if you expect any good of them."

But Barnwell did manage to acquire some light artillery, and although they had little powder, the English "contrived several sorts of Ingenious Fireworks, & a mortar to throw them." Day after day, the Carolinians and Yamasees dug trenches, lobbed mortars, and tried a number of times to burn down the walls of the fort, without success. Digging nearer to the palisade, they were stunned to find that the Tuscarora had been tunneling, too, intercepting their lines. "This siege for variety of action, salleys, attempts to be relieved from without, can't I believe be parallelled against Indians," Barnwell told his superiors.

For ten days, just forty-six Tuscarora men and a few "hardy boys that could draw bow" held almost three hundred Yamasee warriors and Carolina militiamen at bay. Short on rations and powder, Barnwell called another truce and got terms from the Indians—a return of captives, the delivery of highborn Tuscarora hostages, and an agreement to turn over Hancock and three other war leaders (who, the Tuscarora said, were in Virginia). The Tuscarora also agreed to "quit all pretentions to planting, Fishing, hunting or ranging to all Lands lying between Neuse River and Cape Fear." Finally, and perhaps to Barnwell most important, they agreed to tear down the bastions at Catechna and "never to build more Forts."

Barnwell raised the English ensign over Catechna, declared victory, and went home, where he was nicknamed "Tuscarora Jack." To a lot of people, though, it looked like an inglorious English retreat. Worse, Barnwell's Yamasees, under a flag of truce, seized hundreds of Tuscaroras to sell as slaves, thus reigniting the war. The fighting went badly for the colony. By the autumn of 1712, North Carolina—rudderless after the death of Governor Hyde in a yellow fever outbreak—was again calling for help.

South Carolina sent another army north, commanded this time by James Moore Jr., son of the former governor, with thirty-three Carolinians and nine hundred Indians—Yamasees, Apalachees, Catawbas, and Cherokees from the western mountains. Moore's forces labored through unusually deep snows, then became trapped for months by more bad weather in Albemarle, where, once again, the North Carolinians offered thin hospitality. "I want words to express the miserable state of this poore Countrey," said Thomas Pollock, the council president who was running the colony. He blamed the lack of preparation and supplies on "the ignorance and obstinacy of our Assembly." Moore's troops and Indians ate whatever they could beg, borrow, or steal from the surrounding countryside. "The destructions his indians make here of our Catle & Corne is intollerable," Pollock said, "so that some of the people here have been semingly more ready to ryse upe against them than march out against the enemy."

The Tuscarora and their allies had plenty of warning. They focused their collective attention on a new stronghold upriver of Catechna, at the town of Neoheroka, located along Contentnea Creek close to present-day Snow Hill, North Carolina. When Moore and his army finally arrived, around the beginning of March, they found themselves arrayed against a fort even more elaborate than the one that had stymied Barnwell. The Tuscarora had learned from their mistakes; they had trenched and covered a route to the creek for water and had dug bunkers for shelter during mortar attacks. They were confident enough this time that instead of sending their women and children to hide in a swamp, they kept them inside with the men.

Like Barnwell, Moore settled in for classic siege warfare, digging a huge, zigzag trench to get safely within range, then erecting block-houses of his own and drawing up artillery. Move and countermove; as Moore's men dug their way closer, the Tuscarora extended their own trenches to block them. Only after weeks of preparation did the English finally launch what turned into a three-day assault. Sappers undermined the Tuscarora walls and set off explosive charges, but the bad powder did little damage. Instead, the attackers managed to bludgeon their way through the defenses at several points, burning parts of the wall and bastions, then fighting vicious, almost urban warfare from trench to trench and house to house. The Tuscarora kept falling back to new, improvised defensive positions, making the colonials and their Indian allies pay dearly for each foot they advanced.

But in the end, the Tuscarora had put too much faith in their walls. Surrounded, and with their entire population trapped within the fort, they suffered a crushing defeat. Moore wrote to Pollock with the news: "Of ye Enemies Destroyed is as follows—Prisoners 392, Scolps 192, out of ye said fort—& att Least 200 Kill'd & Burnt In ye fort—& 166 Kill'd & taken out of ye fort on ye Scout." Almost a thousand Tuscaroras had been killed or captured, the seven hundred or so survivors bound for slavery. (One of the captives proved to be a furious emissary from the Seneca, who, having been ransomed from his Indian captor for a princely sum, was ferried back to New York on an English ship, with fulsome apologies from the nervous Carolinians.)

The war sputtered on for another two years, but never again did the southern Tuscarora and their allies fight behind walls. They depended instead on classic hit-and-run raids and the safety of deep swamps, trying to evade the English and Tom Blount's northern Tuscarora.

For the southern Tuscarora, the only way to peace seemed to be exile. Since their plea to settle in Pennsylvania had been denied, they turned now to their distant Iroquoian cousins in the Five Nations, who had been acting as advocates on their behalf for some time. Village by village, family by family, hundreds of Tuscaroras shouldered their belongings and trudged north on the paths running along the eastern flank of the Appalachians, resettling in New York and along the Susquehanna Valley in Pennsylvania. There, by the early 1720s, they were formally adopted into the Iroquois confederacy, which would thereafter call itself the Six Nations.

With the Tuscarora threat fading after Moore's victory, South Carolina settled back to the business of slaves and deerskins. Although the Yamasees who had accompanied both Tuscarora expeditions enjoyed a windfall from their captives, they and the other so-called settlement Indians were having a harder and harder time making ends meet. Overhunting and the spread of English plantations throughout the coastal plain made deer harder to come by, at a time when Yamasee demand for trade goods was skyrocketing. Traders were happy to extend obscene amounts of credit. By 1711, the four hundred men in the Yamasee villages alone owed a collective debt—much of it for rum—equal to 100,000 deerskins, almost a year's worth of exports from South Carolina as a whole.

Given the value of slaves—a single Indian man was worth more than a season's worth of deerskins—additional raids would have made sense, but as with deer, the supply of captives was drying up. Yamasee raids into Florida in the previous decade had driven away most of the Spanish Indians, including the Apalachee and Timucua, leaving the Yamasee without a ready source for new slaves. There was "not now so much as one Village with ten Houses in it, in all Florida, that is subject to the Spanish," said Indian agent Thomas Nairne. English traders were receiving slaves from French-held areas to the west, but they came from the Upper Creek and other inland groups, not from the Yamasee, Savannah, Apalachee, Lower Creek, and other settlement Indians, who found themselves in an economic vise.

The traders, always abusive, were becoming increasingly rapacious. Extended families and villages were being held responsible for repaying individual debts, and failing that, the traders might simply enslave the debtor or his relations. This wasn't strictly legal, and such antics were often overturned by the colony's Commissioners of the Indian Trade, who heard a litany of complaints from the Yamasee and other tribes. Some of the behavior was puerile, such as traders who sent "sum of their Indians 2 or 300 miles with a leter to each other that hath litle in it only to call one another names and full of debauchery." John Wright, whom Nairne had replaced as Indian agent, "when out among the Indians have a great numbers only to wait on and carry his lugage and packs of skins," charged a Virginia trader, "purely out of ostentation[,] saying in my hearing hee would make them honour him as their Governour."

Other traders were considerably more sinister, such as Alexander Long, a trader among the Cherokee, and his friend Eleazar Wiggan, whom the Indians called "the Old Rabbit," after a character in Cherokee folktales. Hard-bitten even by frontier standards—Long once detonated a batch of gunpowder, blowing up an Indian at whom he was angry—they goaded the Cherokee into attacking the Yuchi town of Chestowee, near the mouth of the Hiwassee River. That the Yuchi were friendly trading partners with the English made no difference; they were a small community, and some of them were in debt to Long. The Yuchi were about to learn a cardinal rule of dealing with the Carolinians: never become more valuable as a slave than as a customer.

Long had been lobbying for an attack on Chestowee for years—a Yuchi had thrashed him in a fight, and he wasn't one to forget a grudge—and so he concocted a spurious "order" from the South Carolina government demanding that the Cherokee "cut off the said Town." (In fact, another trader was on his way with instructions expressly forbidding such an assault, but he arrived too late.) The Yuchi, confronted with the prospect of bondage, gathered everyone in the town's central ceremonial house. The men killed their children and wives "to prevent their falling into the Hands of the Cherikees," then committed suicide. To the traders' disgust, only six Yuchis survived to be captured.

There was some outrage, and Long fled to the Cherokee. He and Wiggan had their trading licenses lifted for a time, but even the governor of South Carolina reportedly "seemed pleased with the Euchees being cut off and [said] that he knew no Reason why the Prisoners shoold not be Slaves."

Burdened by debts they couldn't pay, and with examples such as the Yuchi suggesting what might happen to them, the Yamasee and other tribes in the English sphere were deeply worried about the future. A census of Indian allies conducted by the English in early 1715, which found more than 28,000 Indians in 141 towns, looked to many of the freshly tallied subjects suspiciously like a run-up to mass enslavement. Talk of preemptive war was already flying among the villages, and there probably were some firmer plans for a coordinated attack being devised among the Creek. Then, into the middle of this already fraught situation fell a nasty, intensely bitter power struggle between Carolina's two Indian agents, Thomas Nairne and John Wright, which set the whole thing alight.

Since Nairne had replaced Wright in the post in 1712, on allegations that Wright had abused his authority, the two had been at each other's throats in a decidedly public way that split the colony. Traders and politicians formed factions backing one or the other; so did many of the Indian leaders with whom they dealt. As the two men dragged each other through the courts and before the Commons House, oversight of the traders evaporated, and the atmosphere became so toxic that the entire system of regulating Indian trade was on the verge of collapse. On top of that, a stalemate between the governor and the chief justice all but paralyzed the South Carolina government in the first days of April 1715, just as warnings of imminent attack by the Creek and other tribes, brought by breathless traders, reached Charles Town.

Spurred to action, Governor Charles Craven made a fatal mistake. Given the heated factionalism, and in the interest of seeking balance, he dispatched both Nairne and Wright to meet with the Yamasee in Pocotaligo, their principal town, about twenty miles north of Port Royal Sound. Nairne, speaking for the government, assured the Yamasee of English goodwill and their intention to resolve the Yamasee complaints against the traders. They all shook hands, and Nairne retired for the night.

Wright, however, apparently stayed up and conveyed a very different message to the Yamasee—"that the white men would come and fetch . . . the Yamasees in one night and that they would hang four of the head men and take all the rest of them for slaves." Wright told the Yamasee they were acting like women, which "vex'd the great warrier's."

The situation must have been baffling to the Yamasee. Two men with whom they'd worked for years and who, in the past, had both spoken for the Carolina government, now painted starkly different pictures of colonial intentions. Governor Craven, who could have settled the matter, was trying to arrange a parley to reinforce Nairne's message of reconciliation, but the Yamasee had no way of knowing this. They debated through the night, made up their minds, and painted themselves for war.

The next morning—April 15, Good Friday—the Yamasee and other settlement Indians killed all the traders they could find. One of the few survivors, shot through the neck and mouth, managed to swim across the channel to John Barnwell's plantation and raise the alarm. Nairne suffered torture by fire for three days before he died. Wright also was killed, although no one is sure how.

Although the war started with the Yamasee and their neighbors along the Savannah River, it seems plain the Lower Creek were ready to lift the hatchet themselves, and once begun, the fighting spread rapidly throughout the Southeast. Ninety of the nearly one hundred English traders working in the region were killed in the weeks that followed, and plantations were sacked and burned as Carolinians fled for Charles Town. Three hundred of them, unsure of their security even there, crammed into a smuggler's ship docked in the harbor.

Through the spring of 1715, the war rippled out to more distant tribes and confederacies, including the Catawba, Congaree, and other small groups along the Virginia border; the Cherokee in the mountains; the Upper Creek in what is now Alabama; and the Choctaw, whose land extended all the way to the Mississippi River. Each town, each council of leaders, had its own reasons for joining the fight. Although the Yamasee War took its name from the Natives who struck the first blow, it grew into the single largest American Indian uprising against any colonial power. Yet it was neither centrally planned nor centrally led. Even more than King Philip's War forty years earlier, it was the combined expression of many grievances and opportunities that differed dramatically from one tribe or confederacy to another, as each wrestled with their own circumstances, eyed their alliances, weighed their risks, and made their own calculus between war and peace.

For those Indians in closest contact with the Carolinians, debt, the loss of hunting land, trader abuses, and fear of slavery were likely the important sparks. The Yamasee were repaying "Oppressions, Grievances & Provocacons Offer'd to those Indians," said Dr. Francis Le Jau, an Anglican minister in Goose Creek. "In my humble Judgmt the Chief Cause of [what] we suffer are our Crying Sins." But for many of the Indians on the colonial periphery—those insulated from the Carolinians by distance and with access to goods from the French, the Spanish, and Virginia—the war may simply have been an opportunity in which fairly modest risks were outweighed by the potential for plunder and slaves. It was an opportunity, too, for Carolina's opponents; not only did the French and Spanish spur on their Indian allies, but some Carolinians accused Virginia's traders of doing the same.

The Carolina frontier essentially emptied, with the English huddled at Charles Town and a few other places where they had some measure of protection. Leaders throughout the English colonies watched the unfolding events there with unease; no one knew how far the repercussions might travel or which of their own Indian allies or enemies might be pulled into the conflict. As far north as New York, colonial envoys were urgently meeting with the Five Nations to ensure that they remained out of the fight. The Iroquois had long raided the Catawba and Cherokee, and their involvement could exacerbate an already complex situation.

South Carolina mustered its militia, hastily built a series of blockhouses and stockades in which to base their troops, and suffered several humiliating defeats, as well as one or two victories in rare pitched battles. They received some outside help; North Carolina sent a small contingent that included northern Tuscaroras, while Virginia and Massachusetts sent guns and powder. (Virginia also offered to send three hundred soldiers—but only if South Carolina sent an equal number of African slave women to replace them.) Facing stronger colonial defenses, the Indians fell back on their traditional lightning raids, but slowly the tide began to turn, especially after a Carolinian expedition south into Guale Province chased many of the Yamasee back to the protection of St. Augustine.

The Catawba swapped sides several times, first proclaiming their willingness to support the English, then joining in the attacks, and finally—after a walloping ambush by Carolina militia and armed black slaves—offering to fight on the colony's side. The Chickasaw, too, seem to have had second thoughts; they publicly blamed the April murders of Carolina traders in their villages on rogue Choctaws, although it seems more likely they simply changed their minds after striking those initial blows.

Similarly, the Cherokee were divided about the war. Some of the Lower Cherokee towns, those closest to Carolina, had participated in the early raids, while others, especially the Overhill communities across the mountains, argued for peace with the English. Sensing an opening, "the Old Rabbit," Eleazar Wiggan, offered—for the right price—to carry South Carolina's peace overtures to the Cherokee, with whom he successfully brokered a tentative deal.

Such frontier diplomacy was a minefield of overlapping and conflicting interests, as evidenced by the incident at Tugaloo, as the English called the Lower Cherokee town of Dugilu´yĭ. Hoping to ratify Wiggan's deal with the Cherokee, the Carolinians had sent three hundred men under James Moore Jr. to meet with the chiefs. While there, they learned that a contingent of Upper Creeks, invited by the Cherokee, were on their way to discuss peace. But at Tugaloo, the English found eleven dead Creek headmen and a twelfth about to be put to death by the Cherokee. The Cherokee story was that a huge war party of Creeks and Yamasees was nearby, waiting to ambush the English, but no sign of such a force was ever seen.

No one has ever been able to determine the truth of what happened—whether the Creek were really interested in peace and were lured into a trap, or whether they were lobbying the Cherokee for an alliance against the whites. The murders may well have been the result of pro-English factions among the divided Cherokee trying to force their opponents' hands. The Cherokee may even, as they claimed, have foiled what would have been a devastating ambush against Moore's troops. Whatever the underlying cause, the killings at Tugaloo precipitated war between the Creek and Cherokee that lasted for almost forty years, and they forced the Cherokee to ally themselves more closely with the Carolinians.

With no decisive victories, the Yamasee War lurched to a messy, drawn-out conclusion. As there was no unified Indian leadership, there could be no single treaty of peace, although most hostilities were over by 1717. The Cherokee alliance caused many colonists to breathe a sigh of relief, as did subsequent pacts with some of the Creek—whom the Carolinians immediately began to rearm, knowing they would use the guns against the province's ostensible allies, the Cherokee. "How to hold both as our friends . . . and assist them in cutting one another's throats without offending either" was how one colonist described "the game we intend to play if possible."

Almost forgotten today, the Yamasee War nearly accomplished what no other Native war of resistance did—elimination of a colonial threat. Even in the regional conflagration that was King Philip's War, there was never any real danger that New England as a European entity would be extinguished. During the early months of the Yamasee War, however, South Carolina's Indian enemies pushed the settlers to the very edge of the continent, and it's easy to imagine that with a few more lucky strokes, they might have smothered the colony entirely.

Although it survived, the war left South Carolina dramatically changed. The burned plantations and ruined farms could be rebuilt, but the Indians who had once served as a buffer on the colony's southern border, and whose trade fueled its economy, had largely left. Carolinians saw their frontier greatly contracted and their flanks vulnerable to incursions from the French and Spanish, along with their Indian allies. Palatines and other Protestant immigrants, who were beginning to flood into other parts of English America, shied away from the Carolinas, sailing instead for places such as Pennsylvania, which had for decades maintained relative peace with its Native neighbors.

South Carolinians tried to pick up the pieces. John Barnwell reached out to the Yamasee, who had once been his neighbors and comrades-in-arms and who continued to raid the colony. In 1719, he made a dangerous, clandestine visit to Spanish Florida, hoping to arrange a meeting with the Huspaw King, the Yamasee chief who many believe launched the war, but with whom Barnwell had once been friends. His message was carried by three Indians, one of whom was the king's relative.

The Huspaw King arrived at St. Augustine borne in a sedan chair and preceded by drums and trumpets, but "in such a temper, that they durst not deliver their errand." In fact, the Spanish had placed the king at the head of five hundred Indians, who were at that moment moving by land and canoe to attack settlers on the Edisto River. Barnwell scratched out an urgent warning. "Our Southward will be exposed to dreadfull depradations," he wrote to the governor, sending the note ahead by fast boat but knowing he was already too late. "I fear much because I perceive ye smoaks of ye land parties to be a head of me this day."

Barnwell's attempts to broker a rapprochement with the Yamasee were just one string in his fiddle. He saw that South Carolina's safety depended on a populated, defended frontier, which he tried to fill by promoting colonization of "the Golden Islands" south of the Savannah River. (This was part of a larger scheme, dubbed the Margravate of Azilia, to settle indebted English immigrants between the Savannah and Altamaha rivers.) Tuscarora Jack beat the drum as the colony's agent in England, where a royal governor was about to take control of Carolina from the lords proprietors following a colonial revolt, and his observation that the French, who had designs on the Altamaha, used frontier forts to great advantage struck home.

In 1721, Barnwell led an expedition to the Altamaha to build Fort King George—a check on French and Spanish interests and the first step toward founding Georgia twelve years later. Georgia was exactly the kind of barrier colony that Tuscarora Jack had long argued for, but he did not live to see it, having died in 1724, at the age of fifty-three.

The Carolina of 1724 was a very different place from the one Barnwell had first known. Immediately after the Yamasee War, the Native slave trade all but collapsed, and traffic in deerskins, furs, and corn shrank to a ghost of its earlier self. Huddled in the Low Country amid the smoldering ruins of their vast farms, Carolinians turned away from the interior, channeling their money and energy into raising indigo and especially rice.

Indian slaves, who accounted for one-quarter of all the captive labor in the colony before the war, dwindled in number, while imports of African slaves—many of them familiar with rice production in their home countries—soared. The number of black slaves entering South Carolina increased almost tenfold immediately following the Yamasee War and continued to rise steadily through the decade that followed. A war caused in large measure by the trafficking in one group of human beings had, in an unexpected way, caused the proliferation of the trade in a different group—and in a very real sense hastened the birth of the Old South economy of slave-supported plantations that would endure until the Civil War. What we speak of as the "antebellum" South—that is, the South, literally, "before the war"—was nothing of the sort. It was a "between-war" South, lasting a century and a half.

Although many Indians took part in the war, it's perhaps fitting that the Yamasee gave it their name. From their creation out of the splinters of fractured chiefdoms, through their seesawing fortunes between Spanish and English spheres, to the moment when they decided to fight instead of risk enslavement, they are emblematic of the bewildering changes that beset the Natives of the Southeast. But unlike such great Indian confederacies as the Creek, which arose from the chaos of the late sixteenth century with enough unity to withstand the buffeting, or the Tuscarora and Shawnee, who left the Southeast for what they hoped would be better homes in exile, the Yamasee did not find a way through the thickets. Pinned between the failing Spanish and the resurgent English and their Indian allies, open once again to attacks and assaults, the Yamasee were whittled down by death and captivity. When the Spanish finally withdrew from Florida in the 1760s, it appears that the last handful of Indians still claiming the Yamasee name went with them—sailing on Spanish ships for Cuba and Mexico, and out of history.

PART III

"WE THAT CAME OUT OF THIS GROUND"

Chapter 8

"One Head One Mouth,
and One Heart"

W HEN THE WINTER ice and snow melted, the runoff
from thousands of square miles of forest wilderness in
the Allegheny Mountains flowed down countless steep-
walled valleys, feeding the monster known, to the handful of English
traders who had seen it, as the Ohio River. At one particular bend, the
river piled up the scarred, waterlogged trunks of immense trees, bark-
less and silvered with age and abrasion. The village that sat on a high
bluff above that bend was, appropriately, called Logstown.

Now it was early autumn, and the Ohio was running smooth and
placid, the kind of forgiving river on which travelers could paddle
to the center current at dusk, lash their canoes together, and sleep
through the night without fear of rocks or rapids. The air in Logstown
was cool; humid with weeks of rain; and heavy with smoke and the
odors of simmering corn mush and bear fat, damp old blankets and
deerskins, and wet camp dogs. Logstown wasn't much to look at—a
collection of sixty or seventy log or bark cabins clustered near a more
substantial trading house—but it was a wonder to listen to. The chatter that filled the air was a babel of tongues reflecting the town's accidental nature—a place where the exiles of many Indian nations (from
the Carolinas to the Great Lakes, Delaware Bay to the mountains

of New England and the edges of the buffalo prairies) had created a community of necessity.

It was September 1748, and Conrad Weiser was feeling weak. Strong stomach cramps had kept him confined to his bed for some time, and he'd allowed himself to be bled in the hope it would purge his bad humors. He finally felt well enough to proceed with the task that had brought him from his quiet farm, a day's ride from Philadelphia, hundreds of miles across the hazy mountains to the far edge of what was known simply as the "back parts."

The lives and histories of Weiser and the three men with whom he sat, in the rough trading house at Logstown, around a smoldering fire, knotted up many of the threads of frontier reality in the mid-eighteenth century. Perhaps no one in the so-called Middle Colonies was as adept at navigating the overlapping worlds of Natives and colonists as Conrad Weiser, a man in his early fifties who spoke English and the Iroquois dialects with a German accent, having been adopted by the Mohawk as a teenager and applying himself thereafter to the study of Indian cultures. He was Pennsylvania's provincial interpreter—its primary diplomat to the Six Nations, who had given him the name Tarachiawagon, meaning "Holder of the Heavens." More recently, he had been known as Brother Enoch, having for some years dropped out of secular life to become a monk in the strange, pietist sect of a charismatic preacher, but Indian politics had pulled him out of the cloister and back into the woods.

Sitting near Weiser was George Croghan, who was as far from a religious ascetic as one could imagine. A Dubliner of around age thirty who had fled a famine seven years earlier, he was open, boisterous, and willing to bet on a weak hand; he also loved to dip snuff and knock back a tot or two (or three) of rum. Starting with nothing, he had in a few years become so successful a backwoods businessman that he was already known as the "king of the Indian traders," while the Mingo of the Ohio country fondly called him "the Buck."

Croghan's long lines of packhorses, laden with cloth, knives, powder, and shot, picked their careful way through the narrow gaps in the Alleghenies, across swollen rivers and west as far as the Wabash and Great Miami, just shy of the Mississippi itself, trading with tribes whose existence was little more than a rumor to most English colonists. Those horses returned laden with pelts and deerskins, taking more and more business away from the angry French. It was dangerous work, about which Croghan tended to make light. Some years later, having survived a bloody tomahawk attack, he would quip, "So You may see a thick Scull is of Service on some Occasions." Among his other achievements, Croghan and his traders would be among the first whites to enter the land of "Kantucqui," at a time when Daniel Boone, with whom Kentucky is most closely linked, was barely out of adolescence.

Facing Weiser and Croghan was an Indian named Tanaghrisson, a leader of the so-called Mingo—Iroquois who had moved to the rich hunting grounds of the upper Ohio country, which they shared with Lenape and Shawnee refugees fleeing the rising tide of settlement farther east. Although he was a respected sachem among the Iroquois, Tanaghrisson had been born a Catawba in the Carolina piedmont, captured as a child, and adopted by the Seneca. His hatred of the French was implacable, because, Tanaghrisson said, they had boiled and eaten his father.

The English called Tanaghrisson "the Half-King" and viewed him as an Iroquois-appointed viceroy over the Ohio Indians. But the idea of Iroquois supervision was a political fiction. In practice, the Ohio Indians went their own way, with little regard for the councils of the Six Nations at Onondaga, hundreds of miles to the northeast—which is why Weiser and Croghan were at Logstown in the first place, anxious to strike an independent deal on behalf of Pennsylvania.

The fourth man around the fire at Logstown was perhaps the most intriguing of all. Andrew Montour was a member of a proud line of interpreters who straddled the uneasy boundaries between Indians and whites, French and English. His uncle, a *métis* (mixed-blood) fur trader who had abandoned the French to work for the English in New York, had been assassinated by order of the governor-general of New France. His aunt was a go-between for the French-leaning Miami of the lower Ohio country, one of whom she'd married. His mother, the famous interpreter Madame Montour, had been born in New France, captured by the Iroquois, and ransomed back to Canada, then returned to Iroquoia to marry an Oneida war chief, who was killed when Andrew, his son, was young. Stepping into her murdered brother's shoes, she then became New York's trusted interpreter.

Andrew Montour was as comfortable in war paint, going to raid the Catawba in Carolina, as he was speaking with provincial officials in Albany or Philadelphia. Several years earlier, a German count, meeting him along the Susquehanna River, had found that his "countenance [was] decidedly European, and had not his face been encircled with a broad band of paint, applied with bear's fat, I would certainly have taken him for one." Montour's English coat, breeches, and scarlet waistcoat were complemented, the count said, by decidedly Iroquois earrings "of brass and other wires plaited together like the handles of a basket."

In the years to come, the Six Nations, who knew Montour as Sattelihu or Oughsara, would metaphorically "set a Horn on [his] Head," making him "one of their Counsellors and a great Man among them." Yet the colonists for whom he long worked would never entirely trust him or his family. His mother was "a French woman," after all, and his cousin, with whom he was so close many thought they were siblings, went by the name French Margaret. With his kaleidoscopic heritage, he felt free to change names the way some men change clothes. Depending on the place, the audience, or his whim, he answered to Sattelihu, André Montour, Andrew Sattelihu, Henry Montour, French Andrew, and simply Monsieur Montour. Some of these names were guaranteed to raise English hackles, whatever his actions.

At times, he was too French, at other times too Indian. In the early days of the Seven Years' War, he barely escaped a Pennsylvania mob demanding any Indian's blood, and he carried a French bounty on his scalp. At still other times, Montour was too white, not wholly accepted by the Indians among whom he lived and moved, suspected (sometimes with reason) of toeing the colonial line to their detriment, despite his ties of blood and marriage to the Six Nations, Lenape, and Conoy, among others.

Unlike Croghan and Weiser, who left bundles of letters and documents written in their own hand, or even Native leaders such as Tanaghrisson, whose words at council meetings and treaty negotiations were transcribed by fast-scribbling secretaries, Montour left almost nothing in his own voice. He was apparently illiterate, and when he spoke in public, it was usually for someone else.

Andrew Montour was a complicated mess. He was, according to the governor of Maryland, "of all the Traders Interpreters or Woodsmen without Comparison the most promising & honest," but his weakness for drink and his tendency toward debt vexed even his staunchest admirers, including Conrad Weiser. "He abused me very much, Corsed & swore, and asked pardon when he got sober, did the same again when he was drunk, again damned me more than [a] hundred times," Weiser told a correspondent. Unable to rouse the half-dressed Montour from a sodden coma, Weiser abandoned him—only to have the interpreter meet him later on the trail, clear-eyed and contrite, having ridden two days and two nights to make the rendezvous.

Weiser, who probably knew Montour better than anyone else, on the one hand praised the young man as "faithful, knowing, & prudent," but on the other hand cautioned that he was "a French man in his heart." In the end, Weiser could only scratch his head and pronounce himself "at a lost what to say of him."

In ways that would haunt him all his life, Andrew Montour was the living embodiment of the patchwork human frontier, a shadow of the physical borderlands. And when he tried to create a place where he could find peace—a place for all the other in-betweens and castoffs, half-bloods and immigrants, refugees and wanderers—both of Montour's worlds, the Indian and the European, conspired to crush his dream. No wonder he drank a lot.

In comparison with the rest of the eastern frontier, the Middle Colonies were oases of relative peace, and none more so than Pennsylvania. This was partly a function of demographics and geography. Through most of the seventeenth century, while New England was bursting at the seams with farmers hungry for land, Virginia was pushing tobacco plantations deeper and deeper into the interior, and Carolina was enslaving Natives left and right, the valleys of the Delaware and Susquehanna rivers were largely devoid of immigrants. The Swedes landed only a few hundred settlers along the lower Delaware, and the Dutch, who wrested away control in 1665, were, as usual, more concerned with creating trading outposts than in colonizing the wider landscape. Too far north to seriously worry Spain and too far south and east to attract much French attention, Pennsylvania was able to avoid involvement in most of the imperial dustups as well.

By the time the Dutch ceded their interest in New Netherland to England and William Penn wrangled a royal grant in 1681 for 45,000 square miles of land he'd never seen, there were perhaps a thousand Europeans along the lower Delaware and its bay. That soon changed. Penn's savvy promotion of his "holy experiment" brought waves of new colonists, and by 1700 some 21,000 Europeans—English Quakers, Germans, Finns, Swedes, Welsh, Dutch, Scots-Irish, French Huguenots, and black slaves—were living along the Delaware.

Their land had belonged to the Lenape, a collection of loosely related groups living, in small bands of about twenty-five or thirty people, from the lower Hudson River to the mouth of Delaware Bay. They may never have been numerous, and by the late seventeenth century, the Unami, or "downriver people," who lived along the lower Delaware, probably numbered fewer than five hundred. Unlike many Algonquian peoples along the eastern seaboard, the Lenape were not serious farmers. Although they planted small maize plots at their summer camps of bark-covered lodges, they lived primarily by hunting, fishing, and foraging.

And unlike the Susquehannock and Five Nations to their west and north, or the Algonquians of the Chesapeake region, the Lenape did not usually fortify their villages with palisades, nor is there much archaeological evidence to suggest frequent warfare in pre-contact days. It's unlikely the Lenape were innately peace loving; they were just lucky enough to live in a fairly safe neighborhood, and the lack of ready access to northern furs kept them out of the bloodiest conflicts of the seventeenth-century Beaver Wars, which all but obliterated the Susquehannock.

By the time Penn arrived, life was already changing for the Lenape. They had briefly taken up intensive corn production, selling their surplus to the hungry Swedes, and by the 1690s many of the Unalachtigo, the Lenape who lived east of the Delaware in what had become the English province of West Jersey, were moving out to avoid the growing numbers of Quaker immigrants there.

Migrating across the river, they found themselves dealing with still more Quakers, most notably William Penn. Because his behavior stands in such vivid contrast to the freewheeling deceit and chicanery that attended many dealings between colonists and Indians, Penn may get more credit than he deserves for scrupulous honesty. After all, he sold the first 1,250 acres of his new empire, pocketing £25, three days before his charter was even signed by the king. More startling, Penn

William Penn's Treaty with the Indians
An engraving (reversed from the original) of Benjamin West's iconic 1771–72 image of William Penn meeting with Lenape sachems along the Delaware River. Although Penn's religious principles compelled him to deal fairly with the Indians, he never dreamed of treating them as legal equals and never imagined that they would not sell him as much land as he and his heirs wanted. (Library of Congress LC-DIG-pga-01451)

sold half a million acres before he arrived in the colony and started talking to the local sachems, from whom he still needed to buy the land. That this might be wrong clearly never occurred to him, and he obviously viewed his negotiations with the Lenape as a formality.

That's the paradox about William Penn—Quaker visionary, famous friend to the Indians, and brash land speculator. On the one hand, his religious principles compelled him to deal fairly with the Indians, to whom he pledged—in all apparent sincerity—that the English "must live in love, as long as the sun gave light." Penn instructed his purchasers that "noe man shall in any wayes or meanes in word or Deed, affront or wrong any Indian" and that any grievance between colonists and Natives would be decided by a panel of six of each. One modern historian (normally no fan of European colonizers) called Penn a "unique blend of the practical, the pious, and the decent." On the other hand, Penn's sweeping rights under English law were wholly feudal, and he never dreamed of treating the Indians as legal equals or imagined that they would not sell him as much land as he and his heirs wanted.

When Penn finally arrived in the colony in 1682 and sat down to treat with a dozen Lenape "sachemakers," as he called them, he struck what he considered a very fair deal for the first parcel he bought—tens of thousands of acres below the Falls of the Delaware, where he would build his home. The payment amounted to more than £240 in goods, the equivalent of about $30,000 in today's currency, including 600 fathoms of wampum (half of it prized black beads); 40 white blankets; 40 fathoms of red and blue stroud cloth; another 40 fathoms of gray duffel; 40 each of muskets, kettles, and woolen coats; 150 pounds of gunpowder; 300 bars of lead; 5 casks of rum, beer, and hard cider; "two handfull ffish hooks"; and shoes, scissors, combs, paint, saws, bottles, hoes, axes, and shirts.

Over the next two years, Penn negotiated a series of additional land deals, acquiring a swath about ten miles deep and running about forty miles along the western shore of the Delaware, as well as parcels on the eastern side of the Susquehanna (land claimed by Maryland, a situation that would later boil over into violence). The venerated sachem Tamanend, exchanging what he undoubtedly thought was the right to use land near Penn's original purchase, made his mark—a small drawing of a snake—on a deed, accepting "so much Wampum so many guns, Shoes, Stockings Looking-glasses, Blanketts and other goods as he ye said William Penn Shall please to give unto me." (A receipt attached to the deed specifies what Penn was "pleased" to pay:

an assortment of fabric, garments, wampum, tools, clay pipes, and "4 hansfull Bells.") To the Lenape, Penn became "Brother Onas," a name that would be handed down to successive colonial leaders who, in Native eyes, embodied Pennsylvania.

When he wrote home to England, Penn could say with a clear conscience that he "bought lands of the natives, [and] treated them largely," meaning generously. The Lenape, of course, considered the guns and blankets and such to be gifts, and as the use of the land for which they were given continued through the years, they pressed for the gifts to be renewed from time to time, to the annoyance of colonial officials, who believed that paying for land once was quite enough.

Yet in all, the relationship was a fairly harmonious one while Penn remained in control. In 1701, a group of sachems from the lower Susquehanna wrote to King William III, saying they expected to live in harmony with Penn and his colonists "as long as the Sun and Moon shall endure, One head One Mouth, and one Heart."

But as English ships unloaded dressed lumber from New York and clapboard houses rose along the muddy streets of Penn's new city of Philadelphia, the Lenape must have realized that the English were a different breed than the Swedes and Dutch who had preceded them. Penn had big plans for Pennsylvania, and like his royal land grant, they extended far beyond the Delaware. In particular, Penn cast his eyes toward the Susquehanna Valley—a focus that would bring his colony into ever closer contact with the Five Nations, the Iroquois from central New York, two hundred miles to the north. In time, these powerful entities—one colonial, one Indian—would recognize the advantages of dealing directly with each other and essentially ignoring the people who actually lived on the land in question.

The League of the Five Nations had come into existence sometime between 1400 and 1600, its origins clouded by a combination of oral history and myth involving the prophetic Great Peacemaker whose message brought the five warring tribes into one: the *haudenosaunee*, "the people of the longhouse." Like the fires of extended families running the length of a traditional bark longhouse, the Five Nations ran the width of greater Iroquoia, from the Mohawk guarding "the eastern door," near the Hudson, to the Seneca in the west. The league's central council fires burned among the Onondaga in the heart of the Finger Lakes, flanked by the Oneida and Cayuga.

Legend aside, the league's evolution was a long and torturous process, marked by lingering hostilities and perennial rivalries, although by the late seventeenth century, the Iroquois were able to present a fairly united front when dealing with New France and the English. Even so, the league often functioned more as a marriage of convenience than a political monolith, no more in lockstep on its policies and aims than were the squabbling, intensely competitive English colonies. Some Iroquois leaders favored the French, others the English. Some had strong connections with the government of New York; others worked closely with Pennsylvania. Personal ambition and historical frictions among the league's members added to the confusion.

Yet despite these handicaps, the League of the Five Nations achieved a level of diplomatic success unmatched anywhere else in colonial North America. By the late seventeenth and early eighteenth centuries, the league had constructed a treaty network, known as the Covenant Chain, which bound the Iroquois, their subsidiary tribes, and the neighboring English colonies into a web of alliance and responsibility, while offering official Iroquois neutrality to the French. Skillfully playing on their position as the Native buffer between two habitually warring European empires, the Five (later Six) Nations managed for generations to hold off England and France, trading on the threat of wholly committing to one side or the other in order to keep both adversaries at bay.

The Susquehanna Valley had largely been depopulated in the wake of the Beaver Wars between the Iroquois and Susquehannock, as well as further conflicts between the Susquehannock and Maryland and Virginia. A few Susquehannocks had, however, managed to resettle in their old homeland near Conestoga Creek, acquiring the creek's English name as their own. The Conestoga were one of many tribes under the Five Nations' "umbrella." Another was the Shawnee, who settled with the Conestoga in 1692 after wandering east from what is now Illinois.

The colonists saw in such arrangements a simple dynamic: the Iroquois as military overlords who ruled the vanquished with absolute authority. The reality was more complex. The Iroquois called the Conestoga "nephews" or "cousins," and the latter in turn called the Iroquois "uncles." At least theoretically subject to Iroquois oversight (with a Five Nations sachem deployed locally to keep an eye on things), the Conestoga, Shawnee, and others were on their own in negotiating treaties and land sales and in deciding whether or not to supply warriors for Iroquois military operations. Although they were not junior members of the league, like the Tuscarora, whom the Iroquois formally adopted around 1723, neither were they defeated slaves, and they enjoyed a significant degree of autonomy.

Still, it was advantageous for the Five Nations to promote the illusion of a vast Iroquois empire, acquired by conquest and rigidly controlled, and for their European neighbors, especially the English colonies, to acknowledge it. This political mirage bolstered the league's always tenuous position with the European powers, and it gave colonial negotiators such as James Logan—the Scots-Irish secretary (and later president) of the Pennsylvania provincial council, who grew wealthy from a near monopoly on Indian trading in the province—a backroom avenue to smooth out troublesome Native affairs, especially land sales. Best of all, because the English considered the Iroquois subjects of their own king (a fallacy to which the Iroquois politely acceded in treaties, while ignoring it in practice), New York and Pennsylvania concluded that all the land within the league's shadow was, by right and law, English as well.

In the early 1700s, Pennsylvania was swelling with immigrants from all directions. As more and more Europeans disembarked at Philadelphia or New Castle, an increasing number of Lenapes forsook their homeland along the lower Delaware and moved north to the land along the fringe of the Appalachian Mountains or west to the Susquehanna Valley. There they mingled with refugees invited by the Iroquois—the Savannah (Shawnee), Tutelo, and Tuscarora from the Carolinas; the Conoy and Nanticoke from the Chesapeake—and by Five Nations members looking to spread south into the good hunting and farming areas along the river. Runaway black slaves from as far away as Virginia found their way to the villages in the Susquehanna Valley, where they were often given sanctuary.

One reason the Long Peace that Pennsylvania enjoyed endured for so many years, even as settlers poured into the province, may be that the Lenape felt they had plenty of room at their backs, especially in the relatively empty country to their north and west. As they sold off tract after tract along the Delaware and lower Schuylkill rivers, they were able to withdraw, especially to two places of extraordinary beauty and fertility.

One was the Forks of the Delaware, about seventy miles upriver from Philadelphia, where the Lehigh River entered the Delaware at the edge of a valley nearly fourteen miles wide and full of deer, elk, and turkeys. The other was about fifty miles southwest of there, along the same valley system—the Great Valley of the Appalachians, which runs from New York to Alabama. On the upper Schuylkill river, several days' long walk from the Forks through majestic hardwood forests of ancient chestnut and oak, this valley was framed by a line of low, jumbled hills along its southern edge and the long, flat-topped ridges of the Appalachians to the north.

The latter valley, like that of the Forks, was watered with clear-running streams that flowed from the mountains or rose out of springs from the limestone bedrock of the lowlands. The hunting was equally rich, the soil deep and good for raising maize and squash. The Lenape called it *tülpewihacki*, "the land abounding with turtles," and shared it with the Conoy, Shawnee, and other exiles. The colonists, mangling the pronunciation, called it the Tulpehocken. That two such idyllic places would soon be snatched from the Indians—one of them through the most infamous land fraud of the colonial era—would become a deep, long-festering wound, with mortal consequences.

The same exodus of Palatines from Germany that brought Christoph von Graffenried's colonists to Carolina, just in time for the Tuscarora War, also produced one of the most enigmatic and confounding figures in the Middle Colonies. No one was more central to Pennsylvania's Indian policy during the mid-eighteenth century than Conrad Weiser, and no one exhibited a stranger divide between his public persona—esteemed interpreter and diplomat, "Holder of the Heavens" to the Iroquois—and the intensely personal (and, to modern eyes, bizarre) spiritual path he followed in his private life.

Weiser was born in 1696 in Groß Aspach, a farming village in the duchy of Württemberg, where his father, a former soldier, worked as a baker. Shortly after losing his wife in 1709, Johann (John) Weiser and eight of his children, including twelve-year-old Conrad, joined the fifteen thousand Palatines who accepted Queen Anne's offer of asylum in England. So many Palatines poured into Rotterdam and from there onto London-bound ships that the British embassy warned its superiors, "You may have half Germany if you please."

What to do with them all became a pressing issue for the British government. Thousands were shipped to northern Ireland, while some joined Graffenried on his venture to Carolina. Robert Hunter, the newly appointed lieutenant governor of New York, floated another proposal. Why not ship several thousand Palatines to the wilderness of his colony—billing them for their own passage—and let them work off the resulting debt by boiling down pine sap to make tar and raising hemp for rope? The Palatine problem could be eased and Britain's crucial naval stores increased.

That sounded good to John Weiser, who signed up. He and his family embarked around Christmas 1709 on what must immediately have seemed a dreadful mistake. The voyage took an unbearable six months, with hundreds, especially children, dying en route. Upon landing in June 1710, two of Conrad's older brothers were trundled off to labor in Long Island, and when the rest finally reached the pine woods where they were to work, winter was coming on fast. Hunter had nothing but tents in which to shelter them, and the man he'd chosen to oversee production was little more than a crook, with no notion of how to make tar. But the undertaking failed for an even more fundamental reason: while upstate New York was full of pine trees, they were mostly white pines—fine for ships' masts, but a poor source of tar and turpentine. Trapped on land unsuitable for farming and unable to produce enough tar to pay off their debts, the Germans came to loathe Hunter and his English colleagues, while the governor, watching them drain his fortune with no tar to show for it, responded in kind.

After several years, both the Palatines and Hunter had had enough. The governor cut them loose, and John Weiser emerged as a leader of the Germans, meeting in the fall of 1713 with the Mohawk to seek permission to settle along Schoharie Creek, a lovely waterway that flows north out of the Catskills to the Mohawk River. This was good land, ideal for crops, the type of land about which the Palatines had dreamed since arriving in the New World. John found many of the "Maquas," including a sachem named Quayant, friendly and welcoming—openhanded in a way that must have stood in contrast with the Germans' treatment by the English—as the Palatines prepared to move the following spring.

The Germans were short on everything. They had to borrow a plow harness and cows, and nine of them pooled their meager cash to buy a horse. Food was so scarce that they dug Indian potatoes and groundnuts, as the Mohawk showed them—mostly with gestures, since neither spoke the other's language. So Conrad, by then sixteen, was packed off with Quayant to learn Mohawk. Whether or not he had a vote in this decision, Weiser never said, but he seems to have been dry-eyed about leaving his family, which had, to all intents, ceased to exist. His father had remarried, and rather than stay with a stepmother they despised, his surviving brothers and sisters had already fled.

Nevertheless, life among the Mohawk was a challenge. "I suffered much from the excessive cold, for I was but badly clothed, and towards spring also from hunger, for the Indians had nothing to eat," he recalled. The Mohawk also drank, and when they were in their cups, they "were so barbarous, that I was frequently obliged to hide from drunken Indians."

Weiser remained with Quayant until July, having "learned the greater part of the Maqua language." (That he managed so much, so quickly, suggests that Weiser was a linguistics wizard, since Iroquoian languages are notoriously complex.) The nearest of the seven German settlements—named Weiserdorf, in honor of his father—was just a mile or two from the Mohawk village. Over the next few years, "there were always Maquas among us hunting, so that there was always something for me to do in interpreting," Conrad said. "I gradually became completely master of the language, so far as my years and other circumstances permitted." Weiser also became so close to the Mohawk that he was adopted—an act that, in their eyes, bound him to them as closely as a blood relative.

But while the Palatines had carefully cultivated friendship with the Mohawk, there was still Robert Hunter to consider. They had escaped the tar works, but not their obligations to him, nor to the men to whom he sold the title to the Schoharie land. When an agent arrived to collect their rent, the Germans mobbed him, trapping him for a time in a building as they exchanged gunfire. An arrest warrant was issued for John Weiser, but when the sheriff tried to execute it, the women of Weiserdorf thrashed him savagely and rode him out of town on a rail. When John and two companions sailed to England in 1718 to plead their case before the new king, George I, they were waylaid by pirates, robbed, and beaten, and upon reaching London, nearly penniless, they were soon thrown in prison for debt.

By the time the elder Weiser limped home in 1723, many of the Palatines had pulled up stakes in the Schoharie, which in the end proved far less than the paradise they had imagined. Cutting a road through the forest to the north branch of the Susquehanna, some had made rafts or hollowed out tree trunks to make crude dugout canoes and paddled south with their possessions, while others had driven their herds of cattle through the mountains. Their destination was a place in the Pennsylvania hinterlands about which they'd heard, a place with good soil and clear streams, far from Robert Hunter and the New Yorkers—a valley called Tulpehocken.

No one formally invited the Palatines to come to the Tulpehocken Valley—especially not outraged Lenapes such as the sachem Sassoonan, the leader of the Tulpehocken band, who "could not believe the Christians had settled on them, till he came & with his own Eyes saw the Houses and Fields they made there." Both the Germans and the Lenape seem to have become mixed up, through no fault of their own, in a testy little power struggle between Pennsylvania's governor, William Keith,* and its longtime council secretary, James Logan, who was the major architect of the province's Indian policy.

While visiting Albany, Keith had quietly mentioned the Tulpehocken Valley to a member of the Palatine community, broadly suggesting that the land was free to settle and sparking the migration from the Schoharie. When the furious Sassoonan objected that the land above South Mountain was supposed to be off-limits to whites,

* Keith was technically the deputy governor. The actual governors of provinces such as Pennsylvania, Maryland, and Virginia often never left England. In their absence, the day-to-day business was conducted by a deputy or lieutenant governor, or by the president of the council, such as James Logan in Pennsylvania.

Logan first tried to smother the Lenape's arguments with paper, producing old deeds purporting to show that Pennsylvania had purchased
the area. When that failed to budge the Lenape sachems, who knew
they'd agreed to no such thing, he shrugged and said that the Germans
had come "without the Consent or Knowledge of any of the Commissioners" and threw the blame onto Keith.

Logan's own hands were scarcely clean. As William Penn's deputy,
he had long since taken to quietly subverting the late founder's stated
principles of fair dealing, furthering his own ends through land speculation and trade. Logan had already cheated the Lenape out of the
fertile Brandywine Valley, which was supposed to have been reserved
for their use, and for all his protestations to the contrary, he would
happily do it again in the Tulpehocken.

The few dozen Palatine families weren't alone in violating Indian
rights in the land abounding with turtles. More and more settlers,
including English Quakers and Welsh, were moving up the Schuylkill
beyond South Mountain. Violence was breaking out between them
and the Indians. When two Welsh brothers went on a short killing
spree, a wider war was averted only by a death sentence for the murderers. When it became clear that no one was going to evict the newcomers, Sassoonan grudgingly accepted the fait accompli — neither
the first nor the last time the Lenape acquiesced, with astonishing
restraint, the usurpation of their land.

When the flotilla of German canoes disappeared down the Susquehanna, bound for the Tulpehocken, Conrad Weiser remained behind
in Weiserdorf. In November 1720, a few weeks after his twenty-fourth
birthday, he had married a young woman named Ann Eva Feck, and
she was pregnant with their first child (of an eventual fourteen) when
the Palatines — including her parents — departed for Pennsylvania.
Whether it was family ties, restlessness, or a building frustration with
the situation in New York, however, Weiser eventually moved his own
fast-growing family to the Tulpehocken. In 1729, he marked off two
hundred acres on the north side of the South Mountain hills and settled down to a farmer's life. To his English and Scots-Irish neighbors,
he and his kind were "Dutch" — from Deutsche, or German. Unused,
Weiser's Mohawk grew rusty.

But as the first winter snows fell in 1731 — after the harvest had been brought in, the potatoes and root crops stored on groaning shelves in the cellar behind the house; the pigs slaughtered, the bacon salting in brine troughs and the hams hanging in the smokehouse; the casks of local cider laid up and the sauerkraut packed in pottery jars; the wood-shed filled with split oak; and all the other, seemingly endless chores of autumn farm life attended to — Weiser's world changed. He appears for the first time in the minutes of the provincial council in Philadel-phia, interpreting for "Shikellima, of the five Nations, appointed to reside among the Shawanese."

Shikellamy was an Oneida sachem, dispatched by the Iroquois League at Onondaga to live at the confluence of the two branches of the Susquehanna and keep an eye on the Shawnee. He may have already known Weiser from the young man's time in the Schoharie, or perhaps he simply met Weiser on his way to Philadelphia, since the main path from the Susquehanna ran practically past Weiser's front door.

However it happened, the encounter was a gift to both the Iro-quois and colonial officials. Here was a Mohawk by adoption and a Pennsylvanian by choice, a man who understood Iroquois culture in a way few whites did, living halfway between Philadelphia and Shikel-lamy's new home along the Susquehanna, and who was happy to serve as a bridge between the two. When the Philadelphia meeting ended, the council voted to pay "forty Shillings to Conrad Weiser the Inter-preter." It was more than the promising start of a new career. James Logan recognized a prodigy, and Weiser was on his way to becoming the indispensable cog in Pennsylvania's Indian policy.

For the moment, though, Logan and the Pennsylvanians were focused on the belief that they had found a way to make an end run around the Lenape and Shawnee. They considered Shikellamy "a trusty good Man & and a great Lover of the English," and a go-between "whose Services had been & may yet further be of great ad-vantage to this Government." Through him, with Weiser as their in-creasingly reliable link, they could work directly with the Six Nations.

For their part, the Iroquois felt a change in the wind, too. Where they had quietly tried to foment an uprising by the Shawnee and Len-ape against the English in 1726 (which fizzled when their supposed tributaries simply ignored the demand), they now saw the advantage of cutting a private deal—the reason Shikellamy had come to Philadelphia in the first place. The Iroquois would claim dominance over the Indians of Pennsylvania based on past military "conquests." That such conquests were largely fictive didn't matter, as long as Philadelphia agreed to recognize them. That Pennsylvania's Indians would be factored out of the equation entirely was bad enough, but there were worse insults to come.

For Conrad Weiser, the hard life he'd endured as a young man was softening. By 1736, Ann Eva had borne ten children, eight of whom lived, and his farm had grown to more than eight hundred prosperous acres. He was active in the community and in the small union church—half the congregation Lutheran, like Weiser, half Reformed. But there were plenty of less traditional avenues for a questioning religious man to follow. It was a time of great Protestant ferment, with pietists who shunned the formal trappings of established churches creating their own denominations, and mystical sects bubbling up in Europe and America.

On the Pennsylvania frontier one might find New Baptists, nicknamed "Dunkers" by their derisive neighbors for their habit of full-body baptism, foot washing, and love feasts; *Neugeborene,* or New-Borns, a celibate sect that rejected most scripture and sacraments and believed they were free of sin; and *Neumondler,* or New Mooners, who timed their worship to the lunar cycles and accompanied their services with blaring trombones (in accordance with the book of Numbers: "Also in the day of your gladness, and in your solemn days, and in the beginnings of your months, ye shall blow with the trumpets over your burnt offerings").

Among the strangest of the lot, well out on the fringe even by Radical Reformation standards, was a Palatine named Johann Conrad Beissel, a short, wild-eyed, self-described prophet, prone to keening an oddly charismatic Rosicrucian philosophy for hours, his eyes closed as though in a trance. Beissel had a magnetic personality, by one account wooing a bride away from her marriage altar, mid-vow. (On a less salacious note, he was also instrumental in publishing a number of important German religious works in America.) He established a hermitage of like-minded followers a dozen miles over the South Mountain hills from Weiser at Ephrata, a community his flock called *der Kloster,* or the Cloister. For someone like Weiser, who had "at a previous period of my life wished that I had never heard of a God," a late-blooming faith can burn strongly, and when he met Beissel in 1735, he fell under the preacher's spell.

While he didn't—yet—become one of the Solitary Brethren, the self-mortifying monks who starved themselves and slept on hard wooden benches, Weiser accepted baptism into the sect as Brother Enoch and burned his books of Lutheran theology (and any others he could find). He later grew a long beard, fasted to the point of gauntness, donned a biblical robe and sandals to preach the new gospel, and tried to hold himself apart from the temptations of the flesh (i.e., his wife). Ann Eva, good hausfrau that she was, gave it a shot for her husband's sake, attempting to live with him at Ephrata, but she soon returned to the farm—as did Conrad, from time to time. If his celibate spirit was willing, his flesh was evidently weak, for Ann Eva's babies kept on coming with regularity.

It's hard to reconcile the journeyman mystic—proselytizing in German communities in the late 1730s, wearing a robe belted with rope like some Old Testament figure—with the pragmatic, tough-as-nails Indian interpreter on whom James Logan and the Iroquois leaned so heavily. Yet these two Conrad Weisers coexisted for years. In February 1737, for example, despite snow and bone-chilling cold, Weiser left the Tulpehocken Valley and set out for Onondaga, where, at the behest of Logan and the governor of Virginia, he was to negotiate a peace treaty between the Iroquois and their perennial enemies, the Catawba and Cherokee. It would be the first of many such diplomatic missions to Onondaga, and very nearly his last.

If Weiser wanted mortification of the flesh, this was far more effective than a hard bed in Ephrata. On the first leg of the journey, Weiser, a German named Stoffel Stump, and a homeward-bound Onondaga named Owisgera rode eighty miles to the Indian village of Shamokin, which they found abandoned. Firing their guns, they finally attracted the attention of a trader on the far shore of the Susquehanna, who ferried them across, "but not without great danger, on account of the smallness of the canoe, and the river being full of floating ice," Weiser noted in his journal. Forced to abandon their horses, they continued upriver on foot to rendezvous with Shikellamy, who was to join them.

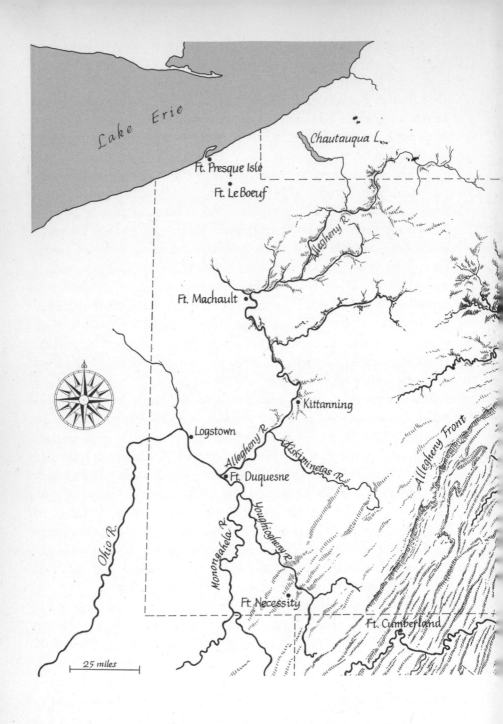

Lake Erie

Chautauqua L.

Ft. Presque Isle
Ft. Le Boeuf

Allegheny R.

Ft. Machault

Kittanning

Logstown

Allegheny R.

Kiskiminetas R.

Ft. Duquesne

Allegheny Front

Ohio R.

Monongahela R.

Youghiogheny R.

Ft. Necessity

Ft. Cumberland

25 miles

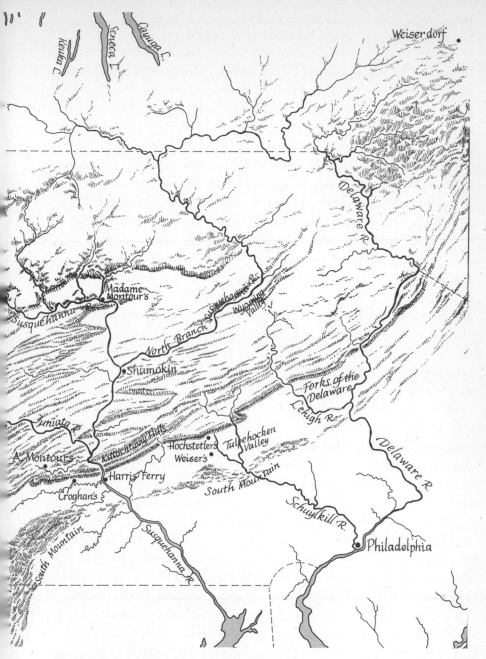

The Pennsylvania Frontier and the Ohio Country

By the early eighteenth century, the Pennsylvania backcountry had become a land of refugees, both Indians and Europeans. For the latter, there were, in effect, two frontiers—one marked by the edge of the Appalachian Mountains, which was for years the limit of agricultural settlement, and the other a commercial frontier hundreds of miles to the west, where traders such as George Croghan operated. (Modern boundaries added.)

What they found was hardly encouraging. Indian hunters said the snow lay waist-deep in the woods, and the rivers were so swollen that two traders, trying to canoe across a nearby stream, had been swept away, one drowning. Weiser intended to resupply at Shikellamy's village, but the Indians themselves were on short commons, and the best he could get was a little cornmeal and a few beans. Still, they pushed on, coming into the village of Ostuaga, along the west branch of the Susquehanna, where Andrew Montour's mother lived. Madame Montour was "a French woman by birth, of a good family, but now in mode of life a complete Indian," Weiser wrote. "She treated us very well according to her means, but had very little to spare, or perhaps dared not let it be seen on account of so many hungry Indians about." Little wonder he was grateful. It had been twenty-four days since Weiser and his companions had left home, 130 miles of hard travel, and this was their first decent meal in a week and a half.

The travelers were barely a third of the way to their destination, however, with the most treacherous terrain still ahead—a trail known as the Sheshequin Path, which crossed the divide between the two branches of the Susquehanna, almost fifty miles of rugged, essentially uninhabited mountains. "We came to a thick forest where the snow was three feet deep, but not frozen so hard, which made our journey fatiguing," Weiser scribbled in his journal. The mountainsides, he said, were "frightful," nearly vertical rocky walls plunging straight into streams flush with snowmelt, so that there was no dry place to walk.

"The passage through here seemed to me altogether impossible, and I at once advised to turn back," Weiser admitted. Shikellamy and the others urged him to push on, all of them clambering on hands and knees over the frozen ground, Weiser using a hatchet to chop footholds in the ice. Their feet were soaked and numb from constantly wading in the frigid water, and after three hours of brutal work, they'd progressed only a mile. Exhausted, they fell into a deep sleep despite the cold, waking stiff and with nothing for breakfast except some boiled meal and beans.

The way didn't get any easier. They passed through "terrible mountains," and in trying to avoid the torture of wading back and forth across icy streams, they attempted to stay high above the creek, inching along the sides of ridges, until a rock shifted under Shikellamy's feet. Unable to stop his fall, the sachem slid toward a hundred-foot cliff over jagged rocks, saved only when a strap on his pack snagged a sapling, leaving him dangling and badly shaken. After Weiser and the others carefully pulled him to safety, Shikellamy "stood still in aston-ishment and said, *I thank the great Lord and Creator of the world that he had mercy on me, and wished me to continue to live longer.*"

The journey became so harsh that Stump said he wished he were dead, and Weiser admitted that, had he not been so close to his goal, he would have turned back. They bickered among themselves, suf-fering more floods, more storms, more deep snow, more dangerous mountains, and near-constant hunger. They crossed the plateau to the north branch of the Susquehanna, where there were a few small camps, but the Indians had scant provisions to share. At one point, Weiser used a stash of needles and "Indian shoe strings" to buy freshly made maple sugar, "on which we sustained life, but it did not agree with us; we became quite ill from much drinking to quench the thirst caused by the sweetness of the sugar."

In villages that had been rich with corn when he'd last visited in 1726, Weiser found famine stalking the Iroquois: "now their children looked like dead persons and suffered much from hunger." A shaman explained that the Onondaga had themselves to blame. In a vision, the Creator had told the shaman, "*You kill [the game] for the sake of the skins, which you give for strong liquor and drown your senses, and kill one another, and carry on a dreadful debauchery. Therefore I have driven the wild animals out of the country, for they are mine. If you do good . . . I will bring them back. If not, I will destroy you from off the earth.*"

Walking on "through a dreadful thick wilderness, such as I have never seen before," Weiser simply gave up. "I stepped aside, and sat down under a tree to die, which I hoped would be hastened by the cold approaching night." Only "the sensible reasoning" of Shikellamy prod-ded him along. The party covered forty miles on their final day and stumbled into Onondaga like ragged scarecrows on April 10, forty-two days after Weiser had left home. Completing his mission in a week, Weiser turned right around and headed back, floating most of the way down the Susquehanna in a chestnut-bark canoe, a journey of just two relatively easy weeks. Quietly thankful, Weiser wrote the verses to a German hymn on the final page of his journal.

The Indian policy that Conrad Weiser was helping to shape in the 1730s and 1740s was mostly about land—particularly, how to convey title to vast tracts of it to the proprietors, who could then offer it for sale to settlers free of any lingering legal claims by Natives. The degree to which these niceties were observed was changing, however. As far as a seventeenth-century Englishman's innate prejudices allowed, William Penn had always tried to deal honorably with the Indians from whom he bought land. But Penn died in 1718, and his sons—whom he had admonished to "let justice have its impartial course" when dealing with colonial matters—cheerfully traded honesty for avarice. The result was the notorious Walking Purchase, which, like the dispossession of the Tulpehocken Valley, would come back to haunt the Pennsylvania frontier.

Many early Indian land agreements used a simple, direct method of determining the extent of the sale—the amount of land a man could walk across (or in some cases, ride a horse across) in, say, a day. Representatives of the buyer and seller would hike the tract together at an easy pace, having agreed beforehand on the route (usually along a river or stream), breaking for a midday meal and ending when the sun reached a predetermined hour.

In 1734, Penn's sons Thomas, John, and Richard, the proprietors of Pennsylvania, were in a bind. They were crushed by debt, and between a boundary dispute with Maryland to the south and land to the north that was yet unbought from the Indians, they had little free property with which they could satisfy their creditors. So they began selling real estate, thousands of acres at a swat, in a place they did not yet own—the Forks of the Delaware, the land between the Delaware and Lehigh rivers, which was, as the Tulpehocken had been, a refuge for displaced Lenapes and Shawnees.

The Penns' justification for this was a piece of paper—they called it a deed, although no one knows just what it was, because the original conveniently vanished—dating to 1687, allegedly giving their father rights to the land as far as a man could walk up the Delaware in a day and a half. Old Lenape sachems dismissed the claim, saying they had walked the river with Penn, and the land in question ended at the mouth of Tohickon Creek, where the authority of the signatory chiefs ended. This was about twenty-five miles above the starting point and barely halfway to the Forks. Badgered by Penns' sons, however, they agreed to allow the boundary to be walked a second time to settle the matter.

Whereas the Indians expected a companionable stroll, stopping for a long lunch and a leisurely smoke along the way, the Penn boys had something very different in mind—and they left nothing to chance. They made their preparations in absolute secrecy. A route was mapped out—not zigzagging up the riverside, but arrowing straight through the forest toward the Lehigh Gap, designed to plunge as deep into Lenape territory as possible—surreptitiously surveyed, marked, and cleared. In May 1737, a discreet rehearsal was run to test the course, with supplies placed at intervals and a support team following on horseback with food and water. To provide legal cover, a map of the sale area was sketched out and shown to the Lenape for their approval, with accurate portrayals of streams (which the Indians of course recognized), but with intentionally false labels (which they could not read and which may even have been added after the fact). Stepping into the trap, the sachems made their marks on a legal release.

On September 19, 1737, a small party of Lenapes met a much more organized assembly of Pennsylvanians beneath a big chestnut tree about ten miles from the mouth of Tohickon Creek. Of the three English walkers, one—Edward Marshall, a young "chain carrier" for the surveyor who had marked the route—probably had participated in the secret rehearsal that spring. The other two, James Yates and Solomon Jennings, were experienced woodsmen who could walk the legs off a horse. All were motivated by the Penns' offer of five hundred acres of land at the Forks to the man who got there first.

At sunrise, the three English and three Lenape walkers set off—easily at first, but as the strides lengthened and the pace built, the Indians cried foul, with little effect. Jennings dropped out before lunch—a quick, gulped meal sent on ahead for the colonials—and the two remaining English walkers were off again, following Indian paths or blazed trees marked months earlier by the surveyors. No one ran, exactly, but Jennings's failure suggests that the pace must have been de-

manding, even for rugged hunters used to endless foot travel. Two of
the Indians quit entirely, complaining of fraud. When the timekeeper
called a halt for the night, twelve hours after they'd started, Marshall
had to sag against a tree to keep from falling.

The Lenape knew a scam when they saw one. The next morning,
one of the sachems who had signed off on the deal told the Pennsylvanians to go to hell, since they had "all the best of the land" already.
They would not continue with what they knew to be a farce. Unen-
cumbered by the need to play-act for the Indians, Marshall and Yates
set off again as fast as possible, jogging wearily through the Lehigh
Gap in the Kittatinny Ridge, across the river, over Second Mountain,
and then another seven or eight miles beyond—almost sixty-five miles
from the starting point. Yates collapsed shy of the finish, and Marshall
won the land the proprietors had promised.

But the Pennsylvanians weren't done. With Marshall nursing his
sore and swollen feet, a team of surveyors immediately started marking a second line to determine the northern boundary of the sale—not
due east to the Delaware, as the Lenape had expected, but northeast
another sixty-five miles, through pine forests so wild and deep it took
the team another four days to reach the river. In all, the Penn brothers
had snookered the Lenape out of nearly twelve hundred square miles
of real estate.

The Lenape threatened to drive out the colonists by main force, but
if they were expecting their "uncles," the Six Nations, to back them
up, they were again disappointed. The league had already sold out the
Conestoga and other Susquehanna bands, deeding both sides of that
river to Pennsylvania in 1736. At that time, they had declined to wade
in on the thornier issue of the Forks of the Delaware, saying it was
none of their business. Now they had no such compunction. James
Logan had been busy, dispatching Weiser to quietly buy Six Nations
assent, aided by ten horse loads of gifts and—most temptingly—a
promise to intercede on the league's behalf in a land dispute with
Maryland and Virginia.

At a dramatic meeting in Philadelphia in July 1742, the Iroquois
reinforced Pennsylvania's demand that the "Fork Indians," as the Lenape from that area were called, leave their homes along the Lehigh
and upper Delaware rivers. With Weiser translating into English and

Sassoonan's nephew Pisquetomen thence into Lenape, the Onondaga chief Canasatego gave a tongue-lashing to Nutimus, the leader of the assembled Delaware, as the Lenape were usually called.

"The other Day you informed Us of the Misbeaviour of our Cousins the Delawares with respect to their continuing to Claim and refusing to remove from some Land on the River Delaware, notwithstanding their Ancestors had sold it by Deed . . . upwards of fifty Years ago," Canasatego told the provincial council and the grimly dignified Lenape sachems, who could not let a trace of anger show. "We see with our own Eyes that they have been a very unruly People, and are altogether in the wrong in their Dealings with You. We have concluded to remove them."

Holding out a wampum belt, Canasatego turned to Nutimus. "You ought to be taken by the Hair of the Head and shak'd severely till you recover your Senses and become Sober; you don't know what Ground you stand on, nor what you are doing . . . We conquer'd You, we made Women of you, you know you are Women, and can no more sell Land than Women."

The Fork Indians had nowhere else to turn. Swallowing their bile, they gathered their belongings and moved, some to the Wyoming Valley along the north branch of the Susquehanna, others to Shikellamy's town of Shamokin, at the north branch's confluence with the west branch. Disgusted and dispossessed, many kept right on moving, up the west branch of the river, across the Allegheny Mountains, and into the heavily forested, deeply eroded plateau drained by the Ohio River.

The Ohio River valley was another war-vacant land, home to ragtags and exiles of all tribes—Lenapes from the Forks, the Brandywine, and the lower Susquehanna; Shawnees, Nanticokes, and Conestogas; Mingo (Iroquois), mostly Cayugas and Senecas—the flotsam and jetsam of generations of displacement, looking for a spot to hunt and grow their corn, away from the seats of power. In the Ohio country, the influence of Onondaga, Philadelphia, and Albany were muted by many miles of empty land.

But the Ohio Indians were pleased to find they were not cut off completely. The English traders, on whom they depended for so much, were only too happy to follow them, leading their trains of packhorses carrying bulging, canvas-wrapped panniers. They did a brisk business

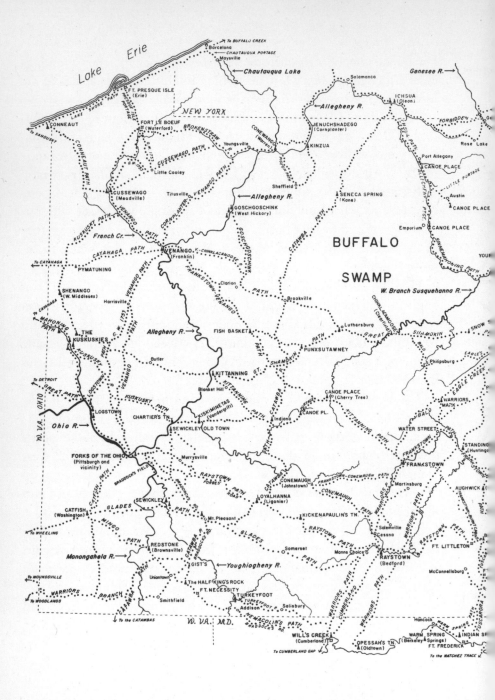

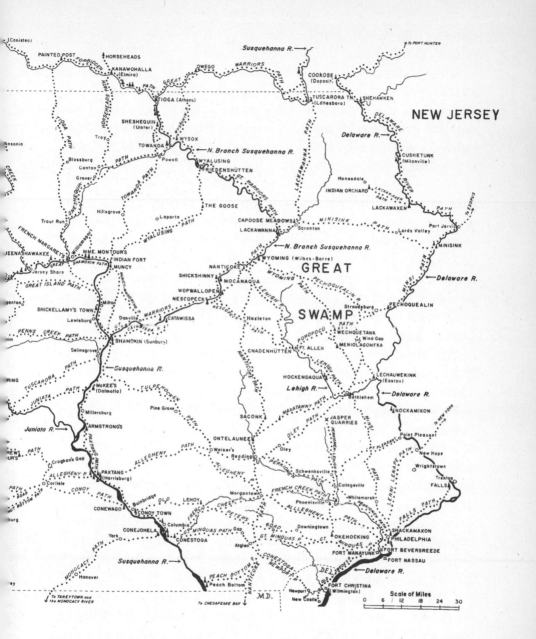

Indian Trails of Pennsylvania

Far from being a trackless wilderness, the frontier was crisscrossed by hundreds of Native footpaths used for trade and warfare. This map shows only the most significant trails through modern Pennsylvania. (Pennsylvania Historical and Museum Commission) not only with the Ohio Indians but also with the tribes much farther west—a fact that drew the increasingly jaundiced eye of New France, which saw those Indians as their own and the traders as an understandably direct threat to French sovereignty and profits. France's policy had been to hem in the English from Canada to Louisiana, holding them east of the Appalachians. Now the English traders—and one especially audacious Irishman in particular—had breached that wall. It was time to deal with them.

Like so many figures on the eastern frontier, George Croghan simply appeared, fully formed, on the historical stage. Nothing is known about his previous life or what brought him to the Pennsylvania borderlands except speculation. None of the many people he met, in the reams of personal letters and official reports in which they describe his actions, even bothered to note his appearance, although he was present at many of the most decisive moments in the long struggle for the Ohio country.

We know he was Scots-Irish, one of the wave of emigrants in 1741 escaping widespread famine caused by poor weather in Ireland. There are suggestions that he grew up in Dublin, and in later years, when some questioned his loyalty to England, suggesting he was a French-loving papist, the governor of Pennsylvania assured his nervous counterpart in Maryland that Croghan "has never been deemed a Roman Catholick." (In fact, Croghan was Episcopalian.) If he had much education, it didn't stick. A biographer pronounced the spelling in Croghan's letters "amusing, provided it is not necessary to decipher many of them," and offered this as an example: "Youl Excuse boath Writing & peper, and guess at my Maining, fer I have this Minnitt 20 Drunken Indians about me, I shall be Ruin'd if ye Taps are nott stopt, it Dose nott cost me less than £3 a day on ye Indians."

Croghan first learned his way around the backwoods by working for an older, experienced trader named Peter Tostee—familiarizing himself with the maze of Indian paths, fords, and mountain gaps that made it possible to reach the hinterlands; coming to understand both the economics of trading and the etiquette of dealing with Indians of many tribal affiliations; starting to build the trust on which he would found his own business. The young Irishman (probably in his early twenties, although his birth date is, like so much else, a mystery) proved a quick study. By 1743 he was pushing out on his own, testing the land speculation waters by buying up more than a thousand acres on the western shore of the Susquehanna, near a popular ferry operated by John Harris. The year after that, Croghan was officially licensed by the province of Pennsylvania as an independent trader.

For almost forty years, Pennsylvania's Indian trade had depended on furs and skins provided by Native hunters within a hundred miles of Philadelphia and collected by dozens of boondocks traders, a system at whose nexus sat the provincial secretary, James Logan. There was almost no way for Logan to lose. He imported manufactured goods from England, provided them to the traders on credit, and had them sold to the Indians at an astounding markup. The pelts and deerskins that flowed east from the frontier were sold in England, with Logan taking a cut on storage fees and a commission on sales, making money coming and going. The best Indian land near the posts, meanwhile, was surveyed and parceled into lots for the trader, roads were cut, and access was improved. Suspicious Indians were told it was all for the good of the trade on which they depended, even as white settlers started showing up, driving cattle and hogs ahead of them, their canvas-covered "Deutsch wagons"—which Logan famously dubbed "Conestoga wagons"—crawling with children.

Best of all, when a trader died or went bankrupt—both common events on the frontier—Logan could seize his property to make good on the trader's debts, thus gaining control of the most valuable land in that region. Given his high and varied government offices, Logan was able to brush away any lingering problems with titles or encumbrances such land might have—and thus did still more avenues of gain present themselves.

By the time Croghan arrived, however, the aging Logan had suffered a stroke (though he still kept a hand in Indian affairs), the Susquehanna had been emptied of its furs, and it was being emptied fast of its Indians, too, squeezed out by the rush of settlers. This created, in effect, two frontiers. To immigrants arriving from Europe, such as Jacob Hochstetler and his Amish kin who settled on the northern edge of the Tulpehocken Valley, the frontier was marked by what they called Blue Mountain, the brooding, two-hundred-mile ridge the Lenape named *keekachtanemin,* "the endless mountains." To the north was "St. Anthony's Wilderness," as the Germans called it, a howling waste of wolves, panthers, and savages.

But traders like Croghan had pushed the frontier into country the German and Scots-Irish farmers had never seen and about which they'd scarcely heard. From his home across the river from Harris's Ferry, along the lazy meanders of Conodoquinet Creek, Croghan and his men could follow the well-worn Indian trail known as the Allegheny Path, which wound 250 miles to the Forks of the Ohio, where the Allegheny and Monongahela rivers met; take other trails to the upper Allegheny River and thence on to Lake Erie; or take still others down the Ohio into Kentucky.

The fifty or so men Croghan employed were constantly coming and going with the seasons—moving through the mountains in autumn, so they could winter with the tribes and get the prime pelts from the cold-weather trapping and hunting, and making another roundtrip to conduct summer trading, exchanging worn-out packhorses at way-station stables in places such as the Penn's Creek valley. Croghan had a major establishment near the Forks of the Ohio, where he maintained a fortified warehouse, as well as smaller posts downriver at Logstown and Beaver Creek, and even one among the Wyandot on Sandusky Bay.

That Croghan worked practically under the noses of the French at Detroit was a testament to his nerve, especially since the two countries were embroiled in another war—King George's War (the North American extension of the War of the Austrian Succession), which had begun in 1744. In a significant way, the war helped English traders; they'd always had an edge with superior goods, but the interdiction of French shipping meant their Canadian counterparts had little to offer the Ohio tribes, and what they did have was outrageously expensive. Conrad Weiser passed on a cautionary tale he'd heard from Shikellamy about one French trader's misfortune: "The French man offering but one charge of Powder & one Bullet for a beaver skin to the Indian; the Indian took up his Hatchet, and knock'd him on the head, and killed him on the Spot."

With his net spread across so immense an area, Croghan's traders bartered for winter furs of fox, fisher, marten, beaver, and otter, but as in the Carolinas, deerskins were the bread and butter of the Indian trade. More and more, the value of goods flowing across the frontier was reckoned by everyone, Native and white alike, in "bucks." And because traders such as Croghan were the only British colonials so deep in the interior, they became vital sources of intelligence about Indian movements and plans, as well as the local mouthpieces and muscle for provincial policy—which often coincided with the traders' own best interests. While most of the fighting had been restricted to Canada, New England, and Iroquoia, no one knew when King George's War might spill over into Pennsylvania. Reinforcing the shaky loyalty of the Ohio tribes made good sense, politically and economically.

Croghan had already been conveying provincial gifts, with which the skids of Indian diplomacy were endlessly greased, to the Shawnee and other Ohio bands.* Because his business took him to places where no other Pennsylvanian ventured, Croghan gradually became the province's de facto diplomat in those far regions. Weiser, anxious to reinforce the progress Croghan had made in flipping the Lake Erie Indians to the British side, suggested sending them a gift by "some Honest Trader, I think George Coughon is fit to perform it. I always took him for an honest man, and have as yet no Reason to think otherwys of him."

In a real coup, Croghan convinced a band of "Twightwees"—Miamis from the Wabash River south of Lake Michigan—to move to a village called Pickawillany, along the Great Miami River in what is now western Ohio, where he established a trading post for them. In the summer of 1748, Croghan shepherded a delegation of Miamis, Shawnees, and Mingos to a "treaty" (as such councils were called, whether or not they resulted in formal agreements) with Pennsylvania officials at Lancaster. During the treaty, one of the Miami chiefs took a piece of chalk and traced out a large map—the Wabash River rising

---

* The importance of gift giving in Indian diplomacy can hardly be overstated. One story, probably apocryphal but accurate in its message, is recounted by Weiser's biographer, Paul A. W. Wallace. Weiser and Shikellamy were walking together along the Susquehanna when the Oneida, admiring the German's flintlock, said, "I have had a dream. I dreamed that Tarachiawagon [Weiser's Iroquois name] gave me a new rifle." Knowing what was required of him, Weiser immediately gave the older man his gun—then turned the tables. "I, too, have had a dream," he said, pointing to an island in the river. "I dreamed that Shikellamy gave me an island." Trapped, the sachem complied—but then said, "Let us never dream again."

just below Lake Erie and flowing south, the villages of the Miami and their allies scattered along its length, the Wabash meeting the Ohio just a few hundred miles from where the Ohio joined the Mississippi. The chief sketched in the locations of two lonely French forts and mentioned that among their towns, the Indians of the Wabash could raise a thousand fighting men.

Laid out like this, it was clear that Croghan had opened a door the British had long sought, and the Pennsylvanians were almost breathless at the thought: "It is manifest that if these *Indians* and the Allies prove faithful to the *English*, the *French* will be depriv'd of the most convenient and nearest communication with their Forts on the *Mississippi*, the ready Road lying thro' their Nations, and that there will be nothing to interrupt an Intercourse between this Province and that great River."

Weiser was translating at Lancaster, as was—for the first time in an official capacity—Andrew Montour, whom Weiser had recommended to the Pennsylvania council just the previous month. No sooner was the treaty concluded than Croghan and the Indians headed back west with gifts for the Ohio tribes, followed a few days later by Weiser, Montour, and Benjamin Franklin's seventeen-year-old son, William, tagging along at the behest of his father, who had a keen interest in Indian policy and the western lands.

Montour and Weiser, two men split between conflicting worlds, would become partners in the go-between life of interpreters. Montour's innate conflicts, racial and ethnic, were deeper and more obvious than Weiser's, whose division was largely spiritual. As a boy, Weiser had envied atheists for their lack of any faith, but as a questing adult, he'd made up for it by throwing himself into the mysticism of Beissel's Cloister at Ephrata. In 1740, four years after his baptism as Brother Enoch, he'd left Ann Eva to live as a monk at Ephrata—a complete embrace that ultimately proved fatal to his faith, revealing what he saw as its hollow, sinful center. After three years, he could no longer ignore the "loathsome idolatry" bordering on messianism; the "spiritual and physical bondage," including Beissel's late-night visits to the quarters of the "Spiritual Virgins"; and the "fulsome self-praise" of Beissel and his cronies. Weiser resigned from Ephrata as emphatically as he had embraced it. "I take leave of your young, but already decrepit sect, and

I desire henceforth to be treated as a stranger," he wrote to the congregation.

Now, in 1748, after sixteen days on the trail to the Ohio—a far easier journey than Weiser's long-ago winter ordeal with Shikellamy, for which the fifty-one-year-old interpreter must have been grateful—the parties reached the river about fifteen miles below the Forks. Here lay Logstown, the largest Indian town in the Ohio country, where a sharp ear like Weiser's or Montour's could pick out almost every Native language and dialect imaginable. Coming into town, Weiser and his companions each pulled a brace of pistols from their belts, firing into the air in salute, and even as the returning volleys from town still echoed, they got to work.

Over the next three weeks, Weiser made the rounds in the laborious, often tedious work of diplomacy at which he excelled—convincing the wavering Wyandot not to return to the French and formally accepting them into Pennsylvania's fold as "brethren"; dispatching Montour to confirm the loyalty of the still angry Lenape; distributing rolls of tobacco, dispensing drams of rum, and punctuating his speeches with gifts of stroud cloth or strings of wampum; and all the while waiting impatiently for the bulk of the gifts, which had been delayed on the trail, to arrive. In each case, Weiser rigorously followed the norms of Native protocol, by which the participants were ritually prepared for the serious work of negotiating. Paramount among these were the "clearing the woods" ceremonies, which, with gifts of wampum, were meant to metaphorically clean a traveler's eyes to see and open his ears to hear, and the condolence ceremonies, in which the exchange of gifts was meant to "wipe the tears" over deaths that had occurred since the last treaty.

Weiser raised the Union Jack in the middle of town as more and more Indians appeared for the council, each band firing off their muskets as they drew near. He met privately with the Seneca, who had recently carried off English prisoners from the Carolinas. "Brethren, you came a great way to visit us, and many sorts of evil might have befallen you by the way," Tanaghrisson told Weiser, handing him a belt of wampum before a gathering of Iroquois leaders. "We give you this string of wampum to clear up your eyes and mind, and to remove all bitterness of your spirits, that you might hear us speak in good cheer."

The sachem apologized for the Carolina raid and, offering Weiser a belt of black wampum, said, "We therefore remove our hatchet, which by the Evil Spirit's order was struck into your body . . . that this thing might be buried into the bottomless pit."

With Croghan's help, Weiser stove in an eight-gallon barrel of whiskey that a white man named Nolling had smuggled into town; growing numbers of "back inhabitants" were getting into their own form of Indian trade, making moonshine and transporting it over the mountains to the Ohio villages. During his time in Logstown, Weiser dealt with complaints from some Indians that the liquor was too plentiful, from others that it was too scarce, and from still others that it was too expensive.

At Weiser's behest, the leaders of each Native band conducted a crude but effective census. For every man of fighting age in their village, they added one twig to a neatly tied bundle. Counting the assembled bundles, Weiser determined that there were almost eight hundred warriors along the Ohio—mostly Lenape, Seneca, Shawnee, and Wyandot, with a few hundred Mohawk, Mahican, Onondaga, Cayuga, and Oneida, and even a few Ojibwa from the far lakes. Even without the Miami farther west, it was an impressive total, more than matching the combined manpower of the Iroquois homeland.

Addressing the council as Tarachiawagon, the name he'd been given by the Six Nations, Weiser said:

> Brethren, some of you were in Philadelphia last fall, and acquainted us that you had taken up the English hatchet, and that you had already made use of it against the French, and that the Frenchmen had very hard heads, and your country afforded nothing but grass and sticks, which were not sufficient to break them. You desired your brethren would assist you with some weapons sufficient to do it; your brethren the President and council promised you then to send something to you next spring by *Tharachiawagon* . . . but other affairs prevented his journey to Ohio.

The council had at last sent "a civil and brotherly present," to which he gestured: five large stacks of goods, one for the Seneca; one to be shared among the Cayuga, Onondaga, Oneida, and Mohawk; one for the Lenape; one for the Shawnee; and one for the Wyandot, Ojibwa, and Mahican.

Despite a temporary peace with France, Weiser had no illusions. This present, Tarachiawagon concluded, "shall serve to strengthen the chain of friendship between us, the English, and the several nations of Indians to which you belong. A French peace is a very uncertain one . . . Our wise people are of [the] opinion that after their belly is full, they will quarrel again and raise a war."

The likelihood of war was also on the minds of Tanaghrisson, the Iroquois "half-king" who held nominal authority over the Ohio Indians, and Scarouady, an Oneida who had been sent by the Six Nations to "oversee" the Shawnee. They met privately with Weiser, Montour, and Croghan, and their message was anything but peaceful. "You will have peace," they told the three men, "but it is certain that the Six Nations and their allies are upon the point of making war against the French." Handing him a belt of wampum, they said, "Let us keep a true correspondence, and let us hear always of one another."

In fact, Tanaghrisson and Scarouady were themselves part of the evolving dynamic of frontier statecraft. As representatives of the Iroquois League, they should not have been happy with the outcome of the just-completed council, and not only because of Britain's temporarily peaceful attitude toward the French. Ever since James Logan and Shikellamy had forged their partnership almost twenty years earlier, Pennsylvania and the Six Nations had sidestepped the "tributary" tribes such as the Shawnee and Lenape. At Logstown, Weiser—and through him, Pennsylvania—had decided to deal directly with allies such as the Wyandot instead of working through Onondaga.

That change would make the leaders of the Six Nations deeply uneasy, but it may not have greatly disturbed Tanaghrisson and Scarouady. They were increasingly independent operators themselves, speaking on behalf of the people they had been sent to oversee. The world was not as it had been. Logan was old and ailing, the aging Shikellamy was in such poor health that he would not live out the year, and the Ohio Indians were a long way from the council fires at Onondaga. The twig-bundle census, as much as anything, showed Weiser that the Native power center had shifted away from Iroquoia.

Weiser wasn't the only one to see what lay ahead. In 1750, Pennsylvania governor James Hamilton wrote to his counterpart in New York, George Clinton:

> You can't be insensible that many of the six Nations have of late left their old Habitations, and settled on the branches of Mississippi, and are become more numerous there than in the Countrys they left At which both the French & the Council at Onondaga are not a little alarmed. The State of Indian Affairs . . . gives me much concern that the Council at Onondaga should not be able to retain their People among them, but by suffering their young Indians to go and settle in those distant parts, give rise to a new Interest that in a little time must give the Law instead of taking it from them.

Weiser had also seen firsthand what astonishing country "those distant parts" were. The Ohio River was so wide and easily navigable that there was "not the least danger of overseting," Weiser wrote after returning home. "The land on both Sides are very good and a great deal of it Extraordinary rich," where "a midling good Hunter among the Indian of Ohio Killes for his Share in one fall 150, 200 dears." The gifts he had just delivered, Weiser said, made "a good start," but the province must move quickly to establish better trade to forestall not only the French—"one thinks it a thousand pity that Such a large and good Country Should be unsetled or fall into the hand of the french"—but also Marylanders and Virginians, who were eagerly eyeing the Ohio country.

To Weiser's mind, the Indians of the Ohio could provide a bulwark against the French, but only if properly courted. That meant leaning on the Iroquois on behalf of the Lenape, persuading "the 6 nations to take off the petticoat from the delewares and give them a Breech Cloath to wear." More and more, Weiser, so long the proponent of Iroquois domination, was pushing for greater independence for the Ohio towns.

Richard Peters, the Anglican clergyman and provincial land clerk who had taken over for James Logan, not only as council secretary but also as proprietary henchman and Machiavellian sculptor of the colony's Indian policy, wrinkled his nose at the idea. Except for the

Wyandot and Miami, he told Thomas Penn, the Lenape and other Ohio Indians were "ye Scum of the Earth . . . [a] mixd dirty sort of people." Even Peters had to admit, though, that they "may be made extremely useful to the Trade & security of this Province"—a fact that the French were about to make very, very clear.

Chapter 9

The Long Peace Ends

I N   T H E   S U M M E R   of 1749, Captain Pierre-Joseph Céloron de
Blainville—former commander of Detroit and Fort Niagara, a
knight who wore the Cross of the Order of Saint-Louis for his
exploits against the Chickasaw—was dispatched to the Ohio, which
the French called La Belle Rivière, "the beautiful river." There he was
ordered to cow the Indians and bring them firmly back into the fold
of "Father Onontio," the traditional name of the governor of New
France; to rid the backcountry of the English traders who were caus-
ing such mischief; and to reestablish the unquestioned rule of His
Most Christian Majesty Louis XV across the vital midsection of the
continent, linking Canada and Louisiana. Céloron intended to accomplish all this with 213 men, 23 canoes, and half a dozen lead plates.

A year after the Logstown council, during which Conrad Weiser
and Andrew Montour shored up the friendship between Pennsylvania and the western Indians, Céloron's expedition was evidence that
the French were finally taking the British threat to the Ohio Valley
seriously—although perhaps not seriously enough. Long practiced in
the nuances of Indian affairs, the French had lately installed a series
of governors who proved singularly ham-handed—stingy with gifts,
insensitive to cultural etiquette, tone-deaf to the effect they were hav-

ing on relations with the Natives inhabiting the *pays d'en haut*—the "upper country" around the Great Lakes and Ohio River from which the furs (and thus the riches) of New France flowed.

Angered by the success George Croghan and other British traders had enjoyed in winning allies in the southern *pays d'en haut,* the latest Father Onontio, Governor Roland-Michel Barrin de La Galissonière, could have sent a large military force into the Ohio country to sweep it clean of Englishmen and fortify the crucial points along the rivers. Instead, he sent Céloron with 20 soldiers and officers and 180 Canadian militia of questionable value. They were accompanied by only 30 Abenakis and French Iroquois; the large contingent of Indians expected to join them from Detroit, knowing a lost cause when they saw one, decided they had better things to do.

Despite this, Céloron pushed ahead, leading his flotilla past Fort Niagara and into eastern Lake Erie. They beached, then cleared a ten-mile road through thick, hilly forest to the south. By stages, they portaged their canoes and a mountain of food, bedding, ammunition, and Indian gifts to a lake the Iroquois called *yjadakoin*—today's Chautauqua Lake. The skies poured incessantly throughout the five-day portage, and Céloron could only comfort himself that all the rain would eventually mean easier canoeing downriver.

Beyond the lake, though, the way grew even worse. Another portage led to streams that would have been easily navigable in spring but were now, despite the downpour, almost dry. While most of the party carted backbreaking loads overland, some of the men dragged their half-empty canoes through a few inches of water, ripping holes in the precious boats. The Ohio country seemed like a cruel mirage. They'd been told to expect a bountiful land teeming with *de boeufs illinois,* but instead of buffalo, they found themselves in a "somber and dismal valley," on guard against *serpents à sonnettes*—rattlesnakes.

Finally, at noon on July 29, the small creek brought them to La Belle Rivière—what the English called the Allegheny River, one of the Ohio's two major tributaries, near the modern town of Warren, Pennsylvania. The company drew up in ranks, arrayed in full uniform with weapons at the ready. Céloron ceremoniously buried a heavy lead plate about eleven inches by seven, pre-engraved with all but the date and place (which had been quickly stamped into the metal that afternoon). The inscription proclaimed the plaque "a monument of the renewal of the possession we have taken of the said Rivière Oyo, and all those which fall into it, and of all the territories on both sides as far as the source of the said rivers, as the preceding Kings of France have possessed or should possess them."

An iron plate bearing the royal coat of arms was nailed to a nearby tree. Finally, a written proclamation known as a *procès-verbal* was prepared and signed by the officers, reiterating the French claim. After a few cheers of "Vive le Roi!" the soldiers and Indians slipped their canoes back into the current and headed south.

Five more times in the next five or six hundred miles, between the upper Allegheny and the mouth of La Rivière à la Roche (the Great Miami, where Ohio, Kentucky, and Indiana meet today), Céloron buried his plates and drew up his written *procès-verbaux*. In Indian villages along the way, he tried to cajole the elders back into friendship with the French, offering a few gifts and occasionally sharing "a cup of the milk of their Father Onontio" (a glass of brandy). But everywhere the French went, they found two things—a distaste bordering on contempt among the Indians, and English traders doing a thriving business.

"Each village, whether large or small, has one or more traders, who have in their employ *engagés* for the transportation of peltries [dressed pelts]," the Jesuit Joseph-Pierre de Bonnécamps wrote in his journal. "Behold, then, the English [are] already far within our territory; and what is worse, they are under the protection of a crowd of savages whom they entice to themselves, and whose number increases every day."

Along the lower Allegheny, the French intercepted five such traders working for George Croghan. "They had some forty packets of peltries," Bonnécamps wrote, "skins of bears, otters, cats, pécans [fishers], and roe-deer with the hair retained—for neither martens nor beavers are seen there." They found the traders "lodged in miserable cabins, and had a storehouse well filled with peltries, which we did not disturb." Instead, Céloron ordered them out of the country and gave them a letter for the governor of Pennsylvania asserting French sovereignty.

The mood at Chiningué (as the French called Logstown)—which

they reached two days later, to find Weiser's faded Union Jack still flying over the village—was so overtly menacing that Céloron ordered
his men to sleep in their clothes, muskets ready, having been warned
of imminent attack. For two days, he parleyed with the Indian leaders,
although his message was so unwelcome that in the middle of one of
his speeches, a blind but still spunky Shawnee chief name Kakowat-
chiky was heard to urge someone—*anyone*—to please shoot Céloron
for him.

No one did, but the captain concluded that Logstown was "a bad
village, which is seduced by the allurements of cheap merchandise fur-
nished by the English, which keeps them in a very bad disposition towards us." But his hands were tied by distance and ineffectuality. The
chief trader there (probably one of Croghan's men), "who saw us ready
to depart, acquiesced in all that was exacted from him," Bonnécamps
said, "firmly resolved, doubtless, to do nothing of the kind, as soon as
our backs were turned."

Céloron had a similarly unpleasant reception at Pickawillany from
the Miami, whose leader, Memeskia, called "La Demoiselle" by the
French, had earned a new nickname: "Old Briton." By the time Céloron returned to Montreal, five months and 3,500 miles after leaving, the weary chevalier seemed to realize that his expedition had had
precisely the opposite effect he'd intended. Instead of strengthening
French rule in the Ohio country, he had shown it to be weak and im-
potent.

The French could do little to peel the Ohio Indians away from
Brother Onas and their British friends. But it was not impossible, as
the next few years would prove—although the British colonies would
largely accomplish the deed themselves, alienating their own allies;
sending them back to the French in anger, frustration, and eventually
sheer desperation; and reawakening grievances stretching back to the
Walking Purchase and beyond. When at last the Ohio Indians lifted
the hatchet, the causes would include British fecklessness and stupid-
ity, along with bad luck, bad politics, and bad judgment. An inexperi-
enced young officer from Virginia would make a military blunder in
a valley called Great Meadows, triggering what has been called the
first truly global war, while a far more senior (and far more arrogant)
commander following in his tracks the next year would err even more

spectacularly, inviting defeat and stampeding the Ohio tribes to the French.

But behind all of this, driving the region toward cataclysm, was the same naked greed for Native land that had underpinned virtually every other frontier war of the previous century and a half. Some things, it seems, never change.

One might think the news of a longtime imperial adversary making a public display of reclaiming the Ohio would snap the British colonies into sharp action. Instead, they seemed barely to notice amid their interprovincial bickering. At times, open war seemed more likely between the colonies than between them and the French.

Far from being a cordial melting pot, the frontier was becoming an increasingly fractious mishmash, although ideas of race and ethnicity were curiously tangled, and the divisions did not always—or even usually—fall out the way one might expect. Animosity between a Presbyterian Ulsterman and a pietist German along the Susquehanna might be sharper than that between either man and a neighboring Delaware. A Protestant New Yorker in the upper Hudson Valley might view a fellow European—a Catholic Frenchman from Montreal—as more vividly different (and thus more ominously threatening) than a Mohawk.

For almost two centuries along the eastern frontier, the sharpest division from a colonial perspective was not racial—Caucasian versus Amerindian—but religious. Whether British, French, or Spanish, colonists almost invariably referred to themselves as Christians and to the Indians they encountered as heathens, pagans, or savages. Before he was torched alive by the Tuscarora, John Lawson advocated an intentional policy of intermarriage with Natives in Carolina to erase that gulf. The colony, he felt, ought to "give Encouragement to the Ordinary People, and those of a lower Rank, that they might marry with these Indians . . . By the Indians Marrying with the Christians, and coming into Plantations with their English Husbands, or Wives, they would become Christians, and their Idolatry would be quite forgotten, and in all probability, a better Worship come in its Stead . . . and the whole body of these people would . . . become as one People with us."

Skin color, the foundation of the modern concept of race, was very much an afterthought. Europeans did not see themselves as a monolithic race; national identity—English, Scots, Welsh, German, Swede—trumped other considerations, and the same was largely true of their Native counterparts. Growing up in Iroquoia, a young man would view himself first as a member of, say, the Wolf Clan; then as an Oneida; and then, to a lesser degree, as part of the Six Nations. A Catawba wasn't a fellow Indian but a traditional enemy, generations of tribal war having carved a deep psychological divide that shared ethnicity was unlikely to bridge.

By the early eighteenth century, however, recognition of racial lines was beginning to crystallize. In Carolina in the late 1600s, the slave-owning immigrants from Barbados already thought in terms of black and white, and it is here that some of the earliest self-references to "white" appear. By the mid-eighteenth century in the Middle Colonies and New England, the term had largely replaced "Christian" as the catch phrase for those of European descent. Most of the official correspondence in Pennsylvania, for example, used the terms "whites" and "white people" when speaking of colonists in general, regardless of their national origin.

Conrad Weiser spoke of the difficulty of judging private quarrels between "the white and the brown People, for the former will out swear the very devil, and the Latter oath is not Good in our Laws," but such references to Indian skin color were still unusual. New Englanders and others had long tended to speak of "tawnies," if they brought up Indian appearance at all, and in 1751 Benjamin Franklin noted that "*Asia* [is] chiefly tawny, *America* . . . wholly so." What he considered to be "purely white People," on the other hand, were rare. Franklin excluded "*Spaniards, Italians, French, Russians* and *Swedes,* [who] are generally of what we call a swarthy Complexion; as are the *Germans* also, the *Saxons* only excepted." Franklin mused that "by excluding all Blacks and Tawneys," the colonies could serve to increase "the lovely White and Red"—by which he apparently meant ruddy English stock.

Ironically, "red" as a description of American Indians, now considered a racist slur, first appeared among some Indians themselves. Sitting in council in 1725, a Chickasaw chief told the Carolinians that his towns "desire always to be at peace with the White people and desire to have their own way and to take revenge of the red people"—one of many references by Cherokee, Creek, and Chickasaw leaders of that period to "red people." As historian Nancy Shoemaker has shown, this term was virtually unknown outside the South, where many Native cultures used red and white as potent symbols of duality, and the term likely made its way into common use in French and English via frontier diplomacy. Among themselves, many Indians adopted some form of "white" to speak of the colonists, usually unkindly. The Cherokee referred to "white nothings," and the Lenape word for settlers, *shuwánakw*, meant "sour people" or "white people."

While racial identity was hardening in the eighteenth century, in other respects the same barriers were breaking down. Intermarriage was a fact of frontier life from the very beginning, whether the unions were formalized or not, and the long tradition of absorbing captives and refugees into Native communities made it especially easy, for example, for freed or escaped African slaves to find refuge and acceptance there. Similarly, subjugated Indian populations, such as the Pequot of southern New England, who were held as slaves or servants after the Pequot War, often married Africans in similar straits.

Traders especially blurred the edges of ethnicity and culture, even as they pushed the boundaries of the frontier. Marrying into an Indian society was simply smart business. It gave a trader immediate access to his wife's many family connections, which often stretched across hundreds of miles and into many villages and towns; it bestowed a degree of legitimacy that fostered trade; and it offered a welcome measure of protection in an inherently dangerous profession. Traders were notoriously lax about the niceties. Many had "country wives" scattered throughout Indian country, in addition to a legally married spouse (white or Indian) back in the settlements. This practice was not confined to lowly traders; Sir William Johnson, New York's powerful Indian agent, maintained a Mohawk "housekeeper" in addition to his wife, although the polygamous Iroquois understood the reality of the situation perfectly.

Cotton Mather disparaged the war parties attacking New England in 1690 as "half one, half t'other, half Indianized French, and half Frenchified Indians," a sarcastic reflection of the fact that intermarriage was both common and frequently encouraged in New France, at least at lower levels of society. Antoine Laumet de La Mothe Cadillac, who in 1701 founded Ville de Troit, "city of the straits," hoped to bring female Indian converts to Detroit as wives for his French soldiers, but most such unions were the result of human nature, not policy, and had been going on since Samuel de Champlain arrived. The métis developed their own distinct culture and, in the western *pays d'en haut,* their own language, known as Michif, which blended French and Plains Cree.

Andrew Montour's family—his mother, uncles, aunts, cousins, and siblings—were a singular example of the way the métis straddled multiple worlds. But while he had been walking the knife-edge between white and Indian, French and English all his life, by the time Montour returned from the Logstown council in 1748, he was looking to make something uniquely his own in a hostile world.

"Andrew Montour has pitched upon a place in the Proprietor's manor, at Canataqueany," a keyed-up Conrad Weiser wrote to council secretary Richard Peters in August of that year. "He Expects that the Government shall build him a house there, and furnish his family with necessarys. In short, I am at a lost what to say of him. He seams to be very hard to please."

The "manor" to which Weiser was referring was not a house, but one of several reservation-like holdings the Penns had set aside, ostensibly for Indian use—in this case, one along Conodoquinet Creek, on the western shore of the Susquehanna, not far from Croghan's establishment. It was initially created "to Accommodate the Shaawna Indians or such others as may think fit to Settle there." (The eviction of the Lenape years earlier from a similar manor along Brandywine Creek, west of Philadelphia, was one of many long-nursed injuries the Delaware held.) The Shawnee shunned it, and white settlers were pouring in. At least officially, though, the land was to "Remain free to ye Indians for Planting & Hunting." Chameleon-like, the multiethnic Montour chose to consider himself if not a Shawnee, at least an Indian.

Peters and Weiser were having none of it. The land was valuable, and Montour was prodded out of the Conodoquinet manor with his home half built. But with the stick, the Pennsylvanians—who saw a way to benefit from the situation—offered an eventual carrot: land north of the Kittatinny Ridge, given because Montour had "earnestly and repeatedly applied for Permission to live in some of the Plantations over the Blue Hills."

Actually, Weiser, Peters, and Pennsylvania governor James Hamilton saw an opportunity to quiet Montour's incessant requests for a home and to use him as a check on squatters who were flooding into land not yet purchased from the Indians. Montour's commission noted that "many Persons are lately gone and continually going over the Kittochtinny Hills to settle . . . notwithstanding repeated Proclimations . . . If it was permitted you to go and reside there you cou'd be very serviceable to both this Government and to the Six Nations in keeping People off."

Montour picked a sweet spot for his home, where the so-called New Path to the Ohio country crossed Sherman's Creek, a deep and clear stream full of trout that flowed into the Juniata River. Perhaps he moved there fully intending to keep his end of the bargain and police the frontier, evicting the white settlers who were drawn by the same good soils and game-rich forests that attracted him. If so, he changed his mind quickly; he was soon encouraging immigrants, white and Native, instead of evicting them. It seems Montour was already envisioning something else: a place where in-betweens, half-bloods, refugees, and the other impoverished riffraff on the colonial margins could make a life for themselves and a living for him—a mini-Penn, a small proprietor of sorts, collecting their rents and overseeing their settlements.

Montour was illiterate, and none of his words and thoughts have come down from his own hand. Historian James Merrell has reconstructed Montour's dream largely from the métis's land deals and from the reactions of those, like Weiser, who were at first puzzled, and later concerned, by his actions in the Juniata country. There is no evidence that Montour was imagining an egalitarian utopia of ethnic equality. More likely, he saw an opportunity to create a community of the kind of people he knew best and among whom he felt most comfortable, a community that would also provide him with the kind of means he'd never enjoyed. Montour may not have thought beyond the rude framework of such a plan, but it's clear that what he and others were doing in the Susquehanna hinterlands was keenly unsettling to those in power.

Richard Peters worried that "the lower sort of People who are exceeding Loose & ungovernable . . . wou'd go over [the mountains] in spite of all measures and probably quarrel with the Indians." Bad enough, Peters admitted, "but Mr. Weiser apprehends a worse Effect, that they will become tributary to the Indians & pay them yearly Sums for their Lycense to be there." The idea of white settlers living under the political and economic control of Indians had been a colonial bugaboo since the earliest days of the Massachusetts Bay Colony, since it would quickly erode the central, determinedly English character of any province. If such a circumstance were to come about, Peters fretted, not only would the Penns lose much of the revenue that now flowed into their coffers, but "the Proprietaries will not only have all the abandon'd People of the Province to deal with but the Indians too & . . . they will mutually support each other & do a vast deal of Mischief."

Mischief to whom? To those who had been pulling the strings in the backcountry for years—not just the Penns and their underlings but the Onondaga council in Iroquoia, which had spent generations building a façade of their supremacy over the Indians of the Middle Colonies. Many of the very people marginalized and dispossessed by the land deals of the Penns and the Iroquois were now gathering under Montour's umbrella. French Andrew was making a lot of powerful people nervous.

The Walking Purchase was hardly the only land fraud of the period, and the tricks cut both ways, even between supposed partners. In 1744, the Iroquois agreed to several large cessions, including one to Virginia for the Shenandoah Valley, which was even then filling up with German immigrants passing down the Great Valley from Pennsylvania. The Iroquois didn't really own the valley, but claimed it by nebulous "right of conquest" over the resident tribes. In council at Lancaster that summer, the Six Nations agreed to renounce their claims to "all the lands within the said colony" of Virginia, for which they gained explicit recognition of supremacy over land they'd never really controlled in the first place.

But the league was itself trumped in a breathtaking piece of legerdemain. Virginia conveniently neglected to point out to the Iroquois that its royal land grant extended all the way to the Pacific coast. By choosing to honor the dubious Iroquois claim of conquest, the British could ignore legitimate Indian claims to a monumental sweep of land. "Virginia resumes its ancient Breadth, and has no other limits to the West . . . to the South Sea, including the Isl'd of California," Virginia governor Robert Dinwiddie would later exult.

California could wait; the province had just acquired an open door into the Ohio country. To exploit it, a number of influential Virginia investors, including "the Maryland Monster," Thomas Cresap,* formed the Ohio Company. In 1749, the company received a royal grant of half a million acres there — provided it settled a hundred families, with a well-built fort for protection, within seven years. Instead, the entry of the company into the delicate balance of conflicting interests and powers in the Ohio would, in a few years, result in one of the worst wars the world had ever seen, brought on by "the heedless greed of a few headstrong men," historian Francis Jennings said. "That they lost money in the process is only a crowning irony."

To navigate the understandable opposition of the French, the Pennsylvanians, the hoodwinked Iroquois, and the Ohio Indians who actually lived on the frontier — and who would resent any attempt to settle it — the company turned to an experienced backwoods hand, Christopher Gist. For seven months beginning in the fall of 1750 and for five months again the next winter, Gist scouted the Ohio country as far west as Kentucky, where he saw "beautiful natural Meadows, covered with . . . blue Grass and Clover, and abound[ing] with Turkeys, Deer, Elks and most Sorts of Game particularly Buffaloes."

He was keeping his eyes open for possible settlement sites, such as the valley of "good level farming Land, with fine Meadows, the Timber white Oak and Hiccory," along the Monongahela River in western

---

* Cresap's reputation depended on which side of the Maryland-Pennsylvania border you lived on. During the 1730s, boundary tensions between the two colonies flared almost into open war, exacerbated by hostility between Scots-Irish immigrants mostly from Maryland and Germans from Pennsylvania. Cresap was the leader of those Marylanders who, legally or not, occupied land west of the Susquehanna claimed by Pennsylvania, earning him the nickname "the Maryland Monster." In 1736, Pennsylvania forces besieged his cabin, burning it and taking Cresap into custody. Arriving in Philadelphia, Cresap famously told his jailer, "Damn it, Aston, this is one of the Prettyest Towns in Maryland."

Pennsylvania, which Gist found so appealing he later established a farm near there himself. Just to the east was a place known as Great Meadows, or the Flats, even more extensive relict prairies nestled within the mountains. Gist crossed paths with Croghan, who made light of the fact that the French were offering an enormous bounty for his scalp, and with Montour, who frequently served as Gist's interpreter among the Lenape and Shawnee, to whom he promised large gifts.

Gist didn't reveal his purpose to the Indians, but Croghan knew what Virginia was planning. The French, no longer relying on Céloron's buried plates, had started seizing English traders and their pelts. Croghan saw that the security of his business depended on a stronger military presence in the Ohio and pressed Pennsylvania hard to build a fort there, despite the reluctance of its Quaker-dominated assembly. (The assembly's actions have long been cast as those of naive, blinkered pacifists, when in fact much of the friction between the assembly and Pennsylvania's governors had more to do with the Friends' principled opposition to an erosion of their strong governing role, as laid out in William Penn's founding charter. The Quakers would eventually prove the only force pragmatic enough to end the coming frontier war.)

Croghan brought the issue of a fort to a head in the spring of 1751, carrying a message from the Ohio tribes directly asking for a fort to provide them with protection against the increasingly potent French threat. The Quakers wavered, despite their pacifist beliefs. Yet when the assembly called Montour to corroborate Croghan's report, the métis shocked everyone by contradicting his longtime friend and collaborator, saying that the fort was Croghan's idea and the tribes would never go along with it. (Subsequent events suggest that Croghan was right. Why Montour blind-sided him in such a public, damaging way remains a mystery.)

The door for Virginia opened even wider. In fact, Hamilton quietly shifted Montour's services to Dinwiddie and encouraged the Virginia governor to pursue the course that Hamilton's own assembly had refused. But Virginia would need explicit permission from both the Six Nations and the Ohio tribes to bring in settlers. Gist was expected to deliver a confirmation of the fraudulent 1744 treaty at Lancaster—with the help of Croghan, Montour, and a mountain of gifts—to a council at Logstown in June 1752.

The Virginia commissioners called the treaty an opportunity for "polishing and strengthening the Chain of Friendship subsisting between us," but historians suspect that backroom dealing led Tanaghrisson and other representatives of the Six Nations to sell out the Lenape once again at Logstown, since the formal agreement permitting Virginia to settle the southeast side of the Ohio bore the names of no Delaware sachems. At least some of the Indians saw Virginia's interest for what it was. "An Indian who spoke good English came to me and desired to know where the Indian's Land lay, for that the French claimed all the Land on one Side of the River Ohio & the English on the other side," Gist said. "I was at a Loss to answer Him."

While the Logstown treaty was taking place, more than two hundred French, Ottawas, and Ojibwas swept into the Miami town of Pickawillany, killing Croghan's resident trader there and thirty others. A special fate was reserved for Memeskia, the Miami leader known as "Old Briton," who was boiled and ceremoniously eaten. The French were snapping up traders and their goods, leaving many Indians destitute and without supplies. Croghan and his partners, their business already foundering, went bankrupt as the entire frontier trading system collapsed. The Ohio tribes clamored for arms and ammunition, the military support Brother Onas had long promised, but they got almost nothing.

A year later, meeting with the Iroquois in Albany in August, Conrad Weiser heard rumors of "a very numerous Army of Men well armed and some great Guns," with which the French "would build Strong Houses . . . and so take Possession quite down [the Ohio] till they meet the French coming from below." He had blunt advice for Virginia: "Raise about 2000 men and take possession of ohio by force[.] Build [forts] . . . and if in the time of peeace will admit knock every french man on ohio that won't run to the head . . . If we don't do it now we never again shall so be able to do it. And our posterity will Condemn us for our neglect."

In fact, with an alacrity the British couldn't muster, the French displayed a military commitment suggesting they'd learned a few lessons from Céloron's feeble bluster. The newly arrived governor-general, Ange Duquesne de Menneville, the Marquis Duquesne, whipped the lax troops into shape, tightening discipline. Mustering two thousand soldiers in the winter of 1752–53 despite strident complaints about the cost, Duquesne used them the next summer to build three forts in rapid succession: one on Lake Erie at Presque Isle Bay; another, Fort LeBoeuf, fifteen miles inland at the headwaters of the Rivière au Boeuf (French Creek to the British; the Venango to the Ohio Indians); and the third, Fort Machault, where French Creek meets the Allegheny River.

Plans called for a fourth fort at Logstown, but the French would have to overcome the animosity of the Indians there if they were to force the issue. Tanaghrisson issued a tense warning to the intruders: "Therefore now I come to forbid You . . . I tell you in plain Words You must go off this Land." He repeated it three times, as required by tradition. The French dismissed him in a "very contemptuous manner," and seeing them move with such determination made many of the Ohio Indians question their already shaky loyalty to the British and to Tanaghrisson.

Both sides recognized that the Forks of the Ohio was the linchpin of the west. The French were dug in about seventy miles to the north, and the Virginians were eighty miles to the southeast at Will's Creek. Dinwiddie was authorized to build a British fort, but unwilling to confront the French without explicit instructions from the Crown, he dispatched an urgent request to London. ("Urgent," that is to say, within the realities of eighteenth-century communications. The messenger sailed in June, Dinwiddie's letter was received in August and responded to immediately, and the governor had his firm instructions from the king in mid-November.) "You are to require of Them peaceably to depart," he was told, but if the French chose to ignore the warning, "We do hereby strictly charge, & command You, to drive them off by Force of Arms."

To deliver that ultimatum, Dinwiddie accepted the services of an ambitious but untested twenty-one-year-old who volunteered to confront the French. On the surface, George Washington had few qualifications. His commission as a major in the newly formed local militia was less than a year old and probably had more to do with family and

George Washington as Colonel in the First Virginia Regiment
This Charles Willson Peale portrait, painted in 1772, shows the middle-aged Washington in the uniform he wore in 1758, when, as a colonel in the First Virginia Regiment, he again marched on Fort Duquesne. (Washington-Custis-Lee Collection, Washington and Lee University, Lexington, Virginia)

social connections than any evident ability. But Dinwiddie knew the young man had worked as a land surveyor in the wilds of the Shenandoah Valley and western mountains, and at nearly six feet four, he was a physically imposing presence—not an insignificant consideration in a military envoy.

The ink barely dry on his instructions, Washington left Williamsburg on October 31, traveling north to Will's Creek, the Ohio Company's station in the mountains of what is now Cumberland, Maryland. Along the way, he hired a Dutch-born, French-speaking soldier named Jacob van Braam to act as his interpreter, and at Will's Creek he hired Christopher Gist and "four others as Servitors," including an Indian trader named Currin who had worked for Croghan.

Despite heavy rain and snow, and having to traverse what he called "extream good & bad Land" and to swim their horses across the freezing Allegheny River, Washington's party reached Logstown in late November. There he met with Shingas, the leader of the Delaware, and with Tanaghrisson, who recounted his humiliating reception at Presque Isle that summer. Although the French wanted to fortify Logstown, Washington, casting an eye over the village, found it wanting from a military perspective. A fort upriver at the Forks of the Ohio would be cheaper and more effective, he felt, "equally well situated on Ohio, & have the entire Command of Monongahela."

Washington and his men labored farther upriver to Fort Machault—probably still just the trading stronghold the French had commandeered from Croghan—where the French garrison uncorked wine and "dos'd themselves pretty plentifully with it." The captain was Philippe-Thomas de Joncaire, an influential interpreter who had grown up among the Seneca, had been with Céloron in 1749, and was now the leading French voice in the Ohio country trying to win over the Indians. He must have been charming; Washington and Gist both said that he treated them "with the greatest Complaisance" for three days before passing them up the chain of command to Fort LeBoeuf to meet the French commander, Jacques Legardeur de Saint-Pierre. A weathered military hand whom Washington called "an elderly Gentleman" (he was fifty-three), Legardeur had arrived at LeBoeuf only days earlier and held a sufficiently lofty rank to accept Dinwiddie's floral demand that the French get lost.

"Sir, in Obedience to my Instructions it becomes my Duty to require Your peaceable Departure, & that You wou'd forbear prosecut-
ing a Purpose so interruptive of the Harmony & good Understanding
which his Majesty is desirous to continue," Dinwiddie wrote. Drafting
a reply took Legardeur and his staff some days, during which Washington suspected him of "ploting every Scheme that the Devil & Man
cou'd invent" to keep his Indian escort behind and leave Gist and
Washington vulnerable. The young officer responded in kind, quietly
assessing the fortifications, counting up the number of soldiers, evaluating the weight of the cannons, even tallying how many canoes lay
along the river. He wrote nothing down, however; to be caught actively spying would be a grievous and dangerous breach of his embassy.

At last, Legardeur handed a sealed reply to the young man he must
have found insufferably callow and arrogant. "As to the Summons you
send me to retire, I don't think myself obliged to obey it," Legardeur
informed the Virginia governor. "Whatever may be your Instructions,
I am here by Virtue of the Orders of my General; and I intreat you, Sir,
not to doubt one Moment that I am determined to conform myself to
them."

Now all Washington had to do was get the letter back to Virginia.
He and his companions perilously threaded their way downriver by
canoe, breaking through thickening ice, until they reclaimed their
horses. The Indian hunters kept them well supplied with bear meat,
but the cold, snow, and lack of forage had weakened their mounts to
such an extent that Washington elected to abandon them and the rest
of the party and to continue on foot with Gist. For his part, the guide
was deeply skeptical of the plan. In fact, Gist's opinion of Washington
may not have been much higher than Legardeur's. "I was unwilling
that he should undertake such a travel, who had never been used to
walking before this time," Gist said, "but as he insisted on it, I set out
with our packs, like Indians, and travelled eighteen miles."

"The Cold incres'd very fast, & the Roads were geting much worse
by a deep Snow continually Freezing," Washington wrote in his journal. "I took my necessary Papers, pull'd off my Cloths; tied My Self
up in a Match Coat; & with my Pack at my back, with my Papers &
Provisions in it, & a Gun, set out with Mr. Gist, fitted in the same
Manner."

It was an exhausting trip even for Gist, who was used to rough travel, while Washington was "much fatigued." Nor was the weather the only danger. A suspiciously ingratiating Indian who fell in with them along the way, trying to steer them back to the north toward French territory, suddenly turned and fired his musket at the men, just missing Gist. The next day, while crossing the river on a jury-rigged raft, Washington was flung into deep water amid the ice floes, and both men were trapped that night on an island, Gist's fingers and toes frostbitten and Washington shivering in his frozen clothes.

The young major straggled into Williamsburg on January 16, presenting Dinwiddie with Legardeur's letter and a copy of his own journal, which the governor then passed on to the newspapers. Soon the journal was being reprinted up and down the colonies—the first time anyone outside of tidewater Virginia heard the name George Washington. One reader in particular—a Glasgow-born shopkeeper named Robert Stobo, who ran a store in Petersburg—was on such fire to see the frontier that he cadged a captain's commission from his friend Dinwiddie and bought a uniform, a sword, and a copy of *A Treatise of Military Discipline,* a handbook for newly minted officers. He loaded hundreds of pounds of supplies, including a cask of wine, into a specially built wagon. Robert Stobo—"a slender man five feet and ten inches tall . . . with a dark complexion, a penetrating eye, an aquiline nose, and a cheerful, round face," in the words of a biographer—was ready to go to war.

No sooner was Washington home, with his recommendation for a fort at the Forks of the Ohio, than Dinwiddie ordered George Croghan's one-time business partner William Trent to build it. Trent set out immediately, while Washington, promoted to lieutenant colonel, was tasked with raising one hundred militiamen, then following Trent's party to man the fort until reinforcements could arrive.

Croghan, wintering in the Ohio country, was still pushing the idea of a Pennsylvania fortification—"If ye government wold Build a Strong Log'd house & Stockead itt Round, itt wold Do," he wrote to Richard Peters in early February—but even as he was dashing off a similarly misspelled note to James Hamilton the same day, Trent and his Ohio Company workmen appeared on the scene. Croghan, who was about to leave, decided instead to stay, along with Montour, and translate, as Trent "Can't talk ye Indian Languidge." A few weeks later, Croghan departed, informing Hamilton that Trent "had enlisted about Seventy Men before I left Ohio. I left him and his Men at the Mouth of Mohongialo building a Fort, which seemed to give the Indians great Pleasure and put them in high Spirits." Tanaghrisson helped lay the first log.

Everyone knew that such a direct provocation, on the doorstep of the French outposts at Machault and LeBoeuf, was sure to draw a response. Shortly after Croghan and Montour left Logstown, a French officer arrived there with a stern warning. "I am convinced that You have thrown away your Fathers and taken to your Brothers the English," he told the Indian leaders. "I tell You now that You have but a short Time to see the Sun, for in Twenty Days You and your Brothers the English shall all die."

By mid-April, Fort Prince George, as Trent called it, still wasn't much to look at—a partially completed log stockade enclosing a single building. Washington and his 150 ill-clothed, ill-equipped colonial militia were en route from Virginia when a message from Trent reached them at Will's Creek "demanding a Reinforcement with all Speed, as he hourly expected a Body of Eight Hundred *French*." It was already too late. A few days earlier, on April 17, with Trent off fetching more supplies and the garrison under the command of Croghan's half brother, Edward Ward, a veritable French armada had rounded the bend of the river.

Sixty flatboats, which the French called *bateaux,* and three hundred canoes beached at the Forks of the Ohio, disgorging more than a thousand soldiers, Canadian militia, and Indians, as well as eighteen pieces of artillery, including heavy nine-pounders. Ward had just forty-one men, most of them laborers, and nothing more potent than rifles. After hemming and hawing for a day, he accepted the inevitable. "Mr. Ward . . . having no Cannon to make a proper defense, was obliged to surrender," Washington wrote. "They suffered him to draw off his Men, Arms, and Working Tools, and gave Leave that he might retreat to the Inhabitants." Demolishing the half-finished walls, the French began building the wide, heavy earth-and-wood breastworks of what they dubbed Fort Duquesne.

The frontier was smoldering, and the Pennsylvania assembly still did little but dither, questioning whether the Allegheny and Monongahela were even in the old Penn land grant. Hamilton couldn't budge the Quaker bloc from its refusal to arm the Indians, which mystified Dinwiddie. "Whatever may be their religious Scruples, I think they should consider the first Laws of Nature Self-Preservation, and not remain inactive when likely to be invaded by the common Enemy," he told Hamilton.

Croghan was also driven to distraction by Philadelphia's inertia. Tanaghrisson was pressuring him for arms and ammunition, without which the Shawnee were likely to go over to the French. Indeed, the whole situation appeared increasingly sinister to the Ohio sachems, who saw the Virginians and French as dogs eyeing the same bone—their land.

"The whole of ye Ohio Indians Does Nott No what to think, they Imagine by this Government Doing Nothing towards ye Expedition that ye Virginians and ye French Intend to Divide ye Land of Ohio between them," Croghan warned Hamilton. Without a "hearty Concurrence" by Pennsylvania's council, which they still trusted, the Indians will believe that Virginia "is only atacking ye French, on Account of Setling ye Lands." The Iroquois League was increasingly fractious and ineffective, Croghan said, and whatever control they once exercised was gone: "I ashure your honour [the Ohio Indians] will actt for themselves att this time without consulting ye Onondago Councel." The promises of British assistance that he and Andrew Montour were carrying to the Ohio villages, Croghan said, "I am a fread will Come to leatt."

Washington didn't have enough men or heavy guns for an assault on Fort Duquesne, whose breastworks were already chest-high and rising. But he continued with part of his original mission, cutting a road through the forest from Will's Creek, following an Indian path about sixty-five miles to the Youghiogheny River, where with reinforcements he could float down the Ohio to the Forks and confront the French. It was slow going—three or four miles a day at most, clearing a path for horses and the rest of the expected artillery. Axes rang through the thick woods of oak and chestnut, although here and there Washington's men emerged from the shadows into open grasslands between the mountains—one place called Little Meadows and, thirty miles deeper into the Ohio country, the long valley of Great Meadows, or the Flats, whose marshy grasslands were a welcome respite from the gloom of the forest and good forage for the horses and cattle.

There was little rest. Gist arrived with news that a French detachment under René-Hippolyte Laforce—one of those who had wined and dined Washington at Fort Machault the previous winter—had rifled Gist's plantation not far away, while Indian scouts dispatched by Tanaghrisson reported a possible French ambush. Washington secured his ammunition and supplies and, hoping to one-up the French, marched forty of his men through a nightlong downpour, in woods so stygian that seven of the militiamen got lost along the way. The forest swallowed up another detachment of forty men, under Captain Peter Hogg, who stumbled off in the wrong direction. After five sopping hours, finding their way almost by feel, Washington's party reached Tanaghrisson's camp at daybreak on May 28.

The French had passed a slightly more comfortable night. Laforce and thirty men under the command of Joseph Coulon de Villiers, Sieur de Jumonville, had bivouacked for several days in a hollow just a few miles away, making crude bark shelters to keep off the rain. Jumonville, who came from a military family—his father and all five of his brothers were officers in the colonial regulars—had been sent from Fort Duquesne to find the British column and, if he believed them to be on French soil, present a letter from his commander demanding they leave. Perhaps because he considered himself an envoy—or perhaps from simple carelessness—the thirty-five-year-old ensign posted no sentries and so did not realize that two British scouts had observed the camp and slipped away.

As the rain continued, the British soldiers and a few of Tanaghrisson's Indians encircled the French, many of whom were still asleep, the smoke from their breakfast fires filling the air. The British troops tried to dry their muskets and replace their wet priming powder as Washington formed his ranks in the murky half-light. Someone among the French spotted the intruders and shouted warnings, and a gun may have gone off. Washington barked a command, and two volleys of British musket fire ripped into the disorganized French troops.

"A smart Action ensu'd; their Arms and Ammunition were dry . . .

[and] we could not depend on ours," reported Captain Adam Stephen, who commanded one of Washington's companies. "Therefore keeping up our Fire, [we] advanced as near we could with fixt Bayonets, and received their Fire." The whole thing was over in a few minutes. Jumonville and several other Frenchmen fell in the initial barrage, and those who lay injured were quickly killed and scalped by the Indians. Tanaghrisson walked up to the wounded Jumonville and said, "Thou are not dead yet, Father?" He tomahawked the ensign's skull, dipping his hands into the warm brains before scalping him.

Washington had just ignited the Seven Years' War (known in the United States as the French and Indian War, even though it was the last of four "French and Indian wars" dating back to King William's War in 1689). Spreading from the Ohio country, the conflict would consume France, Britain, and their imperial allies to become the first truly "world war," scorching Europe, the Caribbean, India, western Africa, and the Philippines, among other theaters of operation. Washington's initial reaction, though, appears to have been a bit giddy. He wrote a quick letter to his brother John, bragging, "I heard the bullets whistle, and believe me, there is something charming in the sound." As often happened in those days, the letter was printed widely in newspapers. When King George II, who had led British troops in battle, read it, his reaction dripped scorn: "He would not say so, if he had been used to hear many."

Green though he still was, Washington must have realized he was in serious trouble. He had fewer than 170 men and supply lines that snaked hundreds of miles to Virginia, while the French had several thousand soldiers and Indian allies solidly based at three—now four—forts not far away. Sending his French captives, including Laforce, back to Virginia under guard, Washington in three days hurriedly raised what defenses he could, entrenchments and a crude palisade, in the midst of the marshy meadow that he optimistically described to Dinwiddie as "a charming field for an Encounter."

The French commander on the Ohio, Claude-Pierre Pécaudy, Sieur de Contrecoeur, was livid. A Canadian survivor of the attack, hobbling into Fort Duquesne on bleeding, shoeless feet, as well as some of the Indians who had been there, said that Jumonville had been cut down while trying, through an interpreter, to read the French demands that the British leave. (Washington called the claim "an absolute False-hood . . . When they first saw us, they ran to their Arms.") To Con-trecoeur, it was little more than an assassination. Believing the British were marching six thousand men into the Ohio, he readied a retalia-tory force to oust Washington and secure the Monongahela Valley before British reinforcements arrived. Just before the French columns marched, Jumonville's elder brother, Captain Louis Coulon de Vil-liers, arrived at Duquesne. Shocked by the news of his brother's death, he asked for, and was given, command of the attack.

French fears aside, the British had just fifty-five men coming to Washington's aid, including Captain Robert Stobo and his well-stocked wagon of equipment and provisions. Worse, they had been stuck in Will's Creek for weeks while their commander shilly-shal-lied. Only a blistering message from Dinwiddie and direct orders to abandon the wagons and march "with the utmost expedition" finally prodded him into action. Traveling with the tardy column as an inter-preter was George Croghan, who before leaving Virginia had agreed to supply fifty thousand pounds of flour and two hundred horses to the expedition. (He had, unfortunately, left the job in the hands of the hapless Edward Ward, who made a botch of it.) With them as well was Andrew Montour, now with a captain's commission from Vir-ginia, accompanied by eighteen Indian scouts.

Washington had asked Dinwiddie specifically for Montour, who "would be of singular use to me here at this present, in conversing with the Indians." Tanaghrisson had come not with a war party, but with twenty-five or thirty Native families, including women and chil-dren. The Ohio was becoming a dangerous place for anyone as vo-cally pro-British as the Iroquois leader. Not only were there rumors that the French were after him, but Tanaghrisson's standing among the very Shawnee and Lenape he was supposed to be governing had fallen so far as to make the security of the British camp attractive. Tanaghrisson had sent Scarouady, with wampum and fresh French scalps, to Logstown to raise more fighters and to ask for help from the Six Nations, but the Ohio Indians were increasingly chary of British connections.

Feeding everyone was an even bigger concern for Washington than inconstant allies. That missing flour was desperately needed, as were ammunition and other supplies. "We have been six days without flour, and there is none upon the road for our relief that we know of . . . We have not provisions of any sort in camp to serve us two days," Washington wrote to Dinwiddie.

Arriving with the reinforcements on June 9, Croghan and Montour found themselves in another crisis—the erosion of Indian support. It was clear to the sachems that war was building, and as Scarouady would later observe, it was impossible to live in the woods and remain neutral. After years of begging for the protection of a British fort, the Indians had seen the French bring an overwhelming force to the Ohio and the British squander their chances. The bitterness over land frauds such as the Walking Purchase had never really disappeared, and remaining in the British camp now meant abandoning the Ohio country and moving to the crowded settlements farther east. It was easier to back the likely winner—besides, few Natives trusted the Virginians' motives, believing the sudden interest in the Ohio was merely another ruse to dispossess them of their land. Croghan and Montour, lobbying the Lenapes and Shawnees gathered at Gist's farm, tried to get them to side with Washington, but their responses were cool and noncommittal. Washington came to suspect that the Indians' "Intentions toward us were evil," and after the council, most just melted away. Many would soon be fighting openly with the French.

Even Tanaghrisson, far more fiercely anti-French than most, sized up Washington and found him wanting. The Virginian was "a good-natured man but had no Experience," the Seneca leader concluded. He "took upon him[self] to command the Indians as his Slaves," while accepting no advice from those who actually lived, hunted, and fought on the frontier. "He lay at one Place from one full Moon to the other and made no Fortifications at all but that little thing upon the Meadow," Tanaghrisson later complained to Conrad Weiser.

Still chopping away at a road to Gist's plantation, debating whether to move his main encampment there and create yet another ad hoc fort, and squabbling over command with the captain of the newly arrived company of British regulars from South Carolina, Washington ran out of time. Rangers reported that the French under de Villiers were closing in. Washington tried to retreat, hoping to reach Will's Creek, but his tired, hungry men moved too slowly, so he fell back

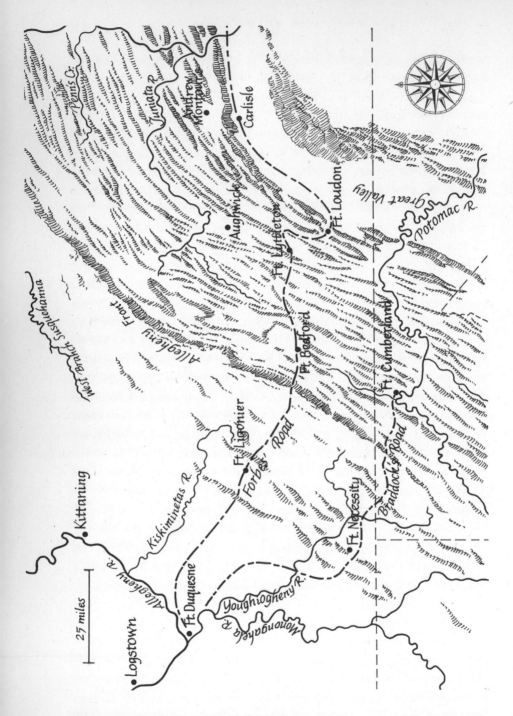

Washington's, Braddock's, and Forbes's Campaigns, 1754–1758
(Modern boundaries added.)

to "that little thing upon the Meadow," the ramshackle stockade that Captain Stobo dubbed Fort Necessity.

In reality, the "fort" was nothing more than a bark-and-skin-covered shelter about fourteen feet square, surrounded by a small circular palisade of logs, sitting in the middle of meadows that were, under the best of circumstances, marshy underfoot. In all, it was perhaps thirty feet in diameter, barely large enough to contain the 284 men. Here they would have to make a stand. Croghan's promised food had never arrived, just a few loads of meager supplies from Will's Creek. Weak with hunger, the men followed orders from Stobo, the regimental engineer, digging trenches beyond the stockade walls and felling trees that could provide cover to attackers. Montour, leading his rangers, was away scouting.

Two days later, late on the morning of July 3, de Villiers reached Great Meadows, having just passed the scavenged bones of his brother and the other dead Frenchmen, killed more than a month earlier and left unburied for the ravens. Once again, it was raining. The British soldiers crouched in cold water in Stobo's ditches, finding what protection they could, while six hundred French and Canadians, along with about a hundred Indians, pinned them down in a crossfire. Choosing his moment, Washington had two of their swivel guns—small, minimally effective cannons—sweep the field with grapeshot, after which the attackers simply kept more closely to the safety of the trees and picked off the defenders from there. The rain fell harder still, turning the fort into a quagmire.

There was no way out. "At Night the ffrench Desired to Parley with our People, But Col. Washington refused," recalled John Shaw, an Irish seaman who had signed on with the Virginia company. Washington smelled a trick; there was no reason for the French to negotiate what they could very easily force. But de Villiers—whose men were themselves short on food and powder, and who feared British reinforcements at any time—wanted the matter closed. The only British officer to speak French was Jacob van Braam, the Dutchman who had accompanied Washington to LeBoeuf the previous winter and now a captain in the Virginia regiment. He met under a flag of truce with de Villiers, who pushed the point, telling Van Braam he was "to be Reinforced in the Morning by four hundred Indians who lay about twelve

Miles off; And then it would not be in their power to give Quarters."
The British were asked to think it over during the night, and if they
wanted to avoid a massacre, they should not hoist the Union Jack in
the morning.

Van Braam took the written terms of surrender back to his commander; by candlelight, Washington and his officers thrashed out the
alternatives. They had more than thirty dead and seventy wounded,
a quarter of their strength, and they'd killed only a handful of the
French. Their options were few. At dawn, the flagpole was bare and
the document had been signed in Washington's forceful hand.

By agreeing to the capitulation, Washington was explicitly ac-
knowledging what the document called *"la assasain qui a été fait sur
on nos officier"*—the assassination of Jumonville, a charge that would
darken his reputation for years. It seems plausible that, as Washington
always maintained, Van Braam (whose French may have been fine but
whose English was shaky) orally translated the passages as "the killing
of" or "the death of" Jumonville. Be that as it may, Washington's journal, which the French discovered in the aftermath of the fight, betrays
a remarkably defensive conscience about Jumonville's death. He called
the ensign's mission "a plausible Pretence to discover our Camp, and to
obtain the Knowledge of our Forces and our Situation!" and through
tortured reasoning he dismissed any notion that the ensign was a formal envoy: "Instead of coming as an Embassador, publickly, and in an
open Manner, they came secretly, and sought after the most hidden
Retreats, more like Deserters than Embassadors . . . Besides, an Embassador has princely Attendants; whereas this was only a simple petty
French Officer."

Of course, much the same case could have been made against a
simple militia officer appearing at a French fort, in the middle of the
winter, accompanied by a few guides, and quietly counting the cannons. But if Washington was self-aware enough to see the similarities
between Jumonville's mission and his own the previous year, he wasn't
going to admit it, even to himself.

And while Washington still thought of himself as British and de
Villiers as French, they and many of those locked in battle—white and
Indian, pureblood and métis, colonial militia and *troupes de la marine,*

Virginians and Carolinians, Pennsylvanians and Canadians, Mingos and Delawares—had been born on the continent on which they were fighting. The motives for looming war may have been international, the goals propelling the armies imperial, but for most of the combatants at Great Meadows, this was home.

In the morning, drums beating and their muskets shouldered, Washington and his men left Fort Necessity. The ranks of the French-allied Indians had swollen overnight, including many whom the British had lately considered their friends. Private Shaw, marching out among the columns, said that the Indians "could hardly be Restrained . . . from falling on our People," and the warriors had already pillaged most of the officers' belongings and killed several of the wounded. The British gave their word—their "parole of Honour"—not to return to the Ohio Valley for one year from that date. It was a far better outcome than the badly outmanned and outgunned colonials had any right to expect.

Captains Stobo and Van Braam watched the columns disappear, trying to maintain a look of professional detachment. "As the English have in their Custody an officer, two Cadets and other prisoners," the signed capitulation had noted, "and which they promise to send with a Safe guard to Fort de Quesne . . . Mr. Jacob Vanbram & Robert Stobo, two Captains, are to be left with us as Hostages until the arrival of our said Canadians & Frenchmen." The officers had turned themselves in the night before, Van Braam wearing a uniform coat he'd bought from Washington so as not to appear shabby, and Stobo having given his new sword to his lieutenant, Will Poulson. Torching Fort Necessity, de Villiers turned toward Fort Duquesne, his hostages in tow. Whatever English homesteads they encountered along the way, including Gist's farm, were burned.

The shock waves from Great Meadows reverberated well beyond the Ohio Valley. Across the colonies, political leaders were stunned by the news but, riven by rivalries and hamstrung by ingrained timidity, did little but wring their hands, despite Dinwiddie's pleas for concrete assistance. Recriminations filled the air. "The Misfortune attending our Expedition is entirely owing to the delay in your Forces," Dinwiddie wrote to Colonel James Innes, whose reinforcements never reached Washington, "and more particularly to the two Independent Companies from N.Y.; how they can answer their disobedience to His Majesty's Commands I know not."

The response was more forceful in England, where a grandly hawkish plan was devised to sweep the French from the Ohio, the Great Lakes, and the Maritimes, overseen by a single, supreme commander in chief—Major General Edward Braddock, a man chosen less for his experience than for his connections and politics. No sooner was the supposedly secret strategy drawn up than word leaked to the French, who rushed to strengthen their defenses and garrisons against the anticipated British attacks the next spring. For hardliners on both sides, unsatisfied with the *status quo ante bellum* conclusion of King George's War in 1748 and itching for a fresh imperial confrontation, things were going splendidly.

In Williamsburg, the notoriously tightfisted House of Burgesses feted the returning army by awarding a gold coin "for each Man as a reward for their bravery in a late engagement with the French." Captain Stobo was promoted, in his absence, to major, while his fellow hostage, Van Braam, already under a cloud for the "assassination" faux pas, was pointedly ignored. Washington, too, came in for public criticism for his leadership. As summer faded to autumn, Washington found himself increasingly frustrated—by the lack of funds to clothe and feed his men, by Dinwiddie's grandiose plans for attacking the French, and ultimately by the governor's restructuring of the regiment into ten companies, which would have effectively demoted Washington from colonel to captain. In October, he resigned his commission and returned to the private life of a planter.

The effects of Great Meadows were, of course, felt most directly on the Ohio frontier, where the defeat was an enormous setback for British interests. Croghan, like most of the western traders, had been bled dry by French depredations over the previous two years, and the loss at Fort Necessity ended any hope that he could recover his business there. He avoided Philadelphia, afraid he'd be imprisoned for his debts, and moved deep into the mountains to a small farm at Aughwick, along the upper Juniata River on the main foot trail to the Ohio, where his workers and a few black slaves stockaded a log house and tilled fields of squash and corn.

Pro-British Mingos from west of the Alleghenies gathered where they could to find safety and food. Among them was Tanaghrisson, now living at Harris's Ferry along the Susquehanna, who in September 1754 accompanied Conrad Weiser on a trip west to Aughwick. Along the way, they found dozens of Lenapes and Shawnees living at Andrew Montour's house, where the métis's wife had to kill their sheep to feed the Indians. There were far more Indians at Croghan's, including Scarouady and Tamaqua, a Lenape chief also known as "the Beaver." "I counted above twenty Cabbins about his House," Weiser reported to Hamilton, having delivered £300 to Croghan for supplies, "and in them at least two hundred Indians . . . and a great many more are scattered thereabouts."

Croghan begged for help from the province. "There is such a number of Women and Children and more coming; they have almost destroyed thirty acres of Indian Corn for me, exclusive of other Provisions, which are very dear and hard to be got," he said. But amid his worries and petitions for assistance, the trader also forwarded to Hamilton two remarkable pieces of correspondence smuggled out of Fort Duquesne.

"I here enclose the Copy of a Letter from Captain Stobo," he wrote, "which I received two days ago by an Indian named Moses." It was the second of two copies, both carried into Aughwick by pro-British Natives. The mere existence of the letters, addressed to Washington, must have raised Croghan's suspicions about their importance. When he broke the seals and read their contents, his eyes undoubtedly widened, and he immediately drafted copies for Hamilton and Dinwiddie, sending them by separate runners.

Stobo was making the best use he could of what he expected to be his short time among the French. Under the conventions of war, both the British and French hostages seized in the two battles could expect a rapid exchange and a return to their companies, probably within a matter of months. Although the defeat at Fort Necessity was clearly a blow, Stobo doubtless took comfort in the words he'd read many times in his thumb-worn copy of *A Treatise of Military Discipline:* "When an Officer has had the misfortune of being beat, his honour will not suffer by it, provided he has done his duty, and acted like a Soldier."

After the surrender, Stobo was escorted with Van Braam to Fort

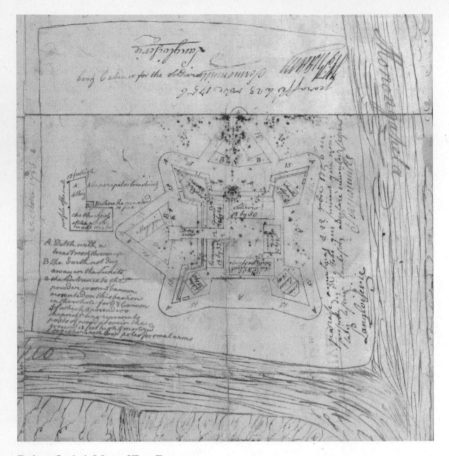

Robert Stobo's Map of Fort Duquesne

Drawn to meticulous scale, Stobo's map—smuggled out in the breechclout of Conrad Weiser's adoptive Mohawk brother, Moses the Song, and forwarded by George Croghan to colonial authorities—laid out the details of the French fortification. It also placed the captive Stobo's head squarely in the noose when his espionage was discovered. (Montreal Archives)

Duquesne. There they were treated cordially, billeted in one of the newly constructed cabins, and given the freedom to wander as they wished within its confines. They were, naturally, expected to obey the unwritten rules of parole, under which a captured officer would, in return for certain liberties, neither seek to escape nor work against his captors.

There was nothing to prevent an active young captain from using his eyes, however, and Stobo applied himself to a thorough examination of the fort. The French had done wonders since dislodging the Virginians, raising walls twelve feet high and more than ten feet thick, with four corner bastions and a six-pointed entrenchment with breastworks. The interior of Duquesne was jammed with barracks, officers' quarters, a smithy, a powder room, and a kitchen, while most of the thousand soldiers, joined by a growing number of Indians, slept outside the walls in bark-covered cabins. Day after day, Stobo the regimental engineer casually but carefully paced off dimensions, memorizing every detail.

No one could fault his behavior were he to carry this information himself back to Virginia, his captivity over and his parole fulfilled. But Stobo decided to take a far more dangerous step—to send it back to Washington by secret courier. Stobo later argued that he had seen prisoners from his own company held by the Indians at Fort Duquesne in violation of the surrender terms and thus considered himself under no obligation to observe those terms either. Whatever the justification, he decided to cross the line from hostage officer to active spy—a choice that could put his head squarely in the noose if he were caught. He was also ignoring another line from *A Treatise of Military Discipline*—one regarding the tendency of young, untried officers "being hurried on, by . . . the impetuosity of their temper, to do something that is great and noble, without considering the consequences that may attend it."

Working quickly but efficiently, he sketched out a remarkably detailed plan of the fort, complete with cross sections of the walls, the layout and function of each building, distances to the river and forest edge, positions of cannons, and more. He added notes on the disposition of the Ohio Indians, many of whom were coming to the French side, in part because of rumors that Scarouady had been murdered by the British and his family given to the hated Catawba and Cherokee. He cautioned that Laforce, the French officer captured by Washington, "is greatly wanted here; [there is] no scouting now, he certainly must be an extraordinary Man among them, he is so much regretted and wished for."

The fort, Stobo said, was weakly garrisoned. "Strike this Fall, as soon as possible . . . One hundred trusty Indians might surprize this Fort." One such trusty Indian was a Mohawk named Moses the Song, who had approached Stobo a couple of days earlier and offered to carry a document through French lines. Besides being Scarouady's brother-in-law, he was Conrad Weiser's adoptive brother, their friendship extending back to their boyhoods together in New York.

Stobo's final lines were a mix of bravado and naiveté. "When We engaged to serve the Country it was expected We were to do it with our Lives, let them not be disappointed, consider the Good of the Expedition without the least Regard to Us; for my part I wou'd die ten thousand deaths to have the Pleasure of possessing this Fort but one day, they are so vain of their Success at the Meadows, 'tis worse than Death to hear them." New as he was to military matters, it's likely that Stobo also underestimated the consequences if he was found out. "If they should know I wrote I should at least lose the little Liberty I have," he said. And with that, he signed his name to the document.

Moses slipped off through the forest, Stobo's letter and map hidden in his breechclout. The next day, the captain smuggled out a second copy with a sympathetic Lenape named Dillaway (or Delaware) George. Then he sat back waiting to be exchanged, satisfied that he'd done his duty.

Actually, Robert Stobo had gotten himself into a pickle. When his letter arrived in Williamsburg, the French prisoners were on their way to be exchanged. But when Dinwiddie read Stobo's remarks about Laforce's value, he recalled them. Under a flag of truce, a British officer met with Contrecoeur, the French commander, offering to swap three junior French officers, but not Laforce or the rest of the men, for Stobo and Van Braam. Contrecoeur refused; he had already learned that information regarding the fort's defenses had leaked to the British — had even been printed in Virginia newspapers — and he suspected his prisoners of cheating on their parole. By late September, Stobo and Van Braam were heading for Montreal.

At the same time Stobo was plunging into espionage, Conrad Weiser was elbow-deep in sordid dealmaking at a great congress in Albany, New York — a gathering of provincial officials and Indian leaders meant to shore up the flagging alliances with the Iroquois and other Natives that the events at Great Meadows made even more critical.

The Covenant Chain of treaties, forged in the late seventeenth century, was in tatters. The League of the Six Nations was fracturing, its long policy of neutrality collapsing as the Seneca sided more and more openly with France, the Mohawk and Oneida aligned with New York, and the Ohio Indians followed their own path. Yet even the Mohawk were growing weary of Britain's ceaseless hunger for land. In June 1753, the aging Mohawk sachem Hendrick announced to a stunned meeting, "The Covenant Chain is broken between you and us. So brother you are not to expect to hear of me any more, and Brother we desire to hear no more of you."

Hendrick may have been playing to the crowd, but his threat of war got Britain's attention. King George himself—not his fractious colonies—would summon the league's sachems and provincial commissioners to Albany in the summer of 1754 to heal the rift in the chain of friendship. At the same time, those advising the king hoped the provinces would find some measure of the unity that had so long evaded them.

That seemed unlikely. Pennsylvania, facing French intrusions to the west, also confronted the prospect of a second invasion from a wholly unexpected quarter—Connecticut. Pennsylvania had already been squabbling with New York for years over the upper Susquehanna, but because of still another pair of overlapping royal grants, the land between the forty-first and forty-second parallels (essentially, the northern third of the modern state of Pennsylvania) was claimed by Connecticut as well, and settlers from New England were preparing to move into the Wyoming Valley along the north branch of the Susquehanna. "Tho' I can scarce persuade myself that any considerable Number would engage in so rash and unjust a Proceeding," Hamilton told Connecticut governor Roger Wolcott, he warned that anyone doing so would be considered "Enemies to their Country," who risked "raising a Civil War."

"I beseech your Honour further to consider that the Six Nations will be highly offended if these Lands . . . be overrun with White People, for they are their favorite Lands and reserved for their Hunting," Hamilton concluded. It wasn't the Iroquois who would be furious, though; it was the Lenape, who had been forced out of the Forks of the Delaware following the Walking Purchase, and who had settled in the Wyoming Valley with a promise that it would remain Indian land. Hamilton was worried enough to urge Weiser to make another winter trip to Onondaga to discuss Connecticut's inroads with the Iroquois League. Weiser refused, recalling how, on his winter march with Shikellamy in 1737, he "almost starved by misrys and Famine I suffered," even though he was then a young man. Now "I am old and Infirm," protested Weiser, who had turned fifty-seven a few months earlier. "I could Expect nothing . . . now than my Grave, or feeding the Wolves and Bears with my Body."

Instead, he traveled to Albany that summer, gliding up the Hudson on a sloop with the Pennsylvania delegates, including Benjamin Franklin, who was hoping to hammer out plans for a "United Colonies," much as New England had tried to forge in the 1600s. A month earlier, Franklin had published a cartoon in the *Pennsylvania Gazette* showing a snake, representing the colonies, chopped into pieces; the legend read "Join, or Die." The verdict seemed to be die. Franklin's proposed Albany Plan of Union failed, rejected by both the colonies and the Crown because, he said, the provincial assemblies "all thought there was too much [royal] *prerogative* in it, and in England it was judg'd to have too much of the *democratic*."

Pennsylvania, New York, Maryland, Massachusetts, Connecticut, New Hampshire, and Rhode Island were represented at the Albany Congress, while others, notably Virginia and New Jersey, stayed away. Sachems of the Six Nations were there, wearing a colorful mix of Native and European clothing, wrapped in red or blue blankets, silver earrings, and ceremonial gorgets flashing in the sun, their finery masking the very deep divisions among their members. The Ohio Indians were, as usual, not invited; the league would speak for them. In fact, the direct negotiations Pennsylvania and Virginia had been conducting with the Ohio tribes were a sore point among the Iroquois, as their sachems made clear.

Weiser had two big jobs—to forestall Connecticut's land grab in the Wyoming Valley, and to try to engineer another on behalf of the Penns. Two months earlier, Weiser had sent a message to friendly elements in the Iroquois League, carried by the sons of his dead friend Shikellamy. The plan laid the groundwork for a sweeping land purchase "for the whole Province"—meaning, as the commissioners explained in Albany, from the Susquehanna "as far Westward as the Province extends, and as far North of the Kittochtinny Hills . . . as

You shall think proper to part with, and the larger the Tract the more agreeable to Us, and the larger our Present to You will be."

Weiser was jockeying with the Connecticut Susquehanna Company's hired man at Albany, a Dutch-born trader and speculator named John Henry Lydius. Tall and lanky, the fifty-year-old Lydius had been skirting the edge of French, British, and Mohawk law for years, developing a decidedly checkered reputation. One Oneida later accused him of being "a Devil [who] has stole our Lands," someone who takes "Indians slyly by the Blanket one at a time, and when they are drunk, puts some money in their Bosoms, and perswades them to sign deeds."

Sly he was, but also effective. "He went in the following clandestine manner to work," reported one witness, "and with tempting the Indians he could prevail upon, with Plenty of Dollars." Moving among the Mohawk, "by many false persuasions, with the offer of 20 Dollars each, [he] brought them to sign their Names." Visiting the Oneida, "Lidius treated them plentifully with Victuals and Drink, & then laid 300 Dollars before them." Employing such rum-and-lucre methods, Lydius got more than a dozen sachems to sign a deed to the Wyoming Valley—few, if any, of them with the authority to do so.

The land swindle was also an invitation to war, which Weiser predicted would erupt should the New Englanders actually move to settle the north branch of the Susquehanna. Hendrick, the Mohawk sachem, told the Pennsylvanians his people reserved the Wyoming Valley "for our hunting Ground," issuing "Orders not to suffer either Onas' People nor the New Englanders to settle any of those Lands." Repeatedly, he drove home the point: "We will never part with the Land at Shamokin and Wyomink; our Bones are scattered there, and on this Land there has always been a great Council Fire."

Hendrick was the lead signatory to Pennsylvania's own deed negotiated at Albany, a document "that reeked somewhat less strongly of fraud" than Connecticut's, but that was still proved to be a profound cheat. The pious Weiser was not above using underhanded methods of his own. "I may perhaps fall in with some greedy fellows for Money, that will undertake to bring things about to Our wishes," he advised Pennsylvania council secretary Richard Peters before going to the congress. Weiser did just that. In return for a thousand golden pieces of eight and the promise of a thousand more to follow—£800, or about $100,000 in modern currency—Hendrick and twenty-two other sachems cheerfully signed away an enormous swath of the central Appalachians.

As with its deal with Virginia years earlier, and with the Lenape's Walking Purchase bargain before that, what the league thought it was selling and what the lines on the deed map showed were two different things. The purchase included about seven million acres west to the current border of Ohio and Pennsylvania and including the west branch of the Susquehanna—a prime and treasured hunting ground that the Iroquois thought they had specifically excluded from the negotiations. When, a few months later, Weiser marked off the line and they saw where it ran, Shikellamy's sons were outraged, and other Indians along the west branch pledged violence if any whites tried to settle the region. (To his credit, though with little immediate effect, Weiser backed their protests over the boundary.)

And once again, those most directly affected by the sale—the Lenape and other Ohio Indians—had been shut out of the negotiations. The province and the league had once more ignored the western villages like the "children" Iroquois sachems often called them in council. All this makes Hendrick's address to the Pennsylvania commissioners at Albany—one famous today for its prescience and its poignancy—a speech soaked in unintended irony.

With seventy of his fellow sachems listening, the old Mohawk spoke:

> What We are now going to say is a Matter of great Moment, which We desire you to remember as long as the Sun and Moon lasts . . . After We have sold our Land We in a little time have nothing to Shew for it; but it is not so with You, Your Grandchildren will get something from it as long as the World stands; our Grandchildren will . . . say We were Fools for selling so much land for so small a matter, and curse Us; therefore let it be a Part of the present Agreement that We shall treat one another as Brethren to the latest Generation, even after We shall not have left a Foot of Land.

In the baroque language of frontier diplomacy, Hendrick was speaking of some distant future, but to the Ohio Indians—especially the

Lenape, who had been down this road a few times before—the moment when they would "not have left a Foot of Land" loomed closer than ever.

While they were busily selling the Ohio tribes down the river, the province and the league also used the Pennsylvania deed to pull a mutually bothersome thorn—Andrew Montour and his Juniata settlements. "Andrew Montour had already sold some of the Land on Sherman's Creek or Juniata . . . and had settled several Others upon it as his Tenants," Weiser had informed the Iroquois that spring, to their shared displeasure. The white and Indian tenants who were flocking to Montour's incipient fiefdom west of the Susquehanna would be effectively beyond the authority of both the Six Nations and the province—but not if Pennsylvania bought the land from the league and sold it, free of legal encumbrance, to white farmers who owed their allegiance (and their payments) to the Penns.

Each for its own reasons, Pennsylvania and the Iroquois League had good reasons to fear the kind of mixed society Montour was forming in the Pennsylvania backwoods, beholden to no one. Better to keep the line bright and sharp—Indian as Indian, white as white. Even as Montour and his eighteen scouts were slipping through the woods along the Monongahela, shadowing the French columns marching toward Fort Necessity, Weiser and the Iroquois quietly but efficiently crushed his dream of becoming lord of a polyglot, polyethnic frontier sanctuary for "the lower sort of People."

When he found out what his former mentor Conrad Weiser had done, French Andrew got murderously and piteously drunk. "He is vexed at the new purchase," Weiser confided to his old colleague Richard Peters, having met Montour at Aughwick two months after the congress.

He abused me very much, Corsed & swore, and asked pardon when he got sober, did the same again when he was drunk, again damned me more than [a] hundred times . . . Told me I cheated the Indians. He says he will now kill any white men that will pretend to setle on his Creek . . . saying that *he was a Warrior—how could he suffer the Irish to encroach upon him*—he would now act according to advice, and kill some of them. I reprimanded him when sober. He begged pardon, desired me not to mention it to you, but did the same again at another drunken frolik. I left him drunk at Achwick, on one legg he had a stocking and no shoe, on the other a shoe and no stocking.

But Weiser had other things to worry about. Tanaghrisson, now ailing,* and Scarouady had not taken news of the Pennsylvania deed to the western lands as well as he'd hoped, and only by invoking the specter of the French coming from the west and New Englanders from the east was he able to cast the purchase in anything like a positive light. Weiser knew that everything hinged on what the Ohio Indians — punching bags for the province and the league for so many years — would do in the months ahead. "What the french and the ohio and other Indians would be to us in time of Warr I leave to you and other Gentlemen to Judge," he wrote to one correspondent. "I can not think on Such a time but with Terror."

---

* Tanaghrisson fell ill while accompanying Weiser on the trip to Aughwick in September 1754, and although Scarouady managed to get him back to Harris's Ferry, Tanaghrisson died there of an unknown illness. "The Indians here blame the French for his death, by bewitching him," John Harris Jr. wrote to Hamilton, "for which they seem to threaten immediate Revenge." Weiser blamed it on the sachem's heavy drinking.

Chapter 10

War Chief, Peace Chief

FOR A PRISONER, Robert Stobo was doing rather well for himself. Moving easily through the social circles of Quebec and Montreal, he spoke with a basic but ever-growing command of French, showing an elegant leg—his suits edged in gold trim and exhibiting the latest Parisian fashion, his lacy shirts closed at the neck with a diamond hasp—to the coquettish ladies with whom he flirted in turn. He had livres in his pocket and—ever the merchant—had launched a profitable side business of Indian trading, through which he had so ingratiated himself with the Mississauga band of the Ojibwa that they tattooed a tribal symbol on both his thighs, pricked into his skin with needlelike fish bones and Native ink. It was a long way from the mud of Fort Necessity, in many respects.

After dispatching his surreptitious letters in September 1754, Stobo and Jacob van Braam had been handed up the French chain of forts to Lake Erie, resigned to the fact that they were not going to be exchanged anytime soon. From Presque Isle, they were bundled off in huge cargo canoes known as "Montreals," each propelled by more than a dozen paddlers, along the lakeshore to Niagara, into Lake Ontario, and down the St. Lawrence to Quebec City. The more than eight-hundred-mile journey took about a month, and at its end the hostages were received by the governor-general, the Marquis Duquesne, who renewed their paroles—either having not heard or choosing to dismiss the suspicions regarding Stobo's role in the leaked map and letter.

Permitted to move freely between cities, Stobo threw himself into the social whirl. He was, by all accounts, a fairly dashing man, and being the only British officer around must have made him an irresistible curiosity to the young mademoiselles and more worldly madames. Weeks of wilderness travel had rendered his uniform a grimy mess, but he was able to draw on credit to outfit himself with panache, a plumed hat and beaver coat being among his purchases. Once more suitably clothed, he set about learning French.

"To acquire it was his first study, in which pursuit he was greatly assisted by the ladies, who took great pleasure in hearing him again a child . . . His manner was still open, free and easy, which gained him ready access into all their company," according to the *Memoirs of Major Robert Stobo of the Virginia Regiment*, an anonymously written, rather turgid account of Stobo's adventures, which despite its fawning tone appears to be fairly accurate regarding Stobo's time in New France.*

By early 1755, however, Stobo was being identified in the French press as the source of the leak regarding Fort Duquesne. The leash tightened, but still, they were only suspicions, and Stobo was able to continue many of his extracurricular activities ("he still preserved his credit with the ladies")—until his comfortable confinement ended with a bang.

Duquesne had been recalled to France, replaced by a new governor, Pierre de Rigaud de Vaudreuil, the Marquis de Vaudreuil, who was incensed by reports in English newspapers crowing about Stobo's letter. Called on the carpet, Stobo admitted what he referred to as a "thoughtless stunt." Although he was held only briefly in prison before Vaudreuil accepted his apology and restored to him a degree of freedom, his days of ranging the St. Lawrence Valley and the salons of the capital at will were over.

---

* The *Memoirs* is an intriguing little puzzle, apparently written in the 1760s but not published until 1800 in England. Because it is so unlike Stobo's straightforward letters, he was probably not the author, but it was obviously penned by someone with access to firsthand information about him. Stobo biographer Robert C. Alberts considered it an overwrought but basically trustworthy account of Stobo's exploits.

Worse was in store. Word reached Montreal of a stunning victory over the British along the Ohio River. A general named Braddock and the army he led had been smashed to pieces on the very doorstep of Fort Duquesne. In the rout that followed, the dead general's own secret papers had been abandoned, and among them was found a map of the fort and the damningly detailed descriptions of its armaments, garrison, and vulnerabilities—in the hand, and under the signature, of one Captain Robert Stobo.

General Edward Braddock was sixty years old, going to gray and running to flesh. Although he was a veteran of the storied Coldstream Guards, his military pedigree looked more impressive than it really was. Despite forty-five years in the service, Braddock had never actually led troops in battle. His rise through the ranks had occurred almost entirely behind a desk, but he knew the right people.

Arriving in Virginia in February 1755, he wielded unprecedented power as commander in chief of all British forces in North America. Braddock mapped out a plan to simultaneously hit French bases in the Ohio country, Champlain Valley, Great Lakes, and Nova Scotia, personally leading the assault against Fort Duquesne.

For the Ohio campaign, he gathered more than two thousand men at Will's Creek, including two regiments of regular British infantry, the Forty-Fourth and Forty-Eighth of Foot, numbering about twelve hundred soldiers—rigid ranks neatly squared off for review in their hot, sweat-stained, red wool jackets, white leggings, and neck stocks, their hair powdered with flour that grew sticky in the humid summer air. Companies of grenadiers, their black miter caps towering above the tricorn hats of the common soldiers, bore themselves with visible élan. Although they no longer carried grenades, grenadiers were the elite foot soldiers, expected to tackle the most dangerous assaults.

The grenadiers looked down on the run-of-the-mill privates, and the British regulars all looked askance at the independent companies, including the Third South Carolina, which had been bloodied at Great Meadows and was seeking a measure of revenge. The red-coated Carolinians held themselves aloof from the blue-coated provincial companies from Virginia, Maryland, and North Carolina, and everyone seemed to despise the Third and Fourth New York, two in-

dependent regular companies wearing green-faced jackets. The Yorkers had been so late in reinforcing Washington the previous year that
Virginia governor Robert Dinwiddie had publicly hung the blame for
the loss on them. They remained poorly equipped and even less well
led. So many of them were "men from sixty to seventy years of age,
lame and in everyway disabled," that it was hard to see them as a serious cog in the military juggernaut Braddock was assembling.

The Yorkers had, like the Carolinians and Marylanders, overwin-
tered at Will's Creek, helping to build Fort Cumberland, a dramatic
expansion over the small Ohio Company compound that had previously occupied the site. Forty yards on a side, with four corner bastions
mounting heavy cannons, the fort sat on a high bluff overlooking the
confluence of the creek and the north fork of the Potomac River, connected to a rectangular stockade two hundred yards long that enclosed
housing for hundreds of troops and masses of provisions and equipment. Virginia pledged to supply eleven hundred cattle and Pennsylvania tons of flour. A thousand barrels of salt-cured beef were unloaded from the troop transports and hauled to Will's Creek, to which
hundreds of hogs were likewise driven for slaughter, the smokehouses
there packed with slabs of bacon. In all, it was reckoned, the promised
supplies were sufficient to feed four thousand men for six months.

Any army is more than fighting men, and Braddock's retinue included not only the soldiers and officers but also companies of so-
called pioneers—axmen, directed by engineers, who would clear the
road for troops and wagons; a guard of light horsemen; a company of
scouts; and even thirty seamen and officers from the HMS *Centurion*,
handy with ropes, to string bridges across the rivers of the Ohio country. There were surgeons, quartermasters, blacksmiths, and a chaplain.
There were muleskinners and drivers to handle the 150 wagons and
500 horses supplied by Benjamin Franklin. Among the wagoners was
a nineteen-year-old from the Yadkin Valley in North Carolina named
Daniel Boone, whose family had moved from the Tulpehocken Valley
just a few years earlier. It was from one of George Croghan's traders,
another Scots-Irishman named John Findley, that young Boone first
heard of Kentucky.

An eighteenth-century army also carried with it an unavoidable
train of women—not only whores and camp followers, but respectable

washerwomen, nurses, wives, and sweethearts. "I do hereby Certifie that the Bearers . . . are wives to Soldiers belonging to the Forces under my Command; And all persons whatsoever are hereby requir'd to suffer 'em to pass without hindrance or molestation," read a pass issued by Braddock to twenty-eight women departing the fort. Braddock was less solicitous of some of the Indian women at Cumberland, who would trade favors for money from "the Officers, who were scandalously fond of them"; he ordered them driven away. He likewise had no patience with English women who refused to serve as nurses in exchange for their meals and sixpence a day, threatening to turn them out of the installation.

Among the women gathering at Fort Cumberland was Charlotte Brown, widowed sister of one of Braddock's officers and matron of the general hospital—the woman who would oversee treatment of the sick and wounded. She and her brother had made the four-month Atlantic crossing the previous winter and by early June were traveling with a detachment of forty men to join the army at Fort Cumberland. She slept rough most nights, confiding to her diary, "My Lodgings not being very clean, I had so many close Companions call'd Ticks that depriv'd me of my Night's Rest." She moved on the army's wearisome schedule: "At 2 in the Morning the Drum beat . . . Being very sleepy we march'd but there is no describing the badness of the Roads." She ate passenger pigeons that her companions shot for dinner; tippled a little peach whiskey, given to her by an elderly Quaker couple; and stayed for the night at "a Rattle snake Colonels nam'd Crisop"—Thomas Cresap, the "Maryland Monster."*

Not until mid-June did Brown reach Fort Cumberland, to find that Braddock had marched his men out a few days earlier. She and the nurses for whom she was responsible would await their return in what Brown described as "the most desolate Place I ever saw . . . For Quarters I was put into a Hole that I could see day light through every Log and 1 port Hole for a Window which is as good a Room as any in the Fort."

Riding with Braddock was George Washington, again in the uni-

* "Rattlesnake colonel" was a dismissive term for a colonial irregular. One of Braddock's men referred to Cresap as "a Rattle Snake Colonel and a D——d Rascal."

form of a Virginia colonel. "I wish for nothing more earnestly than to attain a small degree of knowledge in the Military Art," he had written to Braddock's aide in March, accepting the general's invitation to serve as a volunteer on his personal staff, since "a more favourable opportunity cannot be wished than serving under a Gentleman of his Excellencys known ability and experience."

Actually, Braddock's judgment and temperament had long been questioned. He had a reputation as a harsh commander and severe disciplinarian, all spit-and-polish drill that had never been tested. His frustration with the colonials was searing. He thought their regiments all but useless, considered Cresap and other frontier traders with whom he dealt as little more than thieves, and was unable to safeguard even his men's horses, which wandered off (or were stolen) despite hobbles, pickets, and guards. "None of these were a security against the wildness of the country and the knavery of the people we were obliged to employ," complained Captain Robert Orme, another of Braddock's aides-de-camp.

Braddock considered the provincial assemblies a parcel of short-sighted tightwads. Robert Dinwiddie was inclined to agree, at least to a point. "[I] am sorry to acquaint You that our Assembly will not pay any Money for the Subsistence of any Troops but those on our Establishment," the Virginia governor wrote Braddock in May. "The Backwardness of some and the refusal of others gives me very up hill Work with my Assembly, and I still doubt if they can be bro't to grant any further Supplies."

George Croghan had his own ideas of what constituted backward-ness, and he saw enough of it to feel a growing sense of alarm at how the assault on Fort Duquesne was being executed. Earlier that spring, tasked with marking a road through the mountains along which Brad-dock's army was to march, Croghan arrived at Fort Cumberland to report the survey's completion. There he was castigated by a furious Sir John St. Clair, Braddock's deputy quartermaster general, because the route was not already cut and cleared.

St. Clair "stormed like a Lyon Rampant," an alarmed Croghan wrote to Governor Robert Hunter Morris in Philadelphia. The man was so angry—preparations for the attack were already months late—that he threatened to ignore the French and instead march the troops into the

back settlements of Pennsylvania and "by Fire and Sword oblige the
Inhabitants to [cut the road], and take every Man who refused[,] . . .
that he would kill all kind of Cattle and carry away the Horses, burn
the Houses, & etc., and that if the French defeated them by the Delays
of this Province that he would with his Sword drawn pass thro'. . . and
treat the Inhabitants as a Parcel of Traitors to his Master."

It wasn't only Braddock's underlings who were throwing their
weight around. As the first military commander in chief to be placed
over civilian authority in the colonies, the general was disdainful of
provincial leaders as well. "He shall take due Care to burthen those
Colonies .the most that shew the least Loyalty to his Majesty, and
lets them know that he is determined to obtain by unpleasant Methods what is their Duty to contribute with the utmost Chearfulness,"
warned Edward Shippen IV, writing to his father, a prominent Pennsylvania merchant. While it was true that the provinces seemed inca-
pable of taking any concerted action in their own defense, Braddock's
imperious manner, and the shadow of an ironfisted royal rule that his
words cast, were already rubbing the prickly, independent colonials the
wrong way.

As counterproductive as such highhandedness was with the colonials, Croghan and the other experienced backwoodsmen knew that
Braddock's attitude toward their Indian allies bordered on suicidal.
Croghan and Andrew Montour were on the payroll, their job being to
secure the Native scouts and auxiliaries without whom frontier fighting was impossible. In May, Croghan brought Scarouady and fifty
others down from Aughwick to meet with the general. With Tanaghrisson's death the previous fall, the Oneida was now the "half-king,"
the Iroquois League's main representative among the Ohio Indians.

Tutored by Croghan and Montour, Braddock began the council in
good form, wiping away the Six Nations' tears over the death of Tanaghrisson with the requisite gifts, but neither he nor his staff had the
slightest interest in listening to advice from Scarouady or any other
savage. What's more, they instructed the warriors on proper tactics
and humane behavior; no scalping would be tolerated, for instance.

One of the Ohio Delawares, the sachem named Shingas, asked
Braddock what he intended to do with the land if he drove off the
French. "To which Genl. Braddock replied that the English should

Inhabit & Inherit the Land," Shingas later recalled. Shouldn't the Indians who were friends of the British be able to live there, too, to hunt
and trade? the sachem asked. "On which Genl. Braddock said that No
Savage Shou'd Inherit the Land." Shingas and the other Ohioans were
so baffled by this answer that they returned the next morning, saying
that "if they might not have Liberty To Live on the Land they wou'd
not Fight for it[.] To which Genl. Braddock answered that he did not
need their Help."

The Indians were not happy with what they were hearing. "Mr.
[Richard] Peters related . . . that he found Scarouady, Andrew Montour, and about Forty of our Indians from Aucquick, at the Camp
with their Wives and Families, who were extremely dissatisfied at not
being consulted with by the General," the Pennsylvania council noted.
When Braddock ordered the women and children sent away, most
of the men left, too, so that by the time the army marched a month
later, all but eight or nine of the warriors were gone. Croghan had sent
wampum to the Miami, to no avail, and several hundred Cherokees
and Catawbas, who had actually started north to join the attack, had
been waylaid by an English trader and convinced to go home. The lack
of Indian support was a colossal setback.

William Shirley, Braddock's secretary (and the son of the Massachusetts governor by the same name leading the campaign against
Fort Niagara), saw plenty to trouble him. He had privately shared his
doubts about Braddock's character and abilities, and when the general's staff pulled out of Cumberland with the final companies on June
10, his premonitions of disaster were strong. "The Truth is I have many
things that give me much uneasiness, which I had rather tell you than
write to you, and which put me almost beyond my Stock of patience,"
he admitted to his father's friend Governor Morris. "On Monday we
move from this place with 200 Waggons: How we shall get that Number over Such Roads & thro' such weather as we have had for some
time past I know not."

Neither the heat and summer storms nor the miserable muddy
road—which so slowed the wagons that Braddock ordered all the
heavy baggage and a third of the troops left behind, to catch up as
they might—was the real problem. Moving across the mountains
with thirteen hundred of the best men, Braddock plunged deeper and

deeper into the forests of the Ohio country for the next twenty days with only a handful of Indian and white scouts. His army was all but blind, and French-allied Indians picked off any stragglers.

Of the few Indians they did have, the British almost lost the most important one. When Scarouady and his son were surrounded by a party of Indians and French, the younger man escaped, but the Oneida sachem was taken captive. The French were in favor of killing him on the spot, but their Indian allies—mostly Ohioans who had known Scarouady for years—threatened to go over to the British side if they did. Instead, they tied Scarouady to a tree and left him for Braddock's men to rescue. His son was less fortunate. A few days later, British grenadiers, hunting for Indians who had just killed several soldiers, shot him by mistake. Braddock ordered him buried with military honors, uniformed officers standing by as a salute was fired over his grave.

A few days after that, the broken and demoralized British army was conducting a hurried, secret funeral for Braddock himself, burying him in an unmarked grave over which the retreating wagons rumbled to mask the site. It hadn't been a planned battle or a prepared ambush. The general's advance column of about four hundred men, led by the "pioneer" carpenters clearing the road, had just forded the Monongahela six miles from Fort Duquesne with Croghan and his overextended scouts. Their regimental colors were unfurled, and their drum-and-fife corps was playing, when they simply collided with a party of French regulars, Canadian militia, and Indians hurrying to the river in hopes of slowing the final British advance.

The irony was that a well-led British force of almost any size could have plucked Duquesne like a flower. As Robert Stobo had observed a year earlier, it was undermanned and poorly supplied, its troops now racked by illness. The French commander, Sieur de Contrecoeur, had so little hope of fending off Braddock's regulars that he had, by some reports, already drawn up articles of capitulation to present to the British general.

As the shooting by the ford began—the acrid white smoke drifting on the hot breeze, the officers calling out the drill that would keep the musket fire in crisp salvos: ("Make ready! Present! Fire!")—the French line collapsed. The British shouted "Long live the King!" but the Indians and French quickly regrouped within the cover of the forest. War whoops began ringing through the trees, unnerving the British troops. Croghan, scanning the woods, guessed there were three hundred on the French side. In fact, there were more than three times that many, six hundred of them Hurons, Ottawas, and other northern Indians. The French officers had the presence of mind to flank the confused, milling hordes of redcoats, who acted as though they were on a parade ground.

Braddock's Defeat (ca. 1907)
As Indian and French forces pour fire into British columns on July 9, 1755, General Edward Braddock is hit and Colonel George Washington reaches for his horse's bridle in a twentieth-century depiction of the battle by the American painter Edwin Willard Deming. (Wisconsin Historical Society Image ID 1980)

Instead of taking cover themselves, the British officers held their ranks in the open of the newly cut road, trying to bring up their cannons, while the French poured musket fire into them. It was child's play; the French and their Indian allies could hardly miss. The vaunted grenadiers, by contrast, flung their fire from hundreds of yards back, ineffectual and actually dangerous to their comrades farther ahead.

Braddock famously had four (some said five) horses killed beneath him before he took a musket ball through the arm and into the gut.

He was carried, gasping, to a wagon by Washington, Croghan, and Orme. The general tried to snatch up Croghan's pistols and "die as an old Roman"—by his own hand, on the battlefield—but the trader batted him away.

"I luckily escaped without a wound, tho' I had four bullet holes thro' my Coat, and two Horses shot under me," Washington wrote to his mother, assuring her that he was fine. Croghan likewise came through without a scratch, but in this they were unusual. Two-thirds of the British troops were killed or wounded, including young Shirley, the dead left on the field to be plundered and scalped. One of those killed was Will Poulson, the lieutenant to whom Stobo had given his sword before surrendering himself at Fort Necessity.

Those who kept their feet used them, mostly running in panic—the vaunted regulars worst of all, at least according to the mocking colonials. The troops, "chiefly Regular Soldiers[,] . . . were struck with such a panic that they behaved with more cowardice than it is possible to conceive," Washington wrote.

Carrying little but their own arms—and many had abandoned even those—the retreating British covered almost fifty miles in the next twenty-four hours. Braddock, lingering for four days, died only about a mile from the charred remains of Fort Necessity. The rear-guard commander, Colonel Thomas Dunbar, took charge, ignominiously marching the entire force back to Fort Cumberland. "It is not possible to describe the distraction of the poor Women for their Husbands," Charlotte Brown noted in her diary when word of the defeat reached them. "I pack'd up my Things to send for we expected the Indians every Hour."

Among the cannons, muskets, powder, and food salvaged by the French from the fleeing British troops—all of which went a long way toward shoring up Duquesne's flagging defenses—was Braddock's personal chest. In it, along with his orders from King George and £1,000 in cash, was found Stobo's letter and map, which Contrecoeur forwarded to Montreal. Instead of surrendering in shame, Contrecoeur would later receive the Order of Saint-Louis.

Dunbar, who had fled the field without firing a shot, blamed the defeat on the fact that "General Braddock cou'd not get above eight or nine [Indians] to attend him," a situation he said had grown out of some unspecified "Mismanagement." The sachems, however, knew whom to hold accountable. A month after the rout, a delegation of the Six Nations—including Andrew Montour, not as an interpreter but in his role as Sattelihu, an Iroquois councilor—traveled to Philadelphia. Montour and Scarouady had seen firsthand what had gone wrong at the Monongahela, and they placed the blame squarely on Braddock. "It was the pride and ignorance of that great General that came from England," Scarouady said, with Conrad Weiser translating his words. "He is now dead; but he was a bad man when he was alive; he looked upon us as dogs, and would never hear any thing what was said to him. We often endeavoured to advise him and to tell him of the danger he was in with his Soldiers; but he never appeared pleased with us, & that was the reason that a great many of our Warriors left him & would not be under his Command."

Holding a string of wampum to reinforce his words, the Oneida sachem continued. "But let us unite our Strength. You are very numerous, & all the English Governors along your Sea Shore can raise men enough; don't let those that come from over the great Sea be concerned any more; they are unfit to fight in the Woods. Let us go ourselves, we that came out of this Ground, We may be assured to conquer the French."

If Scarouady thought so, he was one of the few. The Ohio Indians he had for so long tried to "govern," the Lenape and Shawnee in particular, drew from the Braddock debacle a very different lesson. The promises of aid and arms with which to fight the French had proved as empty as William Penn's pledge of faithfulness and honesty in the hands of his sons. It was time not only to abandon the British but to make them pay a price.

The "back inhabitants" of Pennsylvania had been jittery since the battle at Great Meadows, pleading for protection and support. "We are now in most imminent Danger by a powerful Army of cruel, merciless, and inhuman Enemies, by whom our Lives, Liberties, [and] Estates . . . are in the utmost Danger of dreadful Destruction," settlers in the Cumberland Valley wrote to provincial leaders in Philadelphia. Through the summer of 1755, sporadic attacks occurred on the edge of the Ohio country, but the long-feared stroke finally fell in October.

George Croghan had a whiff of the looming trouble. An Indian just back from the Ohio told him that 160 warriors who had been harassing settlements on the upper Potomac were moving north, quietly gathering support among the Indians who lived in the Susquehanna Valley. "He desires me as soon as I see the Indians remove from Sasquehanna back to Ohio to shift my quarters, for he says that the French will, if possible, lay all the back frontiers in ruins this Winter," Croghan wrote from Aughwick, which his men were busily stockading.

Just two weeks later, a series of coordinated attacks hit the Susquehanna and northern Potomac valleys, starting on October 16 at Penn's Creek, a Swiss and German community. "The Enemy came down upon said Creek and killed, scalped & carried away all the Men, Women & Children, amounting to 25 Persons," the Penn's Creek survivors wrote to Morris. "We found but 13 which were men and elderly women, & one Child of two weeks old, the rest being young Women & Children we suppose to be carried away Prisoners; the House . . . we found burnt up, and the man of it named Jacob King, a Swissar, lying just by it . . . barbarously burnt and two Tomhawks sticking in his forehead." For gruesome emphasis, the message to the governor was accompanied by one of the hatchets pulled from Jacob King's head.

War parties seemed to be everywhere, killing and capturing scores. They struck the Juniata and Conococheague valleys. Many settlers fled, while a few tried to fort up with neighbors for protection. (A few even took the opportunity to rifle the homes of their absent fellows.) John Harris Jr. and a burial party—having ignored advice from a war-painted Andrew Montour to avoid the west side of the Susquehanna—were ambushed, losing four men to hostile fire. Once home, Harris cut loopholes in the walls of his trading house through which he could shoot. Montour, Shikellamy's sons, and other Indians living at Shamokin scoured the woods for the attackers. This was shaping up to be a civil war, Lenape and Shawnee against Lenape and Shawnee, yet many were still willing to fight alongside Brother Onas.

"If the white people will come up to Shamokin and assist they will stand the French and fight them," Weiser wrote urgently to Governor Morris, conveying information gathered by his sons. "They said that now they want to see their Brethren's faces, and well armed with smooth Guns . . . they are extremely concerned for the white people's running away, and said they could not stand the French alone."

He continued, "I pray, good Sir, don't slight it. The lives of many thousands are in the utmost Danger. It is no false alarm." Indeed, before the sealing wax had fully hardened on Weiser's letter, a messenger brought him news of still more raids, still more deaths.

Massacre of Conococheague
An engraving from the 1883 book *The Romance and Tragedy of Pioneer Life* depicts an Indian attack on the Pennsylvania frontier during the Seven Years' War. Melodrama aside, the specter of lightning raids by Indian and French war parties had exactly the intended effect—emptying much of the frontier and spreading terror among the remaining inhabitants. (Library of Congress LC-USZ62-727)

The attacks continued through the fall. "There is two-thirds of the Inhabitants of this Valley who hath already fled, leaving their Plantations," the sheriff of the Conococheague Valley wrote to Richard Peters. "Last night I had a Family of upwards of an hundred of Women and Children who fled for Succor. You cannot form no just Idea of the Distressed & Distracted Condition of our Inhabitants." Of ninety-

three people living in the valley known as Great Cove, almost half
were killed or taken.

Behind this sweep into the Susquehanna was the Lenape leader
Shingas—"Shingas the Terrible," as the Pennsylvanians came to call
him. This was the sachem who had listened in disbelief as Braddock
said, "No Savage Shou'd Inherit the Land"—a story Shingas told to
one of his captives that autumn. Shingas, whose name means "bog
meadow" in Lenape, came from a remarkable family. His uncle was
the late Sassoonan, the sachem who had been so angered by the loss
of the Tulpehocken Valley to the Palatines, and his brothers were the
noted chiefs Tamaqua ("the Beaver") and Pisquetomen. All three had
abandoned eastern Pennsylvania in favor of the Ohio country, where
in 1752 at Logstown, Tanaghrisson had raised Shingas to be "king"
of the Ohio Delaware—a bit of self-serving political theater by the
Iroquois League, since Shingas and his brothers were already an acknowledged force among the western Lenape.

Although Shingas was not a physically imposing man, no one questioned his status as a forceful and effective war chief. For the next three
years, he led repeated forays by Lenape, Mingo, and Shawnee fighters,
terrorizing the Pennsylvania backcountry. (The entry of the Delaware
and Mingo into the fight brought fierce satisfaction to the Shawnee,
who had already been fighting the British on their own for more than
a year, avenging a leading chief who had died while in a South Carolina prison.)

Shingas wasn't the only war leader among the Ohio Indians. Pisquetomen struck some of the first blows that October, and Tamaqua
also proved to be a canny commander. So did the Delaware known by
the nickname Captain Jacobs, who had been among those who routed
Braddock. Shingas and Captain Jacobs both soon carried bounties of
350 Pennsylvania dollars on their heads. But Shingas became the infamous symbol of the western raiders, in part because of his curiously
paradoxical nature. He was ruthless in the field, but his captives often
spoke gratefully of his care and solicitation after they were taken.

Charles Stuart, who with his wife and two children were among
dozens of colonists taken prisoner in October 1755, was singled out
with another man to be tortured to death. "First our Fingers were

To be Cut off and we were To Be Forced to eat them, then our eyes pulled out which we were also to Eat, after which we were To Be Put on a Scaffold and Burnt." But Shingas overruled the other raiders, noting that Indians "Frequently Called at [Stuart's] House . . . and had Always been supplied with Provissions and what wanted Both for Themselves & Creatures without Ever Chargeing them Anything."

One often-cited story probably says more about the general Algonquian view of adoption, however, than about Shingas's personal character. The Moravian missionary the Reverend John Heckewelder recalled an occasion when, observing two white boys playing with Shingas's children, he referred to the boys as prisoners. "[Shingas] said, 'When I first took them, they were such; but now they and my children eat their food from the same bowl or dish,' which was the equivalent of saying they were in all respects on an equal footing with his own children, or alike as dear to him."

The Ohio Indian attacks meshed neatly with French strategy, which military commanders today would call asymmetric warfare, and which the French had dubbed *la petite guerre*—hit-and-run attacks by Indians and French *troupes de la marine* designed to terrify and dislodge the inhabitants of the countryside, putting pressure on provincial political leaders, while avoiding battlefield confrontations. On the map, New France dwarfed its British neighbor—the French colony, after all, extended from New Orleans north to the Gaspé and west to the Missouri River and Grand Portage. The British settlements still hugged the Atlantic coast, rarely extending more than a couple of hundred miles inland. But whereas the French were thinly spread out—trading posts and forts isolated amid millions of square miles—the British colonies were jammed with city dwellers and farmers, backed by far more troops than France could muster. Thus Indian allies became critical.

The idea "to carry the war into their country" in Pennsylvania, in the words of Governor Vaudreuil, capitalized on the gnawing bitterness many Delawares felt for being forced off their land. In the end, however, it's too simplistic to say that the frontier war was only about payback for the Walking Purchase and other frauds. Other issues contributed to the Lenape's actions, including very real worries about continuing British inroads on the upper Ohio River. But the decision to make war alongside the French also had much to do with immediate rewards—the chance for booty and prestige in battle—as well as the danger of ignoring French ascendancy and a feeling of betrayal and abandonment by the British, whom they had once considered friends.

You couldn't live in the woods and remain neutral. The Ohio Indians felt they had to make a choice, and for many the choice was Father Onontio in Montreal, not Brother Onas in Philadelphia. Runners from the western villages slipped quietly into such Indian towns as Big Island, along the upper Susquehanna, with strings of black wampum painted red and a message from the governor of Canada: "If Onontio finds any of your Indians among the White people he will kill them, & therefore he gives you this Warning."

Scarouady, with Andrew Montour at his side, confronted the thirty-six-man Pennsylvania assembly, throwing two wampum belts on the table—one black, one white. "I must deal plainly with You," the sachem said, "and tell you if you will not fight with us we will go somewhere else. We never can nor ever will put up [with] the affront. If we cannot be safe where we are we will go somewhere else for protection and take care of ourselves."

But the government of Pennsylvania was deadlocked. Governor Morris was pushing for swift action, but in such a maladroit way that he further alienated the Quaker-dominated Assembly, which knew the Indians had legitimate grievances. "Instead of . . . providing for the safety and defense of the people and Province in this Time of imminent danger," Morris thundered, "You have sent me a Message wherein you talk of regaining the Affections of the Indians now employed in laying waste the Country." His frustration may have been authentic, but his words were also politically calculated. Morris saw a cudgel with which he could bludgeon the Quakers, scoring points against those who had long opposed unbridled proprietary power. With angry settlers, pushing carts in which they carried the scalped bodies of their families through the streets of the city, it was easy and effective to argue that the Quakers were being unreasonably, almost treasonously, sympathetic to the Delaware.

Most Pennsylvanians no longer distinguished friendly Indians from hostiles, and because they recognized some of the faces among the raiding parties—men who in many cases had eaten and drunk at their tables—they no longer felt any reason to try. Weiser, appointed a colonial colonel, came home from Philadelphia to find that war parties had struck for the first time south of the Kittochtinny Hills, and the countryside was in a froth. He was barely able to prevent a mob of five hundred armed men from lynching Scarouady and Montour, whom he sent north under armed escort.

"[They] called me a Traitor of the Country who held with the Indians and must have known this Murder before hand," Weiser wrote to Morris. "I sat in the House by a Lowe Window, some of my Friends came to pull me away from it, telling me some of the People threatened to Shoot me . . . The cry was *The Land was betrayed and sold.*"

Teedyuscung would have found that cry grimly amusing. A big, heavyset man in his fifties, with a gift for oration and an astounding capacity for liquor, Teedyuscung had again and again seen his family pushed off land they thought was theirs. He was born near what is now Trenton, New Jersey, to an increasingly isolated Lenape family surrounded by white-owned farms and towns. For a time, he made a quiet living weaving baskets and making brooms for sale. About 1730, he and his family moved to the Forks of the Delaware, until that sanctuary was lost to them in the Walking Purchase.

Teedyuscung—whose name means "he who makes the earth tremble"—converted for a time to Christianity, becoming a Moravian known as Brother Gideon or Honest John. He was married in a church ceremony to a Delaware woman christened Elizabeth. They lived at Gnadenhütten, a Moravian mission town along the upper Lehigh River whose name means "cabins of grace" in German. Around 1753, at the Six Nations' invitation, Teedyuscung led his family and about seventy other Lenapes and Mahicans from Gnadenhütten north to the Wyoming Valley along the north branch of the Susquehanna. He was at Albany in 1754 and knew the shenanigans by which Connecticut was trying to steal the Wyoming Valley, and he understood that his new community was a line in the sand against New England colonization. But he also saw the Delaware as emerging from the shadow of their "uncles," the Iroquois League, and increasingly able to speak for themselves.

One British officer who knew Teedyuscung described him as "a lusty, rawbon'd Man, haughty, and very desirous of Respect and Command." Most remarkably, the officer observed, "he can drink three Quarts or a Gallon of Rum a Day without being drunk."

Teedyuscung also had a sly, cutting wit, as evidenced by an encounter with a colonist who once hailed him as "cousin." Teedyuscung looked quizzically at the man, who was apparently a bit of a lowlife. Cousin? the Lenape asked. How so? "Oh, we're all cousins from Adam," the fellow replied. "Ah," said Teedyuscung with relief, "then I am glad it is no nearer."

Although the Pennsylvanians usually referred to him as "King Teedyuscung," he was never a sachem in the traditional, generally hereditary Lenape sense. But his flair for language and his knack for being in the right place at the right time won him an outsize role among the eastern Delaware during the Seven Years' War.

In time, he and Shingas would become opposite sides of the Lenape coin—the war chief and the peacemaker—but in the frenzied months after hostilities erupted, Teedyuscung, too, took the red-painted wampum, leading war parties south into the heretofore untouched settlements along the Kittatinny Ridge, within the Forks of the Delaware, and further paralyzing the Pennsylvania frontier.

If the Ohio Indians fought for many different reasons, the goal of Teedyuscung and his followers was fairly direct—to drive the settlers out of the Forks and the Tulpehocken Valley and to reestablish the old line of demarcation that once existed along South Mountain in the days of Sassoonan, three decades earlier. Breaking free of the Onondaga council—and with backing from the Seneca, many of whom were openly supporting the French—Teedyuscung reached out to Father Onontio. The eastern Delaware needed help; they'd entered the war hampered by food shortages, brought on by late frosts and severe drought the previous growing season, and now that the traders had fled, war booty did not make up for the lack of supplies. But when he visited Fort Niagara in the summer of 1756, Teedyuscung was shocked to find the French in equally desperate condition.

Just as Teedyuscung was beginning to rethink the wisdom of war with the British, Pennsylvania took the worst possible action, turning all but the most die-hard, pro-British Natives against the colony. Morris and a majority of the provincial council, thumbing their noses at the Quaker assembly, used their authority in April 1756 "immediately to declare War against the Delawares and all other Enemy Indians" and to pay "the following Rewards to such as shall make captive or put to death any of the said Enemy Indians." It was a scalp bounty—$150 for "every Male Indian Prisoner, above Ten years Old," $130 for a live woman or a man's scalp, $50 for a woman's scalp. It was a death sentence for many friendly Delawares; it was easier to scalp a neighbor than to ambush a warrior.

The Quakers, who though now a minority in the colony William Penn had founded still controlled the Pennsylvania assembly, also found themselves ambushed. Penn's son Thomas—who as provincial proprietor had long abandoned his father's principles, along with most of his Quaker beliefs—was conspiring against the Society of Friends, which posed a roadblock to his power and income. Publicly, the Quakers were smeared as disloyal radicals in collusion with the French, a message that the Reverend William Smith, a Penn lackey and Anglican priest with a talent for libel, happily broadcast in anonymous tracts, pamphlets, and letters to newspapers, urging, among other things, that Quaker pacifists have their throats cut.

Spooked by the bloodshed in the backcountry and urged on by Penn, the British Parliament was poised to require a loyalty oath to the king from Pennsylvania's assemblymen—which, because Quakers refused on religious grounds to swear oaths of any kind, would bar all observant Friends from election or service. Only a last-minute compromise, by which ten pacifist Quakers resigned from the assembly for the duration of the war, derailed the oath bill. If anything, though, it stiffened Quaker determination to find a way to peace.

The war of skirmish, feint, attack, and retreat continued—almost entirely one-sided, since with few exceptions it was a defensive action on Pennsylvania's part. The militia had been raised, what arms and ammunition that could be found had been distributed, and Benjamin Franklin was overseeing construction of a line of forts roughly paralleling the Kittatinny Ridge from the upper Delaware, across the Susquehanna, and into the mountains west of the hamlet of Shippensburg. Typical was Fort Allen, near the Gnadenhütten mission on the Lehigh, where eleven Moravians had been murdered two months earlier. Working in frigid winter rain, Franklin's men cut pines, sharpening the ends to build a circular palisade 150 feet across. They hauled a single swivel gun, the only artillery they had, into place on the rampart and fired it, "to let the Indians know, if any were within hearing, that we had such pieces; and thus our fort, if that name may be given to so miserable a stockade, was finish'd in a week," Franklin said.

The forts, including some private garrisons, gave the nervous settlers an illusion of security, but they did very little to actually curb the ongoing Indian attacks. In fact, French and Indian raiders overran several of the installations, including Fort McCord west of Chambersburg, where Shingas killed or captured more than two dozen people, and Fort Granville along the Juniata, taken by Captain Jacobs and Louis Coulon de Villiers, the victor at Fort Necessity. (Captain Jacobs was later killed during a withering strike on his village at Kittanning, on the Allegheny River—one of the few successful counterattacks the provincials were able to mount into the upper Ohio Valley.)

Having left the government, pacifist Quakers redoubled their efforts to negotiate with the Delaware outside official channels. Via Scarouady, they reached out to the eastern Delaware, and Teedyuscung was ready to listen. In July 1756, wearing a gold-trimmed coat the French had given him, he came to Easton to lay the groundwork for what would eventually be three peace treaties over the next two years, as the Quakers formed the Friendly Association for Regaining and Preserving Peace with the Indians by Pacific Measures and provided funds to make the councils possible.

It was fitting that the treaties took place at Easton, a tumbledown settlement at the very Forks of the Delaware, where the long-concealed deceit of the Walking Purchase, and the Penn family's role in it, would finally be exposed. At the second treaty in November 1756, despite everything that Conrad Weiser and provincial secretary (and Walking Purchase coconspirator) Richard Peters could do to prevent it, the newly installed Governor William Denny asked Teedyuscung and the Delaware, Mahican, and Shawnee leaders present a simple question: "Have we, the Governor or People of Pennsylvania, done you any kind of Injury?"

Denny—a forty-seven-year-old captain, hastily promoted to lieutenant colonel so that he could hold office without embarrassment—has always left historians reaching for adjectives. He was, said one, "venal, lazy and inept, unsteady and self-pitying, boastful but physically timid . . . bewildered, frightened, flighty, irritable." Another described him as "idle, fatuous, greedy, and almost simple-minded." No other Pennsylvania governor, beholden to the Penns for his office, would have dreamed of asking such a loaded question.

According to Quaker witnesses, "the Joy which appear'd in the Countenance of ye Indians canot be express'd." The assembled chiefs punctuated every sentence in the remainder of the governor's speech with loud, approving cries of "Ye-ho!," running to him to shake his hand the moment he finished.

Teedyuscung, seeing his opportunity, asked for a day to consider his reply—and to ratchet up the anticipation. In the morning, he laid a series of wampum belts before the council and invoked the timeless ritual of the "clearing the woods" ceremony.

"The Times are not now as they were in the Days of our Grandfathers; then it was Peace, but now War and Distress," the stocky Delaware told a crowd silent but for the scratch of quill pens; Peters was, as usual, taking the minutes, as were several observers. "I take away the Blood from your Bodys with which they are Sprinkled. I clear the Ground and the Leaves that you may sit down with Quietness. I clear your Eyes that when you see the Day-light you may enjoy it."

What injuries had been done the Lenape? "I have not far to go for an Instance," Teedyuscung said, his voice rising, as he stamped his moccasined foot on the ground for emphasis. "This very Ground that is under me was my Land and Inheritance, and is taken from me by fraud." When he sold land, the sachem said, he meant to sell it: "A bargain is a bargain." But if, after a man's death, "his Children forge a deed like the true one, with the same Indian Names on it, and thereby take the Lands from the Indians they never sold, *this is fraud*."

Just as the sachems and officials were gathering for the second Easton treaty, circumstances took a distinctly ominous turn for Robert Stobo and Jacob van Braam.

In late October, armed with the letter and map found in the ruins of Braddock's retreat, the French brought Stobo and Van Braam to trial for their lives, *pour le crime de haute trahison*. Governor Vaudreuil sat as judge, and Céloron de Blainville—he of the lead plates along the Ohio—served as the king's prosecutor.

The officers' belongings were searched, and the men were questioned separately at length. Van Braam, disavowing any knowledge of the letter, did concede that the handwriting appeared to be Stobo's. For his part, Stobo declined to swear any oaths or sign any papers, although he promised to tell the truth "following my conscience." When confronted with the damning map and letter, however, he refused to confirm anything regarding his past actions, arguing unconvincingly that he did not remember what he had written.

As the trial unfolded, Stobo grasped at two circumstances by which he might extricate himself without perjury. As a newly commissioned officer, Stobo maintained, he had not understood the rules of parole, which none of his captors had explained to him, and to the extent that he *had* understood them, he considered his obligation to abide by the terms of the capitulation void after he saw British soldiers who had surrendered honorably being held as Indian slaves.

It didn't work. When Vaudreuil, Contrecoeur, and the five others on the Council of War met on November 8 to render the verdict, Stobo may have realized the decision was foregone; this was hardly a disinterested jury. For the first time, he admitted the obvious: the map and letter were his. "I believed myself entirely free to do what I pleased for the interest of my country," he told the tribunal, "and to give all the information I could contribute to that end. Therefore I wrote the letter in question and I drew the map that is joined to it."

Van Braam was acquitted; Stobo was not. He was ordered held until a scaffold could be built in Montreal, at which time he was to be publicly beheaded. There were no dissenting votes.

It must have felt a little strange to George Croghan to be back in Philadelphia in the early winter of 1756. For one thing, the fear of being taken up for his enormous debts, which had restricted him to the dangerous backcountry since the fall of Fort Necessity, had just been put in abeyance by a special act of the assembly, granting him and his old business partner William Trent ten years' grace from imprisonment. But more than that, he wasn't there as a frontier trader or a militia captain, but as the deputy superintendent for Indian affairs, the new right-hand man to Sir William Johnson in New York, the Crown's liaison with all the tribes of the northern colonies.

The change had done Croghan good and, given the French bounty on his head, maybe saved his life. But in the mayhem of the ongoing frontier war, Quakers weren't the only publicly reviled religious group; Roman Catholics were automatically suspected of French sympathies. The previous year, Conrad Weiser had been seeing ghosts in every corner, claiming that Catholics were colluding with Indians in his neighborhood. (The provincial council investigated and found "very little foundation for that representation." In fact, there was none.) Croghan's Irish background made him a target of similar rumors—so much so that when a series of letters surfaced in which the anonymous writer, signing himself "Filius Gallicae," proclaimed a secret Catholicism and love of French ideals, took credit for leading British scouts at the Monongahela, and offered sensitive military information, Croghan was an immediate and obvious suspect.

Croghan was undoubtedly innocent, and the letters were almost certainly a plant, likely part of the larger plot to smear Quakers and Catholics. But his fortunes had turned since his salad days on the Ohio. Frontier attacks made life at Aughwick (now garrisoned and known as Fort Shirley) dangerous. Shingas had publicly boasted that he'd take Croghan's scalp and kept a cadre of warriors near the fort at all times, hoping to snare the trader. Croghan's debts were still crushing, even if he was free of the immediate threat of lawsuits. He was blamed in some quarters for Braddock's defeat, and his skills at Indian diplomacy were largely unused.

What Pennsylvania disdained, others appreciated. Croghan accepted an invitation from Johnson, a fellow Irishman with whom he'd been corresponding. He would hold the deputy superintendent's position for the next fifteen years, always in the thick of Indian diplomacy as he was now with the Easton treaty, acting on behalf of the Crown instead of the province.

Nor did Pennsylvania retain much appeal for Andrew Montour, who had once dreamed of creating a new kind of world there. Montour found his life on the Susquehanna increasingly awkward; like Croghan, he was deeply in debt, and his second marriage, to an Oneida woman named Sarah Ainse, had ended in divorce. (So had his first, to one of Sassoonan's granddaughters. Native tradition made such dissolutions relatively simple.) With the coming of war, French Andrew was an object of misgiving and doubt among the colonists. With Scarouady beside him, he left the province and moved to New York, too, in the spring of 1756.

The war in America—now a formal conflict, having been duly declared by London and Paris—was not going well for the British. John Campbell, Earl of Loudoun, had taken control of British forces in the colonies in 1756 at almost the same time the new French commander, Louis-Joseph de Montcalm, was capturing Fort Oswego on Lake Ontario. The next spring Loudoun failed in a bid to capture Louisbourg, on Île Royale (Cape Breton Island), shortly before seeing Fort William Henry on Lake George fall to Montcalm. (At the latter, as had happened at Oswego the year before, Montcalm looked the other way when his Indian allies massacred the surrendering garrison.) Canadian militia, skilled in forest warfare, for the most part fought rings around their British counterparts.

Nor was the situation much better on the Pennsylvania frontier. Although attacks from the eastern Delaware had abated thanks to Teedyuscung's peace overtures, Shingas and the other Ohio war leaders, as well as dissident eastern Delawares and some French-allied Iroquois, continued to apply pressure.

"It is now Come so farr that murder is Comited allmost every day," Conrad Weiser lamented in the fall of 1757. "Five Children have ben carried [off] last Fryday, some days before a sick man killed upon his bed, begged of the Enemy to shoot him through his heart, which the Indian answered, I will, and did so . . . I have neither men nor a sufficient number of officers to defend the Country." In all, hundreds of Pennsylvanians had been killed, hundreds more carried away, and multitudes of farms abandoned—a thousand in Cumberland County alone.

Among the victims that autumn were the family of Jacob Hochstetler—the Amish farmer in the Tulpehocken Valley who stood by his nonviolent principles and wouldn't allow his family to shoot at attacking Indians. His wife, a son, and a daughter dead, Hochstetler and two other boys—Joseph, age thirteen, and Christian, age eleven—were taken away as captives. (Having suffered such a blow, many of the

Amish survivors along Northkill Creek decided to move away from the dangerous edge of the frontier, relocating to Lancaster County and forming the nucleus of the modern Amish community there.)

The Hochstetlers' experience as captives was both typical and unusual. As in frontier wars dating back to King Philip's uprising, most of the abductees in the Pennsylvania backcountry were women, children who were old enough to walk, and teens; these were the easiest groups to integrate into village life and added the most to Indian families decimated by war. Christian and Joseph fit this pattern, and once on the upper Ohio, both were quickly adopted and treated as full members of the community.

Adult men and older boys, by contrast, were often killed on the spot or brought home for torture. Yet for some reason, Jacob Hochstetler was spared. Separated from his sons, he was moved deeper and deeper into the interior, to Presque Isle on Lake Erie, then to Detroit, and eventually to somewhere in what is now northeastern Ohio or northwestern Pennsylvania. His mustacheless beard, the hallmark of a married Amish man, was pulled out—not from cruelty, most likely, but because Native Americans tolerated no facial hair on themselves and considered it grotesque on others. The hair on his head was also yanked out, one painful tuft at a time, until he sported the same four-inch-wide scalp lock, hung with feathers, as any Lenape warrior.

Held for three years, never adopted but permitted to take a gun to go hunting, Hochstetler hoarded pinches of powder and the occasional lead ball until he had enough ammunition to risk an escape. Unfortunately, he had no clear idea of where he was and no notion of where home lay. One day, he saw a warrior drawing lines in the ashes of a campfire, explaining to some boys how his raiding party had crossed the mountains and followed a river called "Sasquehannah." That was enough.

Making his break, Hochstetler traveled by night, wading in streams when he could to throw off tracking dogs, until he came to a river he believed was the Susquehanna. Taking a risk, he kindled a fire and used it to burn a large dead log into six sections, then lashed them together with vines to make a raft. When the river bent to the southwest, he despaired, thinking it must be the Ohio, carrying him farther from home. In fact, however, he was on the north branch of the

Susquehanna, floating through the Wyoming Valley. He survived on whatever he could scrounge, mostly nuts and raw crayfish, and once a maggot-filled opossum "that he ate until his hunger was appeased." When he finally drifted past Harris's Ferry, he was too weak to reach the shore. Only by good chance did a man watering his horse down-river see Hochstetler feebly waving his arms, and he called for a rescue.

In prison, Robert Stobo discovered the pleasures of smoking—about the only diversion available to him following his conviction and death sentence. Taken from Montreal, he and Van Braam (who though ac-quitted of treason was still a hostage) were placed in solitary cells in Quebec. But soon, for reasons they did not fully understand, the con-ditions of their imprisonment eased somewhat. They were permitted to share a comfortable room in "a common jail, where two stout senti-nels were posted at [their] door, and two below [their] window." Some time after that, they were allowed to walk the corridor and entertain a few visitors. Their cleaning and cooking were seen to by three ser-vants, all female, ages eighteen to twenty. Stobo's biographer, Robert C. Alberts, hints at "other services that can only be conjectural."

Up to this point, it's hard not to see Stobo as a little bit of a popin-jay—with his grandiose notions of military valor, his plumed hat, and his romantic tête-à-têtes—who was too clever by half and got caught for it. But it was now, after he'd been condemned to death and his fortunes appeared most desperate, that Robert Stobo began to show a more remarkable side.

On the night of May 1, 1757, Stobo and Van Braam made a break for it. Somehow they were able to unlock the corridor door and slip out, climbing out the back of the building and through a manure pile to freedom. It wasn't until the next morning, when their jailer tried to wake them, that he found the British officers had used that most hackneyed of ruses, the stuffed-clothes-in-the-bed trick.

"I touched Mr. Stobo," the jailer later testified, "and I was very much surprised to find, instead of a man, a beaver great-coat in the bed, with a night cap and a shirt, which was made to look like the head of a man." Van Braam was impersonated by "a valise, likewise a bonnet and cotton shirt."

The escape didn't last long; the men were picked up four or five days later, having crossed the St. Lawrence but gotten no farther. Undeterred, Stobo broke out a second time, in July, by himself, carrying thirty pounds of food he'd carefully dried and squirreled away over the preceding months. This time he used a butter knife to grind away the weak stone around his iron window bars. He was free long enough to make it five or six miles down the north shore of the St. Lawrence, wading neck-deep across the St. Charles River and edging past the thundering cataract at Montmorency Falls, before he was nabbed. Stobo—and Van Braam, too, for good measure—were thrown into harsh confinement, Stobo to await his execution.

What he did not know was that the French had no intention of actually beheading him. Vaudreuil had apparently been under orders to find Stobo guilty, condemn him to death, and then stay the execution. It may be that the French felt their own moral standing was questionable, since even some of their own officers agreed with Stobo that the terms of capitulation had been broken.

As far as Stobo knew, though, he was a dead man unless he did something. He hadn't given up.

Teedyuscung's charge that the Penn family had knowingly cheated the Lenape at the Walking Purchase and other land deals was explosive. Most immediately, it showed that the Delaware had legitimate complaints that had led them to war and strengthened the Quaker push for a negotiated peace. But it also exposed the venality of the Penn family and their associates, including council secretary Richard Peters, and provided powerful ammunition for Benjamin Franklin and those opposed to proprietary power, who were fighting to have Pennsylvania taken away from the Penns and made a royal colony. Nor did it reflect well on the Iroquois League, which had ratified these sweetheart deals. In fact, Teedyuscung had rhetorically stripped off the political "petticoat" that the Iroquois had long imposed on the Delaware and was now insisting that the Lenape would henceforth speak for themselves, not through the Six Nations.

It was, therefore, in the best interests of almost everyone—the proprietary leaders in Pennsylvania; the Iroquois League and Sir William Johnson in New York, for whom the Iroquois alliance was paramount; Conrad Weiser, whose allegiance lay with the province and his adopted Mohawk, and who saw the Quakers as meddlesome fools who were prolonging the war—to discredit "King Teedyuscung." Unfortunately, the Lenape's insatiability for rum and his habit of inflating his own importance, especially when in his cups, gave them handy ammunition. "Mr. Weiser informed the Council that the King and the principle Indians being all yesterday under the Force of Liquor," Peters noted primly in the council minutes, which already included a report that "the King and his wild Company were perpetually drunk . . . The King was very full of himself." Even more damaging, the efforts of the Friends to act as intermediaries with the Indians allowed Peters and the rest of the Penn faction to argue that the claims of fraud were all a Quaker fabrication.

Still, Teedyuscung was able to score some important victories at Easton. In return for peace with the eastern Delaware, he asked for compensation for lost land and for sureties that the Wyoming Valley would remain in Delaware hands. "We would have a Certain Country fixed for our own use & the use of our Children for ever," he said. When he demanded to see the original deeds on which the Walking Purchase and other frauds were based, Peters threw up his hands and said they couldn't be produced—only to have the provincial commissioners produce duplicates.

Teedyuscung kept slogging away, making progress by inches. The second Easton treaty in the fall of 1757 had dramatically reduced attacks by the Susquehanna Indians and returned some of the white captives who had been taken. Now he reached out to the western villages on the upper Ohio, from which most of the remaining violence originated. At Teedyuscung's request, and with Quaker backing, a pious Moravian cabinetmaker named Christian Frederick Post, who had been married to the sachem's daughter-in-law, set out in July 1758 for the Ohio country. He was accompanied by Shingas's older brother Pisquetomen, who had taken extraordinary risks to come to Philadelphia that spring to assess the prospects for a truce.

Long a go-between for the Christian Indians of the Wyoming Valley, Post was an inspired choice if only for his Lenape name: Wallangundowngen, meaning "making a good blessing" and the Delaware term for peace. He also brought a guileless serenity of purpose to what was undoubtedly a very dangerous mission, even with an influential

Lenape like Pisquetomen at his side. "If I died in the undertaking," Post wrote in his journal, "it would be as much for the *Indians* as the *English*." (Weiser, who daily dealt with the aftermath of the settlement attacks, was infuriated by the whole escapade. "The Quakers submit to Barbarians, to the Murderers and Heathens," he seethed.)

Crossing the Allegheny Mountains, Post saw old, dried scalps—one with long white hair—tied to bushes along the path as a warning, as well as the red-painted poles "to which [the Indians] tye the prisoners, when they stop at night, in their return from their excursions." At one village he was surrounded by angry men with drawn knives, and a few days later he was warned "not to stir from the fire; for that the *French* had offered a great reward for my scalp, and that there were several parties out on that purpose. Accordingly I stuck constantly as close to the fire, as if I had been chained there."

But he also met with Shawnees with whom he had been friends in the Wyoming Valley, who greeted him warmly, and with Delaware (or Dillaway) George, one of the two men who had smuggled copies of Stobo's map out of Fort Duquesne and whom Post found to be a forceful advocate for peace with the British. The war was taking a harsh toll on the Ohio villages, with food and supplies running low. Pisquetomen's brothers, Shingas and Tamaqua, both seemed open to peace—though with obvious concerns. Shingas asked "if I did not think, that, if he came to the English, they would hang him, as they had offered a great reward for his head," Post said. "He spoke in a very soft and easy manner. I told him that was a great while ago, it was all forgotten and wiped clean away; that the English would receive him very kindly."

If not kindly, at least grudgingly, if it meant an end to frontier raids and help against the French. After years of defeats, the British effort was turning a corner. With a new administration in London under William Pitt, and with the pitiful James Abercrombie ("Mrs. Nanny Cromby" to his troops) replaced by Jeffery Amherst as commander in chief of North America, the differences were dramatic. Amherst had already taken Louisbourg in Acadia, and he tasked Brigadier General John Forbes with leading a new assault against Fort Duquesne. The hard lessons of the Monongahela had been learned. Forbes embraced the critical importance of scouts and light skirmishers and, unlike

Braddock, understood the need to peel away as many of the western Indians as possible.

To accomplish this, he was authorized to deal directly with the Ohio tribes, and he gave Post's mission his full support, as well as prodding Governor William Denny and the recalcitrant Pennsylvanians to accede to Teedyuscung's demands. As autumn 1758 approached and the Pennsylvania mountains began to flame with color, Forbes—so ill with dysentery he was carried in a litter slung between horses—oversaw the completion of a new road and fortified supply posts through the mountains, while a so-called grand council was convening at Easton, under the auspices of Pennsylvania and New Jersey, to finally cement a comprehensive peace.

Easton was awash in Indians. While the previous treaties had been small affairs, with Teedyuscung and a few followers, this time there were more than five hundred Natives present, representing almost every tribe for hundreds of miles—and a multiplicity of agendas. Besides Teedyuscung and a hundred or so followers from the Delaware and mixed-tribe villages along the upper Susquehanna, there were Mahicans and Shawnees, Wappings from east of the Hudson, and a swollen delegation from the Six Nations proper, including six Tuscaroras. The Nanticoke, Conoy, and Tutelo, all recently adopted as "younger brothers" by the Iroquois, were also represented.

The Six Nations were anything but united. The Francophile Seneca, led by a sachem named Tagashata, and the Anglophile Mohawk, led by Croghan's father-in-law, Nichas, were barely civil to each other, although such divisions were papered over for the council. French Margaret, Andrew Montour's cousin, was there with her husband, a Mohawk known as Peter Quebec. There were more women and children present than sachems and warriors, in fact, which posed its own problems. At an earlier treaty, the angry townspeople of Easton "observed that the Shirts which the Indian Women had on were made of Dutch [German] Table Cloaths, which, it is supposed they took from the People they had murdered on our Frontiers."

Conrad Weiser, his health poor, was in attendance. (Easton was to be Tarachiawagon's last hurrah.) So was George Croghan, Esquire, serving as William Johnson's eyes and ears, and "Captain Henry Montour" (Andrew using one of his many *noms de frontière*), "Interpreter

Indians Giving a Talk to Colonel Bouquet
This engraving, based on a Benjamin West painting, dates from the end of Pontiac's War in 1764, but it shows many of the features typical of a frontier treaty—the covered arbor to shelter some members of the large negotiating parties from sun and rain, an Indian orator holding wampum to reinforce his words, and a colonial scribe furiously recording the proceedings for later publication. (Library of Congress LC-USZ62-104)

in the Six Nations and Delaware Languages." One familiar face was missing, however. Scarouady, the Oneida "half-king" who had been central to frontier statecraft for more than a decade, had died that spring, possibly from smallpox.

The formal council sessions took place outdoors, beneath a shed or bower built for the purpose and open on the sides to accommodate the overflow crowd. Governor Denny launched into the "clearing the woods" formalities, saying as he held the wampum, "Brethren: With this String I wipe the Sweat and Dust out of your Eyes, that you may see your Brethren's Faces and look Cheerful." But Tarachiawagon was having trouble translating. "Mr. Weiser interpreted the Substance of this Speech, [but] saying his Memory did not serve him to remember the Several Ceremonies in [it] . . . he desired Nihas, a Mohock Chief, to do it for him, which he did," the council minutes note. Montour, meanwhile, was translating into Delaware, even though Teedyuscung and many of the Lenape were conversant in English.

Teedyuscung had been ducking the Iroquois for almost a year; he'd stood them up at a treaty in Lancaster the previous spring because the aging Lenape knew what was coming. Almost no one at Easton that autumn had the best interests of his Wyoming Valley followers at heart, and the Six Nations council was determined to get a leash around Teedyuscung's neck again. His leverage was gone; he'd made peace and was no longer needed. Perhaps that got his back up, because in session and out, drunk and sober, "He Who Makes the Earth Tremble" never missed an opportunity to twist the collective Iroquoian nose.

With dramatic timing, Pisquetomen arrived in Easton just as the council was about to begin, carrying Post's journal with word of their success among the Ohio Indians. Pisquetomen pledged, on behalf of his brothers Shingas and Tamaqua, Delaware George, and others on the Ohio, that if the council ended with an agreement, the western tribes would observe it. "When you have settled this Peace and Friendship, and Finished it well," he said, "and you send it to me, I will send it to all the Nations of my Colour."

This was the best possible news. But before anyone could relax, Nichas, the Mohawk, rose to speak, his words coming in a torrent, directed with venomous intensity at Teedyuscung. Weiser blanched.

Nichas spoke at length, but no translation followed. "Mr. Weiser was ordered to interpret it, but he desired to be excused, as it was about Matters purely relating to the Indians themselves, and desired Mr. Montour might interpret it." Montour had no desire to put his foot in that hornets' nest, however. Weiser suggested that the speech be translated "at a private Conference."

In that meeting, Nichas and a parade of Iroquois sachems rose to denounce their "nephew" Teedyuscung. "I tell you we none of us know who has made Teedyuscung such a great Man. Perhaps the French have, or perhaps you have . . . We, for our parts, intirely disown that he has any Authority over us," said Sagughsuniunt, an Oneida known as Tom King. In front of the Lenapes, Denny answered with a theatrical shrug: "We believe[d] what your Nephew told us, and therefore, made him a Counsellor and Agent for us . . . I can only speak for myself, and do assure you that I never made Teedyuscung this great Man." In the end, Teedyuscung was humiliated, forced to acknowledge Iroquois supremacy—the very "petticoat" he'd tried for years to throw off.

Weiser met quietly with the league's principal chiefs, and what emerged was, from certain angles, a political masterstroke. Pennsylvania would cede back to the Iroquois the land beyond the Allegheny Mountains that had been claimed by Britain at the Albany Congress four years earlier, and the loss of which had been such a bitter pill. That, with a vow never to fortify or settle the upper Ohio and recognition of their independence of Onondaga, mollified Pisquetomen and the western Indians and ended three years of horrific frontier war—no small feat.

Further, the Pennsylvanians and the Iroquois would cut Teedyuscung's legs out from under him by appearing to accede to the old Lenape's demands. A promised investigation of the Walking Purchase was simply shunted into a bureaucratic limbo in London from which it would never emerge. His insistence on "a Certain Country fixed for our own use" resulted not in a legal deed for the eastern Delaware, but the patronizing consent of the league that Teedyuscung's people might continue to live in the Wyoming Valley under their sufferance.

When Teedyuscung finally spoke at the end of the Easton conference, it was with the heartsick words of a man who knew he'd been outflanked and hopelessly boxed in. Even so, he refused to give up on a deed for the land in the Wyoming Valley, though he must have known it was fruitless. "I sit here as a Bird upon a Bow [bough]; I look about and do not know where to go; let me therefore come down upon the Ground, and make that my own by a good Deed, and I shall then have a Home for Ever; for if you, my Uncles, or I die, our Brethren, the English, will say they have bought it from you, & so wrong my Posterity out of it."

Post and Pisquetomen were on the road from Easton to the Ohio before Halloween, carrying with them news of the peace treaty and a white wampum belt that depicted two people joined by a black line. "If you take the Belts we just now gave you, in which all here join, English and Indians, as we don't doubt you will," said a speech Post carried from Denny, "then by this Belt I make a Road for you, and invite you to come to Philadelphia to your first Old Council Fire, which was kindled when we first saw one another, which fire we will kindle up a gain and remove all disputes."

Post also carried messages from Forbes to Shingas and the Ohio sachems, even as a strike force of 2,500 hundred men from Forbes's army left their outermost stronghold, Fort Ligonier, to make an unexpected stab at Fort Duquesne. Guiding them, fresh from Easton, were Croghan, Montour, and fifteen Iroquois scouts. No one expected the British to move so quickly, so close to winter, least of all the beleaguered French commander, François-Marie Le Marchand de Lignery, who was short on men and supplies. He learned of Forbes's approach just as his Ohio Indian allies accepted Denny's wampum and returned to the British fold. His position untenable, Lignery emptied the fort, planted enormous quantities of gunpowder beneath its walls, and on November 25, 1758, blew it sky-high.

"Honble. Sir," wrote Colonel George Washington, in command of the Virginia troops, to his province's new governor. "I have the pleasure to inform you, that Fort du Quesne, or the ground rather on which it stood, was possessed by his Majesty's troops . . . The enemy, after letting us get within a day's march of the place, burned the fort, and ran away (by the light of it,) at night, going down the Ohio by water . . . The Delawares are suing for Peace, and I doubt not that other Tribes on the Ohio will follow their example." From the smoking ruins of Duquesne, a new fort—Fort Pitt—began to rise. Pisquetomen and the Delaware looked on in worry; their still-fresh peace treaty called for trading posts, not British forts.

Cheers went up throughout the colonies. But as Teedyuscung left Easton, saying goodbye to Israel Pemberton—leader of the Quaker association and the one steadfast friend he had left—he wept like a child.

Robert Stobo was dying by degrees in his clammy prison cell in Quebec. "His looks grew pale, corroding pensive thought sat brooding on his forehead, and left it all in wrinkles; his long black hair grows, like a badger, grey; his body to a shadow wastes[;] . . . his health was gone," according to the *Memoirs of Major Robert Stobo*.

But again, a woman to the rescue. What the *Memoirs* calls "a lady fair, of chaste renown, of manners sweet, and gentle soul," interceded just in time, using her influence to have him moved to better quarters. Stobo himself simply spoke of his "close confinement, only on account of its long duration being reduced to such a bad state of health that my life was despaired of, [that] I was moved from the common jail to a house of an inhabitant." However it happened, it was a lifesaver.

There, nursed enthusiastically by the homeowner's daughter (Stobo, it seems, could have found a willing woman on the moon), he became acquainted with a number of other British soldiers held in Quebec. Late in 1758, as the Easton council was reaching its climax, Stobo began plotting yet another escape attempt, with a Connecticut lieutenant named Simon Stevens, who had been a member of the famed Rogers' Rangers. One hurdle after another arose. Before they could slip away on snowshoes, brutal winter weather locked Quebec in a vise. Plans to commandeer a French sloop, anchored at the shipyard, were rejected as too risky. Finally, with the connivance of a longtime English captive named Clark, they obtained guns, powder, and a big Montreal cargo canoe. Accompanied by Clark, his wife and their three children, and two other English prisoners, they slipped away.

It was the night of May 1—the second anniversary of Stobo's first abortive escape, and nearly five years since his surrender at Fort Necessity. Stevens, in his journal, noted that "about Ten a Clock that Evening we embark'd in our Birch Canoe . . . there being nine in Number, which were Capt. *Stobo, Lakin, Clark* and his Wife, with three Children, *Denbo* and myself."

Buffeted by storms and soaked by spray, their provisions ruined, they nevertheless made fair progress away from Quebec, until on the fourth day they stumbled upon an Indian man and woman who, mistaking them for French, approached and were captured. "I took hold of the Indian and march'd forward," Stevens said, but the man "sprang from me with a Design to make his Escape, upon which Clark being the next behind me, shot him dead. I then gave Orders to *Denbo* to kill the Squaw . . . We scalped each of them." Out of the cold-blooded murders came half a bushel of dried corn and two hundred pounds of sugar, which the fugitives plundered from the dead Natives' camp.

The St. Lawrence was wide but not empty. They dodged a flotilla of troop transports, but a two-masted shallop approached them, and after a flurry of gunfire Stobo's party seized the boat—and the elderly gentleman at its helm. "I am Monsieur Chev. la Darante," the old man said with a bow—Louis-Joseph Morel de La Durantaye, lord of the manor at Kamouraska, ferrying a load of wheat to Quebec with three servants. Durantaye was the soul of civility, but he hoodwinked the English escapees into anchoring near a French guard post that night, and they barely evaded a pursuing sloop ("a lofty frigate," according to the *Memoirs*). Durantaye and his crew hauled on the shallop's oars as they labored to escape, with Stobo threatening to shoot the first one who slacked off even a fraction. Cannonballs slashed the water nearby. "The Tide being against us very strong," Stevens said, "which gave the Enemy a very great opportunity of firing at us; but by good Fortune no Person was hurt."

A rising wind finally carried the shallop out of range, but not out of danger. Days later, having rounded the Gaspé Peninsula, the boat ran aground during a storm, smashing its hull. For more than a week, their food almost gone, Stobo's party tried to cobble together enough driftwood planks and nails to repair the shallop, stuffing the gaping cracks with handkerchiefs and spruce pitch. About to launch, the party was again spied, this time by an armed but unsuspecting schooner and a sloop. Surprising a three-man shore party from the vessels, whom they left tied up and under the guard of Clark's wife, the four men opted for a desperate gamble.

In the middle of the night, and bailing frantically to keep their battered shallop afloat, they quietly approached the schooner, the larger of the two ships. In for a penny, in for a pound, they figured, and if they were successful with the schooner, they would have less to fear from the smaller sloop. Stevens was poised in the bow with a grappling hook, Clark took the helm, and Stobo stood ready, bristling with a cutlass, a pistol, and a musket. They took the ship quickly, though Stevens was later at pains to note that Stobo lost his sword and pistol over the side while boarding and shot a man to whom Stevens had already granted quarter. ("I hope the Reader will excuse my being so particular," the lieutenant said, "as Capt. *Stobo* has reported he was the first that boarded the Schooner, and the only Instrument in taking her.")

Sailing their new prize alongside the smaller sloop, they fired on it, captured the crew, and burned the ship. Gathering those on the shore, they sailed triumphantly a week later into the harbor at Île-Saint-Jean (Prince Edward Island), "where we all safe arrived (thank GOD)," according to the otherwise unflappable Stevens. It had been almost a month since they'd fled Quebec.

After five years, Captain Robert Stobo was free, and for the first time, he learned he was a major. He also learned that Major General James Wolfe had just launched a long-anticipated attack on Quebec. Stobo and Stevens immediately reversed course and joined Wolfe, who had boldly sailed a fleet up the treacherous St. Lawrence (following surveys made by the young but already gifted James Cook), where he disembarked nine thousand men and besieged the city.

In the two and a half months that followed, a military stalemate in which neither side distinguished itself, Stobo served closely with Wolfe and was injured by the same cannonball that almost killed the general during one French bombardment. One small patrol to seize an arms depot, which Stobo helped lead, resulted instead in more than 150 prisoners—many of them women, many of them familiar faces. "Monsieur Stobo's name is all that's heard for half an hour at least," the *Memoirs* states. On another occasion, Wolfe entertained a French official under a flag of truce—the *Memoirs* doesn't say who, other than that it was one of the judges in Stobo's trial—and as one of Wolfe's staff, Stobo felt required to dine with him. "Ill went the victuals down with Major Stobo, and every mouthful offered fair to choke him, nor yet the glass could cheer."

And though it's unproven, it seems likely that this regimental engineer, who once sized up Fort Duquesne so precisely, had observed Quebec City with a similar eye for detail and described to Wolfe a footpath up a steep slope—the route Wolfe used on September 13, 1759, to slip his troops onto the Plains of Abraham just west of the city and at last bring Montcalm to a battle in which both would die. The victory was just one of many that famously marked 1759 as an annus mirabilis for Great Britain, the turning point in its ultimate victory over France. Within a year, Vaudreuil would surrender all of Canada, and while the war would grind on overseas for three more years, it was essentially finished in North America.

Stobo wasn't there for the joy or the grief. Wolfe had dispatched him to Boston immediately before the fight, with orders to proceed to Crown Point on Lake Champlain, carrying messages to General Amherst. En route, Stobo's sloop was taken by a French privateer off Nova Scotia. He barely had time to change into a common soldier's clothes, sinking his officer's uniform and documents over the side. Set ashore with the others in Halifax, he begged his way to Boston, where, after hearing his nearly unbelievable story, Massachusetts governor Thomas Pownall loaned the now penniless man a few pounds to continue his mission.

Before Stobo left, he accepted a letter of Pownall's for Amherst. It seems unlikely that the major knew it contained one of the great understatements of the war. Stobo, the governor observed to Amherst, was marked "by a fatality that pursues him with repeated misfortunes."

Chapter 11

Endings

THE COLD TAIL of November 1759. Robert Stobo, granted leave by General Jeffery Amherst, had just completed the long journey from Lake Champlain to Virginia, the first time he had been home in five and a half years. In his pocket was a letter from the commander in chief to Governor Francis Fauquier, begging leave to recommend the major "to your particular notice and favor" and promising "as I on my own part, when I have an opportunity, shall be very glad to reward his zeal and services."

Fauquier and the House of Burgesses were as good as Amherst's wishes. They granted Stobo back pay, a bonus of £1,000, and a resolution of thanks "for his steady and inviolable Attachment to the Interest of this Country; for his singular bravery and Courage exerted on all Occasions during this present War, and for the Magnanimity with which he has supported himself during his Confinement in *Canada*."

Robert Stobo was never going back to shopkeeping. He was in the army for life—though his fickle stars scarcely relented. Sailing to London a few months later, he was overtaken again by privateers; again he had to fling his papers over the side of the ship and shed his uniform; again, though, he was able to resume his trip after paying a ransom. He later commanded a company at the surrender of Montreal in 1760; among the liberated prisoners was Jacob van Braam. Beating incalculable odds, which by now must have seemed second nature to him, Stobo even found his old sword, given to his lieutenant at Fort Necessity and plundered from the dead man at the Monongahela.

But then Stobo's luck, which had never thrown him a challenge he couldn't overcome, began to change. From Montreal and Quebec, his company joined a British fleet sailing against the French and Spanish islands of the Caribbean. He survived the taking of Martinique but was seriously wounded during the siege of Havana in 1762, his skull fractured by chunks of masonry falling from the castle walls. His recovery was long; the war was over before he rejoined his company in Quebec, then followed it to England in 1768.

But his was no longer the life of a celebrated hostage and self-made spy. Stobo was snarled in a complicated legal fight over land along Lake Champlain, and the lot of a peacetime captain of infantry afforded none of the zest to which he'd become accustomed. He was drinking heavily. On June 19, 1770, after a long bout with the bottle, he blew out his brains with his pistol. His sisters blamed it on his old head wounds: "He never afterwards got the better of their banefull effects."

Jacob van Braam retired to a farm in Wales before being called to active duty during the American Revolution. He served in Florida and the Caribbean, returned to his farm, and was never heard from again.

At almost the same time that Stobo was thanking the Virginia burgesses for his £1,000 bonus, Conrad Weiser was drawing up his will, a long dense document appropriate for a gentleman farmer and county judge: "In the name of God. Amen. I, Conrad Weiser . . . being of perfect health of body and of sound and disposing mind and memory (blessed be God for the same) . . ." In fact, Weiser was in anything but perfect health. He had just turned sixty-three, his constitution had been failing for years, and some days the "lameness" of which he'd often complained so crippled his right hand that he could not write. So he made his preparations.

To his "beloved wife Ann Eve," he left their home in Reading, an annuity of twenty pounds a year, "two of my best feather beds of her own choice; all my kitchen utensils and the sum of fifty pounds current money." His plantation below South Mountain, nearly nine hundred acres, he divided among four of his sons; his land "beyond the Kittochtany Hills, and all my grants or sights to lands lying beyond the same mountains," were to be split among all seven of his surviving children.

On July 12, 1760, Weiser left his house in Reading with Sir John St. Clair, still the quartermaster general of the army, although no longer threatening to set the indolent scum of the Pennsylvania backwoods to the sword. Weiser may have been planning to travel to the Ohio country with St. Clair—the frontier grapevine was buzzing, and Weiser had intimations of trouble brewing in the west. But they did not get past the Tulpehocken before the old interpreter was struck with severe pain in his gut, and he barely made it to his farm there. From the symptoms, it seems likely he was suffering from renal failure, and he died the next day.

. . .

Aug. 13, 1762

To the Hon'ble James Hamilton, Esqr., Lieutent. Governour of Pennsylvania, &c.

The Humble Petition of Jacob Hockstetler of Berks County.

Humbly Sheweth:

That about five Years ago your Petitioner with 2 Children were taken Prisoners, & his Wife and 2 other Children were kill'd by the Indians[,] that one of the said Children who is still Prisoner is named Joseph, is about 18 Years old, and Christian is about 16 years & a half old, That his House & Improvements were totally ruined & destroyed.

That your Petitioner understands that neither of his Children are brought down, but the Embassadour of King Kastateeloca, who has one of his Children, is now here.

That your Petitioner most humbly prays your Honour to inter-
pose in this Matter, that his Children may be restored to him, or
that he may be put into such a Methods as may be effectual for that
Purpose.

And your Petitioner will ever pray, &c.

his

JACOB   X   HOCKSTETER

mark

. . .

On August 18, 1762, James Hamilton, once again lieutenant governor
of Pennsylvania, tidied up some lingering business. "On the proprietaries' Commissioners producing & reading sundry Writings & papers, Teedyuscung was convinced of his Error, and acknowledged that
he had been mistaken with regard to the charge of Forgery," he said.
Actually, it was the payment of £400, a flurry of (probably doctored)
papers, and some backroom arm-twisting that "convinced" Teedyuscung to drop his claim that the Walking Purchase was a fraud.

In fact, Hamilton, the Pennsylvanians, and the Iroquois League
were in rare agreement on one pressing issue—the Susquehanna
Company of Connecticut, which was again making inroads into the
Wyoming Valley. Hamilton may have thought he'd pulled Teedyuscung's fangs, but the Yankees realized the old Lenape was a still formidable roadblock in the valley—and they may have seen an opportunity
to rid themselves of him. On April 16, 1763, as Teedyuscung slept off a
bottle of rum, he was burned to death when persons unknown torched
his and twenty other Indian cabins in the valley. While some blamed
the Iroquois, Teedyuscung's son, Captain Bull, had no doubt who the
perpetrators were, since the murder occurred just as the first wave of
Connecticut settlers arrived. That autumn, Captain Bull led a series
of bloody raids that destroyed the new colony, with the loss of twenty
lives.

While the ruins of Fort Duquesne were still smoldering in 1758, Shingas's brother Tamaqua gave Christian Frederick Post—and through him, the British—a pointed message dressed in silk. "Brother," the sachem said, "I would tell you, in a most soft, loving, and friendly manner, to go back over the mountain, and stay there." The French were gone, and the Ohio Indians wanted no whites west of the Allegheny hills.

But the British, having promised nothing but trading posts to the Natives in the upper Ohio, had no intention of leaving. By 1761, there were more than a hundred white households at Pittsburgh, as the growing community by Fort Pitt was known. Anger was again simmering among the western tribes, and then General Amherst, his budget slashed by London, made the worst possible move, cutting off the gifts by which Indian alliances had traditionally been maintained.

The trouble that Weiser had suspected erupted in May 1763 and became known as Pontiac's War, named for an Ottawa chief who was among its most visible leaders. But like the Yamasee War half a century earlier, the uprising involved many tribes with many rationales, and the revolt quickly spread across the old *pays d'en haut*. The Ohio tribes and western Iroquois joined the fight, again ravaging frontier settlements in Pennsylvania and Virginia. It was the panic of 1755 all over again. Or worse. Fort Venango (as Machault was now known), Fort LeBoeuf, and Fort Presque Isle all fell to the Seneca and their allies. Fighting raged as far north as Michilimackinac, which was taken by the Ojibwa and Sauk, and south into the buffalo prairies of the Illinois country.

Pontiac's War was an ugly affair, violent on all sides but especially notorious for one particular event. Fort Pitt was besieged immediately, crammed with frightened civilians (some of whom had been escorted there by Shingas and Tamaqua, who provided the garrison with intelligence). Conditions inside were vile, and smallpox was spreading. George Croghan's old business partner William Trent, who ran a trading post at Pitt, was commissioned a captain and put in charge of a militia company.

On June 23, two Delaware emissaries formally sought the fort's surrender, which was refused. The commander boasted that he could hold

Pitt "against all the Indians in the Woods." But he also knew that there
were other ways to fight than with a musket. "Out of our regard to [the
emissaries] we gave them two Blankets and an Handkerchief out of
the Small Pox Hospital," Trent noted in his diary. "I hope it will have
the desired effect." Ever the businessman, he submitted an invoice for
"Sundries got to Replace in kind those which were taken from people
in the Hospital to Convey the Smallpox to the Indians."

This was the first documented attempt at biological warfare, with
uncertain results. Smallpox was already spreading through the Indian
ranks, and it's unclear whether the infectious "gift" spurred the ensuing epidemic.

Not long afterward, Shingas disappears from the historical record;
no one knows what happened to him, but it's possible he died in the
smallpox outbreak. His brother Tamaqua survived and, having converted to Christianity, helped Post's Moravians establish missions in
the Ohio country before his death in 1771.

Pontiac's War ended in stages, with the British forcing the repatriation
of many white captives, some held for more than a decade. In the fall
of 1764 or the spring of 1765, Joseph Hochstetler was brought to Fort
Augusta, along the Susquehanna at Sunbury, Pennsylvania. He had
been well treated by his adoptive family, and the parting was hard. According to Hochstetler family tradition, in later years he "often visited
them and enjoyed himself hunting, fishing, running and jumping."

Some of the captives had been taken at such a young age that they
had lost their English or German and most recollection of their previous lives. The transition was wrenching, as soldiers forced them from
the only families they knew; many had to be physically carried away.
Parents, gathering at transfer points such as Carlisle, where the former
captives were being brought, searched the unfamiliar and often sullen
faces for any sign—the curve of a nose, a dimple, an old scar—of the
children they'd last seen years before.

One of those who searched the crowds at Carlisle in 1765 was a
widow named Leininger, whose family had been attacked in the first
massacre at Penn's Creek a decade earlier. Two of her daughters had
been carried off, but none of the young women looked like either of

Indians Delivering Up English Captives
Children were commonly reluctant to leave their adoptive Indian families, as shown
in this engraving based on a Benjamin West painting. Many colonials, even those kidnapped as teenagers or adults, found Native life difficult to leave. Many were returned
to their birth families only under great duress, and some remained with their adoptive
families for the rest of their lives. (Library of Congress LC-USZ62-103)

them. In desperation, she began singing an old German hymn her daughters had loved:

> *Allein, und doch nicht ganz allein*
> *bin ich in meiner Einsameit . . .*

> (Alone, yet not all alone am I,
> in my loneliness . . .)

One of the girls gasped and rushed forward, singing the hymn, and embraced the woman she now realized was her mother.

One day at the Hochstetler farm, as everyone was sitting down to the evening meal, an Indian peered nervously through the front door, muttered a greeting to the family, and sat on a stump in the yard. No one knew what he wanted; food, probably. When dinner was cleared, Jacob Hochstetler walked outside—with decidedly mixed feelings, one imagines. But then, in halting German, the Indian said, "My name is Christian Hochstetler."

As with many former captives, the decision to leave the upper Ohio and return to the settlements had been an agonizing one for Christian. He had grown close to his adoptive family, and at first Jacob could barely get his son to enter the house to eat. For the longest time, Christian wavered between staying and returning to the west. Then he met a girl, and his eventual marriage settled the matter, although he left the Amish congregation to become a baptizing Dunkard and, with two of his brothers, eventually moved west of the mountains to settle new land.

As Sir William Johnson's deputy for Indian affairs, George Croghan held the ultimate inside track on all manner of business ventures in the west, and he pursued them with gusto. Although his ten-year amnesty from prosecution for debt was overturned, it took five years to do so, and in the meantime he was involved (officially or otherwise—mostly otherwise) in everything from a monopoly on whiskey sales to the Ohio forts to farming and mineral exploration. He bought a four-acre estate in Philadelphia that he named Monckton Hall and set about adding to its splendor.

He also had one of the closest calls of his long, eventful life. In the summer of 1765, he was sent on a mission far down the Ohio to bring peace to the Illinois tribes. His party was attacked by eighty Kickapoo and Mascouten warriors, who killed two of Croghan's men and three accompanying Shawnees. The survivors were captured, including Croghan, groggy from a tomahawk blow to the head. (He later quipped, "A thick Scull is of Service on some Occasions.") Thanks to intervention from some of his old Miami customers from the days at Pickawillany, Croghan and his men were soon freed, and with Pontiac's help, he accomplished his assignment.

It was land speculation, though, that really set Croghan's heart aflutter. He held grants, patents, and claims of varying validity on hundreds of thousands of frontier acres, including forty thousand acres in New York he called Croghan's Forest and partial interest in a breathtaking two-and-a-half-million-acre grant that William Trent and others had obtained in the Ohio country from the Six Nations. Yet the thicket of debts, old and new, grew ever more tangled, and to satisfy them, he began selling off his holdings and even taking multiple mortgages on some parcels before selling them to unsuspecting buyers.

As the Revolutionary War began, Croghan found himself suspected of Tory tendencies, much as he'd been labeled a French sympathizer twenty years earlier—and with as little cause. He successfully fought a charge of treason, but Monckton Hall was toppled and burned by British troops quartered in Philadelphia. When independence finally came, it ruined him by invalidating most of his remaining land claims. Almost penniless, Croghan died on August 31, 1782, outside Philadelphia, expecting almost to the end to find a last jackpot in the west.

And what of Andrew Montour, whose very life was, in many ways, the embodiment of the eastern frontier? Because he wrote no letters and filed no reports of his own, Montour left far fewer traces than Croghan and Weiser. He drifts in and out of the historical record—interpreting for Croghan, guiding an unsuccessful midwinter attempt by Major Robert Rogers and his famed Rangers to reach Fort Michilimackinac during the waning days of the Seven Years' War. He tried to start a trading post at the Forks of the Susquehanna with one of his sons, and during Pontiac's War he led Indian counterattacks under Johnson's direction. In one such action in the winter of 1764, he took a war party of nearly 150 men, mostly Oneidas, on a sweep against Delaware towns in southwestern New York and the western Susquehanna Valley. Among their captives was Captain Bull, Teedyuscung's son.

When the Indian wars were over, though, there was little need for Montour's services as a captain or an interpreter. "I shall send Montour off for Fort Pitt in a day or Two," Johnson informed Croghan in 1766, "where you may dispose of him, or take him with you as you shall judge best. I have done my utmost for these 2 years past to keep him out of Debt, and he goes now pritty clear of ye. World."

Two years later, Montour did interpret for a grand council at Fort Stanwix, New York, in which the Iroquois made still more concessions of western territory. After that, Montour's name shows up in only a few scattered land dealings—for a tract at the mouth of Loyalsock Creek called Montour's Reserve, where his mother's old village of Ostuaga stood; for three hundred acres on the Ohio; for fifteen hundred acres on Penn's Creek. Whether he lived on any of this land, we don't know. In 1771, he was at Fort Pitt, where he may have been living with his niece Mary.

That's also where he died. "Captain Montour the Indian interpreter was killed in his own House the Day before Yesterday by a Seneca Indian who had been intertained by him at his House for some Days," Major Isaac Hamilton wrote from Fort Pitt on January 22, 1772. "He was buried this Day near the Fort." The assumption has always been that it was a drunken quarrel that went bad, but like so much of his life, Montour's death is an unanswered question.

By the time Montour died, much of the frontier he knew had moved on.

The years following Pontiac's War saw great leaps in westward settlement, even as official British policy sought to squelch its advance. In 1763, King George III issued a proclamation that forbade colonial occupation of British land beyond the Appalachians. His ministers drew a meandering line on the map tracing the limit of the eastern seaboard, setting aside "any lands beyond the Heads or Sources of any

Rivers that fall into the *Atlantick* Ocean" and declaring that "the several Nations or Tribes of *Indians,* with whom We are connected . . . should not be molested or disturbed in the Possession of such parts of our Dominions and Territories as, not having been ceded to, or purchased by Us, are reserved to them . . . as their Hunting Grounds."

The so-called Proclamation Line snaked down the White Mountains of New Hampshire, across the waist of Iroquoia, along the windswept crest of the Alleghenies and the Blue Ridge to the Smokies, and then back to bisect Georgia. Like many other treasures the tribes had once held dear, the Susquehanna—the Wyoming Valley where Teedyuscung had plaintively begged for "a Home for Ever," the rich hunting lands on the west branch for which so many had died—was entirely on the colonial side of the divide.

Although the Ohio Indians may have been initially reassured by the proclamation, the British military carved out exceptions for themselves, in the form of military bases such as Fort Pitt, and for British subjects on old French land in the buffalo prairies of Illinois. Even civilians such as George Washington, retired from the army for a decade, saw it for what it was: a minor inconvenience. Writing to his surveyor, Washington admitted, "I can never look upon that Proclamation in any other light (but this I say between ourselves) than as a temporary expedient to quiet the Minds of the Indians."

The few Indians who had remained in the now settled parts of Pennsylvania were increasingly isolated and endangered. The attempt to spread smallpox among the western Indians at Fort Pitt was an especially horrific example of an ever more pervasive view among white colonists and the British military that the Natives were little more than animals and deserving of as much consideration. From "back inhabitants" to army officers and politicians, there was less and less attempt to separate Native allies and neighbors from enemies. Increasingly, all Indians were seen as uniformly subhuman threats.

When the hysteria of Pontiac's War hit the polyglot settlements along the lower Susquehanna, for example, the result was massacres of those who posed the least threat and were least able to defend themselves. In a snowstorm on December 14, 1763, a mob of fifty-seven settlers from Paxton, near Harris's Ferry, butchered half a dozen Conestogas who farmed peacefully on one of the remaining reservation-like "manors" among a sea of white neighbors. Two weeks later, these so-called Paxton Boys—armed with long rifles, tomahawks, and scalping knives—slaughtered six adults and eight children who had been confined to the Lancaster workhouse, where they were supposed to be under government protection. It was essentially the end of what had once been the mighty Susquehannock, with only two elderly Conestogas escaping the violence.

Vigilantes, lynch mobs, and frontier justice ruled in much of the "back parts," which ranted against what many pioneer families saw as highhandedness by provincial and military leaders. James Smith, captured by Indians at Braddock's defeat and held for five years, led a backwoods uprising in the Alleghenies in the late 1760s. Known as the Black Boys, his men painted their faces with red ocher and charcoal. Claiming the precedence of local magistrates over the distant authority of Philadelphia or London, they waylaid trade caravans (burning the goods, shooting the packhorses, and flogging the wagoners), kidnapped military couriers, captured Fort Bedford, and besieged Fort Loudoun.

The glimpse of direct, imperial rule that the overbearing Braddock had attempted to impose in 1755 had left a bad taste in the colonies' mouth, and frontier rebellions were not its only expression. The widening chasm between the primacy of local rule, by which the colonials saw themselves as worthy equals under British law, and the supremacy of a Parliament that viewed them as subjugate to royal command only worsened through such escalating crises as the Stamp Act of 1765 and the hated Townshend Acts beginning two years later. The reaction was so poisonous that British troops occupied Boston in 1768, and many of the wrongs the colonies perceived in these acts—the forced quartering of British troops in times of peace, for example, and taxation without representation—would explode in revolution just a few years later. Perhaps Benjamin Franklin put it best in 1773, when he published a satirical article laying out British missteps titled "Rules by Which a Great Empire May Be Reduced to a Small One."

Andrew Montour's unmarked grave is long gone. Perhaps it is lost somewhere beneath the twenty lanes of freeway that now converge directly above the site of old Fort Pitt (where the fort's two reconstructed bastions peek out from beneath the overpasses), though more likely it lies to the east, farther up the peninsula, in what was once forest and pasture and is now the squared-off blocks of downtown Pittsburgh.

Robert Stobo, a suicide, could not be buried in a churchyard. "Having committed an act that was considered both immoral and illegal," his biographer concludes, "he presumably was buried according to custom in an unmarked grave." George Croghan's pauper's grave is likewise forgotten. Tanaghrisson, Scarouady, Teedyuscung, and Shingas and his brothers all lie anonymously and, if the bulldozers have missed them, one hopes quietly, someplace where the birds sing.

They sing over Conrad Weiser. Twenty-six neatly manicured acres of his Tulpehocken Valley farm are now a state historic site, including the small stone house he built in 1729. Visitors find a monument, erected in 1909, to "Conrad Weiser: Pioneer, Soldier, Diplomat, Magistrate" and a 1930s-era statue of Shikellamy, who with his upraised hand, long braids, and peace pipe looks as if he just stepped out of an old movie. At night in the autumn, you can see the lights of a high school football field two miles to the east, where the Conrad Weiser Scouts play.

Weiser's simple gray tombstone, its epitaph in German, stands on a little knoll near the farmhouse, beside the grave of his wife, Ann Eva. Around them are a number of small, plain slabs poking out of the grass. According to family tradition, they mark the resting places of Indian friends and acquaintances. Even in death, it seems, Tarachiawagon keeps a foot in both worlds.

Notes

Introduction

xii keekachtanemin: Sassoonan (September 7, 1732), in *Pennsylvania Archives*, 1st ser., 1:345.
tülpewihacki: Per Heckewelder and Du Ponceau 1834.
"back parts": By the 1720s, the terms "back parts" and "back settlements" were in common use to describe the Pennsylvania frontier; see *Minutes of the Provincial Council of Pennsylvania*, 3:103 and 3:282, for examples of early appearances.

xiv *"a fleshy woman"*: Harvey Hostetler and William Franklin Hochstetler, *Descendants of Jacob Hochstetler, the Immigrant of 1736* (Elgin, IL: Brethren Publishing, 1912), 30. My account of the attack and the Hochstetlers' captivity is drawn largely from this source, which is in turn based primarily on family tradition. The Hostetler/Hochstetler family website, www.hostetler.jacobhochstetler.com, was also an important source of information, including boundaries of the original farm.

xvi *"not between strangers"*: Jane T. Merritt, *At the Crossroads* (Chapel Hill: University of North Carolina Press, 2003), 8.

xix *"a convenient, if belabored"*: William Burton and Richard Lowenthal, "The First of the Mohegans," *American Ethnologist* 1 (November 1974): 592.

Chapter 1: Mawooshen

3 *along the margins:* No one is sure exactly what ethnic group lived along Muscongus Bay in Maine in the early seventeenth century, much less what dialect they spoke. They certainly used a form of what is now called eastern Abenaki, one branch of the immense Algic language family, of which Algonquian is a major part. Anthropologists assume that they belonged to the wider cultural group that included the Penobscot, Passamaquoddy, and Maliseet (and that, with the western Abenaki and Mi'kmaq, formed the Wapánahki confederacy). There is a contrarian view, however, that cultural affiliations were originally much different from those recorded by Europeans just after the devastating epidemics of the early 1600s and that seventeenth-century records should be viewed with skepticism. The Native words recorded by Rosier in 1605, according to linguist Frank T. Siebert, indicated that Indians of several related groups were involved in interactions with the English.
Although this reconstruction of the events of June 4, 1605, assumes that the Indians involved were a distinct Wapánahki subgroup known as Wawenock, whose home was described by early settlers as the mouth of the St. George River, the Algonquian transliterations used here are primarily Penobscot, based on Speck 1940. In fact, no one knows who or what the

Wawenock were, and most modern historians dismiss "Wawenock" as one of many confusing labels applied to the same coastal people by Europeans of several nationalities and languages. For a discussion of this issue, see Bourque 1989 and, arguing for the Wawenock as a distinct group, Brack 2008.

3 As several specialists have pointed out, trying to draw direct and unaltered connections between pre-contact groups and modern Indian cultures along the Northeast coast over-looks the tectonic changes forced by disease, warfare, and immigration/emigration in the seventeenth and eighteenth centuries. "Instead, the modern Penobscot, Passamaquoddy, and Maliseet tribes are better understood as products of ethnic realignments, shifts in residence, territorial loss, and the Indian policies of New England and New France" (Bourque 1989, 274). The language of the Wapánahki, like all native languages along the eastern frontier, was originally oral. French missionaries and, to a lesser extent, English colonists tried to represent it in the Latin alphabet, but with limited success, and later western Abenakis such as Joseph Laurent and his son Stephen from Quebec, as well as recent scholars of eastern Abenaki, such as Siebert, codified the languages. The orthography of eastern Abenaki uses a variety of con-sonants not found in English, such as k^w, pronounced "kww," while written western Abenaki was filtered through French, which lacks the English *w*. That consonant is represented as *8* in western Abenaki, while *ô* is pronounced as a nasal "ohn." In Penobscot, one dialect of eastern Abenaki that has survived (barely) to the present, the symbol *a* is used for an "uh" sound, like the *e* in "flower."

Two long, high-prowed: Although the coastal Wapánahki used large dugouts for ocean travel, the account by Rosier is quite specific that the canoes were "of the bark of a Birch tree, strengthened within with ribs and hoops of wood." Although I have used the modern western Abenaki word for birch-bark canoe, *wigwaol,* James Rosier recorded a different name, say-ing the Wapánahki called the English ship as "they call their owne boats, a Quiden" (James Rosier [1605], *A true relation of the most prosperous voyage made this present yeere 1605, by Cap-taine* George Waymouth, *in the Discovery of the land of* Virginia, in Quinn and Quinn 1983, 288).

woleskaolakw: Western Abenaki, per Wiseman 2001.

4 mata-we-leh: Penobscot, per Nicholar 2007, 165.

"doer of great magic": Penobscot, per Dana et al. 1989.

5 wαpánahki, *"the people of the east":* Most Native groups referred to themselves with terms that are usually translated as something like "the people" or "the real or original people," which in practice probably meant the most fundamental form of "us." These are almost never the names by which they became known to Europeans, however, and thus to modern history. Arriving in a new cultural territory, the French, English, or Spanish explorers would, in ef-fect, ask their Native guides—who were usually from a previously contacted group—"Who are these people?" Not surprisingly, the names given in response were often scatological or derogatory. *Maliseet,* for example, is probably a Mi'kmaq term meaning "broken talker" or "doesn't speak like us."

heavy, waterproof moccasins: The description of the clothing worn by Ktѳhanѳto and his companions, given by Rosier, corresponds with other eyewitness accounts of contact-period Wapánahki dress.

6 *A week earlier:* May 22, 1605.

"Woodeénit atók-hagen Kѳlóskαpe": Adapted from Leland 1884, 28.

it had not been as grim: Geophysicists have linked the eruption of the Peruvian volcano Huaynaputina in 1600 with profound cooling across much of the globe, including Europe, Asia, and South America. The effects likely would have been severe in North America as well, at least through 1603.

giwakwa *cannibals:* Most accounts of Algonquian legends, such as Leland 1884, use the Mi'kmaq term *chenoo,* but this is western Abenaki, per Day 1994.

7 mardarmeskunteag: The modern Damariscotta River; the Penobscot name is from Nicholar 2007. The Whaleback shell midden, though largely removed in the nineteenth century, was one of the largest shell heaps along the Atlantic coast. Thirty feet deep in places, it was more than a thousand years in the making.

8 k'chi-wump-toqueh: In modern Penobscot usage, this term refers to a Canada goose, but this may be a linguistic evolution. The nineteenth-century Penobscot historian Joseph Nicholar used it to refer to a swan (Nicholar 2007, 166n), and oral traditions of first contact often refer to European ships under sail as looking like swans.

pɑnáwɑhpskek: The origin of "Penobscot," meaning "where the rocks widen, open out, spread apart." The first English record of the name refers to a village near the modern town of Orland, at the head of Penobscot Bay.

Bashabes: Historians have wrangled for centuries over whether "Bashabes" is a name or a title. Snow 1976 argues that Samuel de Champlain and other early explorers referred to every other Indian sagamore they encountered by name rather than rank, making it unlikely that Bashabes was an exception. Pierre Biard, a Jesuit who traveled to Penobscot Bay in 1611, wrote that "the most prominent Sagamore was called Betsabés, a man of great discretion and prudence" (Thwaites 1896, 2:49). Rosier, however, specifically notes that the Indians they met applied the term to Captain George Waymouth, "whom they called our Bashabes" (Rosier, A true relation, 276). John Smith, in his account of his 1614 visit to Maine, indicates that Bashabes and the man he called "Dohannida" (Ktəhɑnəto) were brothers; this is the only reference to a blood tie (John Smith, A Description of New England (1616; repr., Boston: William Veazie, 1865).

wenooch: Derived from the word meaning "Who is that?" (Prins 1994, 114).

the first time anyone: Samuel de Champlain, meeting the Bashabes in 1604, said that he was told it was "the first time they ever had beheld Christians" (Bourque and Whitehead 1994, 135).

square-rigged sailing bark: The exact design of the Archangell is unknown, but it was likely similar to Bartholomew Gosnold's Godspeed and other ships of the day—a squat, two-masted square-rigger that would have been beastly cramped with a full complement of about thirty men aboard.

12 "the most fortunate": Rosier, A true relation, 259.

"above thirty great Cods": Ibid., 260.

"we [arrived] there": Ibid., 262.

13 pinnace: Rosier actually refers to the smaller craft as both a pinnace and a shallop, both small, one- or two-masted vessels that could be sailed or rowed.

"thirty very good": Rosier, A true relation, 264.

"but the crust": Ibid.

"fourteen shot and pikes": Ibid.

"very great egge shelles": Ibid., 262. These were most likely the eggs of great auks, a flightless, goose-size relative of the puffin, which was hunted to extinction in the North Atlantic by 1844, and whose remains have been found on other islands in Muscongus Bay.

the mouth of a river: Although there is no doubt that the Archangell's anchorage lay among the Georges Islands at the mouth of Muscongus Bay, in a natural harbor framed by Allen, Burnt, Benner, and Davis Islands, exactly which river Waymouth explored has been the subject of considerable historical discussion. Waymouth took precise soundings and mapped the area in detail, but this information (since lost) was left out of Rosier's published account, doubtless to throw off competitors. It has had the same effect on historians and geographers.

Two rivers debouch into the sea near the Archangell's anchorage—the St. George River, immediately to the north, and the very much larger Penobscot River, to the northeast. I find myself persuaded by the arguments of Morey 2005 and others that although the St. George lies closer to the presumed anchorage, it would have been harder to detect. The Penobscot, opening to its immense bay, is the more likely candidate and better fits Rosier's descriptions. A modern replica of the Waymouth light horseman, equipped with a sail and manned by enthusiastic if inexpert rowers, made it from Allen Island to the Penobscot and back in less than twenty-four hours, suggesting the trip was well within the capabilities of Waymouth and his trained seamen.

14 *"canoas"*: Rosier, *A true relation,* 267.
"the maine land": Ibid., 260.
"But when we shewed": Ibid., 267–68.
"We found them then": Ibid., 269.
"in this small time": Ibid., 271.
"knives, glasses, combes": Ibid., 273.

15 *"whereat they much marvelled"*: Ibid., 274.
shoggah: Rosier compiled a list of four hundred to five hundred Wapánahki words, but he did not publish it in his 1605 booklet *A true relation.* A truncated list of about one hundred words, possibly with transcription errors, was published by Samuel Purchas in his *Pilgrimes* in 1625 (*Hakluytus Posthumus, or Purchas His Pilgrimes* [Glasgow: James MacLehose & Sons, 1906], 18:358–59) and is all that survives. An annotated version appears in "James Rosier's List of Indian (Eastern Abenaki) Words, Recorded in Samuel Purchas, *Pilgrimes* (1625): A Preliminary Analysis by Philip L. Barbour," in Quinn and Quinn 1983, 481–93.
mawooshen: This word has been the subject of much scholarly debate and little resolution. It does not appear in Rosier's published list of Indian words, but *The Description of the Countrey of Mawooshen, Discovered by the English, in the Yeere 1602* (in Quinn and Quinn 1983, 470–76), published at the behest (and possibly by the authorship) of Ferdinando Gorges shortly after the voyage, makes it clear that the English, at least, believed this was the name given by the Indians to this stretch of coastal Maine, roughly from the Piscataqua River to Mount Desert Island. "Although we may always remain in the dark about the precise meaning of *mawooshen,* it probably refers not to a stretch of land but to the confederacy of allied villages under a regional grand chief known as the Bashabes" (Prins 1994, 100–11).
"behaved themselves": Prins 1994, 275.
Owen Griffin: Griffin is described by Rosier (Purchas, *Hakluytus Posthumus,* 18:344) as "one of them we were to leave in the Countrey, by their agreement with my Lord the Right Honorable Count Arundell (if it be thought needfull or convenient)," suggesting that Arundell and the other backers had considered the possibility of leaving some of the crew to overwinter. Quinn and Quinn 1983 (p. 277) suggest that Griffin, a Welshman, may have been chosen because of the contemporary belief that Indian languages were a form of Welsh.
"the men all together": Rosier, *A true relation,* 278.
"their Bashebas": Ibid., 280.
"their excellent Tabacco": Ibid.

16 *"utterly refused"*: Ibid., 282.
chipping his name: The inscription, "1605 Tho. King," remains today on the bedrock just above the tide line at Davis Point in Knox County, Maine.
"every one his bowe": Rosier, *A true relation,* 282.
"very trecherous": Ibid., 283.
"Wherefore . . . we determined": Ibid.

18 ania *for brother and* adesquide *for friend:* Trade pidgin, from the Basque *anaia* and *adeskide,* per Bakker, "The Language of the Coast Tribes," 1989.

18 tomhikon: Besides "tomahawk," the Algonquian family of languages bequeathed dozens of other words to English, filtered by unschooled ears and altered by awkward tongues: *asquutasquash* would become "squash"; *opasson* (from *op,* "white," and *assom,* "dog") would become "opossum"; and *aroughcun* would become "raccoon." *Cau'cau'as'u,* an adviser or counselor, would become the verb "caucus." Toboggan, chipmunk, moose, hickory, hominy, persimmon,

papoose, quahog, skunk, succotash, terrapin, powwow, wigwam, wampum, and many more Algonquian words would find a place in the language of their invaders.

19 *When Ktэ̀hɑnэto was a boy:* In 1579, the Portuguese mariner Simão Fernandes sailed along the coast of Maine, and in 1580 an Englishman named John Walker raided a camp, probably along the Penobscot, stealing three hundred dried hides.

a great canoe: In September 1604, Samuel de Champlain visited the "Bessabez," as he spelled the sagamore's name or title.

20 skrælings: Though usually translated as "wretches," the word may have a less pejorative meaning. Icelandic scholar Nancy Marie Brown notes that "*skræ-* seems to derive from *skrá,* 'dried skin,' particularly the kind of well-scraped and stretched dried skins on which books were written. This kind of skin is not so dissimilar to the leather used by Native Americans for clothing, making a Skraeling a person dressed in leather clothes, as opposed to the Vikings, who wore woven wool and linen" (Brown 2007, 190).

21 *By the summer of 1578:* Turgeon 1998 notes that such eyewitness accounts would have overlooked ships out on the Grand Banks and inshore boats fishing along thousands of miles of coastline, suggesting that the true number could have been considerably higher.

23 *"a Baske-shallop":* John Brereton, *A Briefe and True Relation of the Discoverie of the North Part of Virginia,* ed. Luther S. Livingston (1602; facs. ed., New York: Dodd, Mead, 1903), 4.

"They spoke divers": Gabriel Archer, *The Relation of Captain Gosnold's Voyage to the North Part of Virginia, 1602,* in *Collections of the Massachusetts Historical Society,* 3rd ser. (Boston: Charles C. Little & James Brown, 1838), 7:73.

Mathieu Da Costa: His name is also spelled Mathieu De Coste in French and Matheus de Cost in Dutch. He's referred to as *"een Swart genumd Matheu"* (a black named Matheu) in one source (A. J. B. Johnston, *Mathieu Da Costa and Early Canada: Possibilities and Probabilities* [Halifax, NS: Parks Canada, 2009], www.canadachannel.ca/HCO/index.php/Mathieu_Da_Costa_and_Early_Canada,_by_A._J._B._Johnston). Da Costa wasn't the only black interpreter along the Atlantic coast during this period; the Dutch also employed Jan Rodriguez in New Amsterdam along the lower Hudson in 1613–14.

"the language of the coast": Marc Lescarbot, quoted in Bakker, "The Language of the Coast Tribes," 1989, 124. Bakker notes, however, that none of the Native words recorded by Rosier are of Basque origin, suggesting that the Wapánahkis Waymouth's crew encountered were not themselves fluent in trade pidgin, which was more common to the north.

reported seeing a "chaloupe": Marc Lescarbot, *History of New France,* 2:309, quoted in Bourque and Whitehead 1994, 333.

"two French Shallops": Henry Hudson, quoted in Robert Juet, "The Third Voyage of Master Henry Hudson," in *Henry Hudson the Navigator,* ed. G. M. Asher (London: Hakluyt Society, 1860), 60.

24 *"thinking thereby":* Rosier, *A true relation,* 284.
"It was as much": Ibid.
"Thus we shipped": Ibid., 285.
"being a matter": Ibid., 284.
"being too suspitiously": Ibid.

25 *"he so much esteemed":* Ibid., 287.
"we had then no will": Ibid., 288.
"the most rich, beautifull": Ibid., 293.
"I will not prefer": Ibid., 292.
"having but little wood": Ibid., 293.
"parching hot": Ibid.
"notable high timber": Ibid., 294.
"many Furres": Ibid.
"shot both to deffend": Ibid., 295.

26 *"great store"*: Ibid., 296.
   "did so ravish": Ibid.
   "killed five savages": Champlain, quoted in Grant 1907, 77.
   whose name is also spelled: Linguist Frank Seibert, who worked for years with the Penobscot, has suggested the man's name was really Ktəhanəto. Tehánedo/Ktəhanəto was not the only one whose name was rendered in confusing variety. Skicowáros was spelled Scikaworrowse, Skettawaroes, or Skitwarres; Maneddo was spelled Maneduck, Maneday, and Manida; and Sassacomit was spelled Satacomoah, Assacomoit, Assecomet, or Assacumet, depending on the writer.
   "a Sagamo or Commander": Rosier, *A true relation*, 309.
   "Gentlemen": Ibid.
   "servant": Ibid.
   "First, although at the time": Ibid., 302.
27 *"a huge, heavy, ugly"*: John Aubrey (1813), quoted in Ibbetson 2004.
   "chiefe lord": The Description of the Countrey of Mawooshen, 472.
28 *"putrified"*: Edward Dodding (1577), quoted in Vaughan 2006, 8.
   "the New Found Island," "clothed in beast skins," and *"apparelled after Englishmen"*: "The Great Chronicle of London," ca. 1513, rendered in modern spelling from Myra Jehlen and Michael Warner, eds., *English Literatures of America, 1500–1800* (Oxford: Routledge, 1996), 42: "Thys yere alsoo were browgth unto the kyng iii men takyn in the Newe Found Ile land . . . clothid In beastis skynnys and ete Rawe Flesh and spak such sech that noo man cowde undyrstand them, and In theyr demeanure lyke to bruyt bestis . . . Of whych upon ii yeris passis afftyr I sawe ii of theym apparaylyd afftyr Inglysh men in Westmynstyr paleys, which at that tyme I cowde not dyscern From Inglysh men tyll I was lernyd what men they were, But as For spech I hard noon of theym uttyr oon word."
29 *"he is rich"*: Sometimes translated as "he who is wise." For other views on the translation, see Mook 1943 and Tooker 1905.
31 *"They wounded Assacomoit"*: John Stoneman, "The Voyage of M. Henry Challons Intended for the North Plantation of Virginia, 1606," in Samuel Purchas, *Hakluytus Posthumus, or Purchas His Pilgrimes* (Glasgow: James MacLehose & Sons, 1906), 19:288.
32 *"much welcomme"*: William Strachey (ca. 1610–12), "The Narrative of the North Virginia Voyage and Colony, 1607–1608," in Quinn and Quinn 1983, 404.
   "some Tobacco": Ibid., 411.
33 *"as if he would light"*: Ibid.
   "ignorant, timorous": Ferdinando Gorges, quoted in Alfred A. Cave, "Why Was the Sagadahoc Colony Abandoned?" *New England Quarterly* 68 (December 1995): 627.
   "The maine assistance": Smith, *A Description of New England*, 63–64.
   "indifferent good English": Thomas Dermer, quoted in George Parker Winship, *Sailors [sic] Narratives of Voyages Along the New England Coast, 1524–1624* (Boston: Houghton Mifflin, 1905), 255.
   "cut [off] his head": William Bradford, *Bradford's History "Of Plimoth Plantation": From the Original Manuscript* (Boston: Wright & Potter, 1901), 118.

Chapter 2: Before Contact

35 *There was only forest*: Adapted from "Koluskap and His People," in Dana et al. 1989, and from Leland 1884. Because Wapánahki traditions were oral, there is no way to know precisely what versions of the stories were told in the early contact period, and those later collected by non-Indian scholars doubtless suffer from varying degrees of cultural ignorance and misunderstanding, from the loss in earlier centuries of oral knowledge, and from a blending of once separate traditions assimilated from neighboring groups during centuries of cultural

upheaval. Even versions coming from Native sources underwent inevitable alteration, such as the retelling of Klose-kur-beh (Klooskape or Kəlóskɑpe) stories by the nineteenth-century Penobscot elder Joseph Nicholar, who freely mixed Christian and Penobscot elements to such a degree that even other Penobscot storytellers of the period remarked on it (Nicholar 2007).

37 *"some peopling the lands":* José de Acosta, *The Natural and Moral History of the Indies,* trans. E. Grimston (1604), ed. C. R. Markham (London: Hakluyt Society, 1880), 61.

39 *a wide, formidable deep-water:* Humans were thus able to breach one of the most dramatic ecological divisions in the world, separating the fauna of Asia to the west from that of Australasia to the east—tigers, orangutans, and rhinos on one side; tree kangaroos, cassowaries, and birds of paradise on the other. First recognized by Alfred Wallace, codiscoverer of the theory of evolution by natural selection, the division is known today as Wallace's Line.

42 *"supporting the hypothesis":* Walter A. Neves and Mark Hubbe, "Cranial Morphology of Early Americans from Lagoa Santa, Brazil: Implications for the Settlement of the New World," *Proceedings of the National Academy of Sciences* 102 (December 20, 2005): 18309.
   "I didn't want to get": Will Thomas, quoted in Anna King, "The Man Who Found Kennewick Man," *Tri-City (WA) Herald,* July 25, 2006.

44 *"biography is written":* Hugh Berryman, quoted in Melanthia Mitchell, "Kennewick Man's Biography Is Written in His Bones," *Tri-City (WA) Herald,* July 10, 2005.
   He was probably: Kennewick Man's age at death was initially estimated to be forty-five to fifty-five, but later study revised the age down to about thirty-eight.
   "It's no wonder": Richard Jantz, quoted in Anna King, "Meet Kennewick Man," *Tri-City (WA) Herald,* February 25, 2006.

45 *"They were from Iberia":* Dennis Stanford, quoted in Mike Lee, "New World Habitation Tricky Issue," *Tri-City (WA) Herald,* December 26, 1999.
   "A Solutrean hunter": Bradley and Stanford 2004, 470.

46 *most of the East was dominated:* Paleoecologists have been able to reconstruct a remarkably detailed picture of past plant communities, thanks to the nearly indestructible nature of wind-blown pollen grains, which form annual layers in the bottom sediments of ponds, lakes, and bogs, and which allow scientists to infer the local climate and plant cover.

48 *haunted the barren:* The relatively scarce Paleolithic sites excavated in the Northeast are consistently placed above and along valleys, lakeshores, and other landforms that would channel big-game migration, most likely caribou. In the Magalloway Valley in northwestern Maine, a number of Paleo-Indian sites have been uncovered, all hugging the east side of the valley, downwind of game moving through the river corridor. Taking modern caribou-hunting cultures as a model, it is likely that Paleo-Indians covered tremendous distances in their annual movements. The territories of caribou-hunting Inuit, for example, often encompassed more than eight thousand square miles.

49 *good stone was as critical:* Paleo-Indians had extensive trade networks (or themselves traveled very widely), since a single site in Maine or Vermont may have tools made from chert or other types of stone from locations as far away as Pennsylvania and Nova Scotia.

56 *"Peoples about whom":* Hann 2003, 3.

58 *gave coffee a run:* "One of the puzzles about black drink," writes anthropologist Charles Hudson, "is why we do not drink it today" (Hudson 2004, 7). Colonial Spaniards did, calling it *té del indio* or *chocolate del indio.* Hudson notes that it also enjoyed favor in England as "South Sea Tea." Yaupon tea's demise as a popular drink, he contends, came from its abundance and social stigma. It was so common and so cheap that it was a standard drink of the poor, and after the Civil War, those higher on the economic ladder avoided it.

60 skarò·rəʔ, *"hemp gatherers":* Etymology of the name is from J. N. B. Hewitt, "Tuscarora," in *Handbook of American Indians North of Mexico,* ed. F. W. Hodge (Washington, DC: Bureau of American Ethnology, 1910), 2:842.

61 *the Lenape, who occupied:* The word *lenape,* meaning "the real people" or "the original people," is from the Unami dialect and appears to have been used to describe members of one's own band or village. *Lenni lenape,* or "real Lenape," came into use in the eighteenth century and "is rejected as redundant by twentieth-century speakers" (Goddard, "Delaware," 1978, 236). Becker 1989 departs from the conventional assumption that the river cultures of the Hudson and Delaware were closely allied. He argues that "Lenape" should apply only to those living along the lower Delaware River, later called Unami, and that the Munsee and what he terms "Jerseys" (Unalachtigo), who migrated from southern New Jersey, were culturally distinct.

haudenosaunee: The name "Iroquois" was first recorded by Samuel de Champlain in 1603 and is presumably of Algonquian origin, although its source and meaning are unknown. There has been a growing tendency to use Haudenosaunee instead of Iroquois when referring to the Five (later Six) Nations.

65 *"spread out as far":* Garcilaso de la Vega, quoted in Denevan, "The Pristine Myth," 1992, 375.

66 *not by an absence of paths:* "When an Indian lost his way in the woods, as he sometimes did, it was as likely as not because there were *too many* tracks and he had taken the wrong one" (Wallace 1965, 8).

"need not be seriously": James Mooney, quoted in Douglas H. Ubelaker, "The Sources and Methodology for Mooney's Estimates of North American Indian Populations," in *The Native Population of the Americas in 1492,* ed. William M. Denevan (Madison: University of Wisconsin Press, 1992), 252.

67 *precolonial population of North America:* Denevan, "The Pristine Myth," 1992, suggests a New World population of 43 million to 65 million, with 3.8 million in North America, although he notes that some specialists have estimated as many as 80 million people in South and Central America alone. "In any event, a population of 40–80 million is sufficient to dispel any notion of 'empty lands.' Moreover, the native impact on the landscape of 1492 reflected not only the population then but the cumulative effects of a growing population over the previous 15,000 years or more," he observed (p. 370).

"open and free": Giovanni da Verrazano, quoted in Gloria L. Main, *Peoples of Spacious Land* (Cambridge, MA: Harvard University Press, 2001), 243.

"could be penetrated": Ibid.

"but little wood": James Rosier (1605), *A true relation of the most prosperous voyage made this present yeere 1605, by Captaine George Waymouth, in the Discovery of the land of* Virginia, in Quinn and Quinn 1983, 293.

"the trees growing there": William Strachey (ca. 1610–12), "The Narrative of the North Virginia Voyage and Colony, 1607–1608," in Quinn and Quinn 1983, 407.

"by reason of their burning": John Smith (1607–09), in *Capt. John Smith: Works, 1608–1631,* ed. Edward Arber (Westminster, UK: Archibald Constable, 1895), 363.

68 *"growne over":* Leonard Calvert to Richard Letchford, May 30, 1634, in *The Calvert Papers,* no. 2, Fund Publication no. 34 (Baltimore: Maryland Historical Society, 1894), 20.

"high timbred Oakes": John Brereton (1602), "A Briefe and True Relation of the Discoverie of the North Part of Virginia," in Quinn and Quinn 1983, 151.

"in the thickest part": Ibid.

"medowes very large": Ibid., 152.

"adorn'd with spacious Medows": Daniel Denton, *A Brief Description of New-York Formerly Called New-Netherlands* (1670; repr., Cleveland: Burrow Brothers, 1902), 54.

"where is grass as high": Ibid., 55.

"The Indian is by nature": Hu Maxwell, "The Use and Abuse of Forests by the Virginia Indians," *William and Mary Quarterly* 19 (October 1910): 86.

69 *"The tribes were burning":* Ibid., 103.

Given that Indians: In fairness, not all ecologists accept the idea that Indian burning was so ubiquitous that it altered the forests on a landscape level. See especially Foster and Motzkin

2003 for evidence suggesting that southern New England had very limited open-country habitat. Also see Russell 1983, who questions the extent of intentionally set fires in the Northeast.

70 *"In the year 1153"*: Antonio Galvano, *Discoveries of the World,* ed. C. R. Drinkwater Bethune (1563; repr. London: Hakluyt Society, 1862), 56n1. Bethune's translation, used here, is at odds with the more smoothly edited 1601 English edition, which he also reprinted: "In the yeere 1153 . . . it is written, there came to Lubec, a citie in Germanie, one canoe with certain Indians, like unto a large barge: which seemed to have come from the coast of Baccalaos, which standeth in the same latitude that Germanie doth."

Stories about Indian merchants: One of the more recent books to explore the possibility of Native voyages to Europe is Jack D. Forbes, *The American Discovery of Europe* (Chicago: University of Chicago Press, 2007), which traces the origins of the Indian voyage to Germany cited here. Although some anthropologists have praised Forbes for shining a light on the neglected subject of pre-Columbian contact between the hemispheres, others have excoriated him for sloppy scholarship and faulty logic.

"Florida" was a nebulous concept: Turgeon 1998 also notes that French maps and fishermen still referred to the entire North American coast as "Florida" as late as the mid-sixteenth century.

"The Germans greatly": Galvano, *Discoveries of the World,* 56n1.

"nevertheless it is quite": Ibid.

Chapter 3: Stumbling onto a Frontier

73 *"remote heathen":* Quoted in Rory Rapple, "Gilbert, Sir Humphrey (1537–1583)," in *Oxford Dictionary of National Biography,* ed. H. C. G. Matthew and Brian Harrison (Oxford: Oxford University Press, 2004). www.oxforddnb.com/view/article/10690.

"never heard nor read": Walter Raleigh to Francis Walsingham, 1581, quoted in W. Noel Sainsbury, ed., *Calendar of State Papers* (London: Royal Stationery Office, 1893), vi.

"a demi-Moor": Quoted in Ronald Pollitt, "John Hawkins's Troublesome Voyages: Merchants, Bureaucrats, and the Origin of the Slave Trade," *Journal of British Studies* 12 (May 1973): 26.

74 *"The Judith . . . forsooke us":* John Hawkins (1569), "The Unfortunate Voyage Made with the Jesus, the Minion and Foure Other Shippes," in Hakluyt 1589, 556.

"Hides were thought": Ibid.

"Our people being forced": Ibid.

"It would have caused": Miles Phillips (1582), "A Discourse Written by One Miles Phillips Englishman," in Hakluyt 1589, 566–67.

75 *"about 140 leages west":* David Ingram (1583), "The Relation of David Ingram of Barking, in the Countie of Essex," in Hakluyt 1589, 557.

76 *"great pearles":* Ibid., 557. Here Ingram may have been only stretching the truth, not breaking it entirely. The rivers of the southern Appalachians, especially in the Tennessee Valley, have the richest diversity of freshwater mussels in the world. They are the source of abundant (if smallish and low-grade) pearls, and a traveler through the region would undoubtedly have encountered them.

"the maine sea": Ibid., 562.

"which thing especially proveth": Ibid.

"this Ingram": Ibid., 559.

"the examinate": Ibid., 560.

"whitenes of their skins": Ibid., 557.

"yet alive, and marryed": Phillips, "Discourse Written," 568.

"buffes": Ingram, "Relation of David Ingram," 560.

"a bird called": Ibid.

"most excellent, fertile": Ibid., 559.

77   *"how easily they may":* Ibid., 562.
     "pillers of massie silver": Ibid., 558.
     "verie beautifull to beholde": Ibid., 560.
     "the reward for lying": Samuel Purchas, quoted in Bromber 2002, 124.
     One possibility: Charlton Ogburn, "The Longest Walk: David Ingram's Amazing Journey,"
     American Heritage, April/May 1979, www.americanheritage.com/articles/magazine/ah/1979/
     3/1979_3_4.shtml.
78   *"penguins":* Ingram, "Relation of David Ingram," 560.
     "the shape and bigness": Ibid.
     "the end product": Quinn 1974, 217.
     "Wee are as neere": Edward Hayes (1589), "The Voyages and Discoveries of Sir Humphrey
     Gilbert," in Hakluyt 1589, 695.
     In 1497: Quinn 1974 (pp. 93–94) notes that William Adams's *Pettie Chronicle* from 1625 incor-
     rectly gives the date as 1496, a mistake frequently repeated in other sources.
79   *"Understanding by reason":* John Cabot, quoted in "A Discourse of Sebastian Cabot," in Hak-
     luyt 1589, 512.
     "I began therefore": Ibid.
     "dispairing to find": Ibid.
80   *"hugged him":* Giovanni da Verrazano (1524), in George Parker Winship, *Sailors Narratives of
     Voyages Along the New England Coast, 1524–1624* (Boston: Houghton Mifflin, 1905), 9.
     "We took the little boy": Ibid., 10.
     "the men at our departure": Ibid., 22.
     "the land of the Great Khan": Ginés Navarro (ca. 1527), quoted in Reginald Poole, ed., *The
     English Historical Review* (London: Longmans, Green, 1905), 20:118. Navarro was a Spanish
     sailor who encountered an English ship, likely Rut's bound for home, in Santo Domingo.
     "eleven saile of Normans": Ibid., 20:122.
81   *"giving them to understand":* Jacques Cartier (1545), in Cook 1993, 70.
82   *"Hardly had he fallen":* Garcilaso de la Vega, *The Florida of the Inca,* ed. and trans. John G.
     Varner and Jeannette J. Varner (Austin: University of Texas Press, 1951), 257.
83   *ancient and incontrovertible:* The best overviews of the subject are Axtell and Sturtevant 1980
     and Richard J. Chacon and David H. Dye, eds., *The Taking and Displaying of Human Body
     Parts as Trophies by Amerindians* (New York: Springer-Verlag, 2007).
     Entire heads were often: Ron Williamson, "'Otinontsiskiaj ondaon' ('The House of Cut-Off
     Heads'): The History and Archaeology of Northern Iroquoian Trophy Taking," in *The Taking
     and Displaying of Human Body Parts as Trophies by Amerindians,* ed. Richard J. Chacon and
     David H. Dye (New York: Springer-Verlag, 2007).
     "scalping [became] as Anglo-American": James Axtell, "Scalping: The Ethnohistory of a Moral
     Question," in *The European and the Indian* (New York: Oxford University Press, 1981), 232.
     "the skins of five": Quoted and translated in Axtell and Sturtevant 1980, 456. This translation,
     which the authors believe is more accurate, differs from the traditional translation of "five
     scalps" in H. P. Biggar, *The Voyages of Jacques Cartier* (Ottawa: King's Printer, 1924), 177.
     repartimiento: The *repartimiento* system partially replaced an earlier approach, the *enco-
     mienda,* the abuses of which horrified many Spaniards, most notably Bartolomé de Las Casas,
     a priest who went from being one of the *encomenderos* who benefited from the system to its
     fiercest critic.
84   *"We set ourselves":* Jean Ribault (1562), "Jean Ribault's First Voyage to Florida," in *Historical
     Collections of Louisiana and Florida,* ed. B. F. French (New York: J. Sabin & Sons, 1869), 184.
     "fragrant odor": Ibid.
     "during his lifetime served": J. Leitch Wright Jr., "Sixteenth Century English-Spanish Rivalry
     in La Florida," *Florida Historical Quarterly* 38 (April 1960): 267.

84 *"the rakehell"*: William Cecil (1583), quoted in Richard Simpson, ed., *The School of Shakespeare* (London: Chatto & Windus, 1878), 1:41. Simpson's volume remains the best source on Stucley's checkered career.

85 *captured and clapped in irons:* That was hardly the end of Thomas Stucley, who had more lives than a cat. Tried but acquitted, he became embroiled in Irish revolt plots, was imprisoned again, escaped to Spain, and in 1578 led, with a papal blessing, what was to be a Catholic invasion force aimed at Ireland. Instead, he became involved in a misbegotten Portuguese offensive against the Moors in North Africa, where he fought despite his belief that the odds were against them. He was killed in the Battle of Alcazar. The life and death of "lusty Stucley" provided ample fodder for Elizabethan poets and playwrights.

86 *"some of our companie"*: Thomas Hariot, *A Briefe and True Report of the New Found Land of Virginia*, introduction by Luther S. Livingston (1588; facs. ed., New York: Dodd, Mead, 1903), F2r.

87 *"Some would likewise"*: Ibid., F2v.
"plant [the colony]": "Instructions Given by Way of Advice" (1606), quoted in Haile 1998, 21.
"poore Gentlemen": John Smith, *The Generall Historie of Virginia, New England & the Summer Isles* (1624; repr., Glasgow: James MacLehose & Sons, 1907), 1:196.
"Ten good workemen": Ibid.
"have great care": "Instructions Given," 21.

89 *"halfe rotten"*: Smith, *Generall Historie*, 1:180. Smith wrote of himself in the third person.

90 *"snatched the King"*: Ibid., 1:166.
"neare dead with feare": Ibid.
"You promised to fraught": Ibid., 167.
"Men may thinke": Ibid., 1:170–71.

92 *"Some doubt I have"*: Ibid., 1:157.
"the poorer sort": Ibid., 1:204.
"did kill his wife": Ibid., 1:204–5.
"Now whether shee": Ibid., 1:205. Smith had, of course, already left Virginia in October 1610, before the worst of "the starving time," but his *Generall Historie of Virginia* continues through 1623, based on reports and conversations with those who remained in the colony.

93 *so now had the English:* Ironically, a second, smaller English fort thirty miles downriver at the mouth of the James, near present-day Hampton, Virginia, was essentially ignored by the Indians, and its thirty-man garrison survived the winter in relative comfort, eating hogs and crabs. It, too, was to be abandoned during the planned pullout.
"Of all the foure": Smith, *Generall Historie*, 2:14.
"high craggy clifty": Ibid., 2:24.
"well inhabited with many people": Ibid., 2:23.
"well inhabited with a goodly": Ibid., 2:14.
"country of the Massachusits": Ibid., 2:25.
"Paradice of all those": Ibid.
"not much inferior": Ibid. Emphasis added.

94 *as few as 900,000:* After Smithsonian ethnographer James Mooney made his first estimate of pre-contact Indian populations in 1928, his figure of 1.5 million in North America was revised down to 900,000 by anthropologist Alfred L. Kroeber, who calculated that the entire hemisphere was home to 8.4 million Natives, mostly in Mesoamerica and the Andean highlands.

95 *numbered between 300,000:* Las Casas and other early Spanish accounts estimated the pre-contact population at up to 4 million, while some modern scholars believe the island held many more. Livi-Bacci 2003 concluded the total was closer to 300,000 or 400,000 based on agricultural carrying capacity, Taino social structure, and other clues.

- 95 *the greatest loss:* W. George Lovell, "'Heavy Shadows and Black Night': Disease and Depopulation in Colonial Spanish America," *Annals of the Association of American Geographers* 82 (September 1992): 426–43.

97 *the Black Death:* Not all specialists have agreed that the Black Death was a rat-borne human plague caused by the bacterium *Yersinia pestis,* with some arguing that it was a different disease, possibly a viral hemorrhagic fever. In 2010, however, DNA analysis of the remains of Black Death victims confirmed that *Y. pestis* was the cause. See Haensch, Bianucci, and Signoli et al., 2010.

98 *no evidence that they spread:* See Dobyns 1993 for a contrary view suggesting that sixteenth-century epidemics ranged as far north as Canada.

    "rare and strange accident": Thomas Hariot, "A Briefe and True Report of the New Found Land of Virginia," in Richard Hakluyt, *The Principal Navigations, Voyages, Traffiques and Discoveries of the English Nation* (Edinburgh: E. & G. Goldsmid, 1889), 13:352.

    "Within a few dayes": Ibid.

    "This happened": Ibid.

    "all that space": Ibid., 8:381.

    "subtile devise": Ibid., 8:380.

99 *the relative scarcity:* Not everyone is convinced that epidemics were rare or unknown north of Mexico in the sixteenth century. Dobyns 1993 and others have postulated that Native-to-Native transmission fanned smallpox outbreaks that reached as far north as Canada long before Europeans were present to chronicle the event, although evidence for this is scant.

    along the St. Lawrence River: The culture affinities of the groups Jacques Cartier and other French explorers encountered in the upper St. Lawrence Valley from 1535 to 1543 are unclear and controversial. Usually referred to as St. Lawrence (or Laurentian) Iroquois, because of perceived similarities in language, they apparently shared no common heritage with the Five Nations Iroquois to the south. By the time Champlain arrived in 1603, they had vanished, perhaps scattered to the south and west.

    "the strangest sort": "A Short and Briefe Narration (Cartier's Second Voyage 1535–1536)," in Henry S. Burrage, ed., *Early English and French Voyages, Chiefly from Hakluyt, 1534–1608* (New York: Charles Scribner's Sons, 1906), 73.

    "ripped"—crudely dissected—"to see": Ibid., 74.

100 *"God of his infinite":* Ibid.

    bark and needles: Cartier assumed he'd found a magic potion and dubbed the plant "tree of life." Although no one knows what species was used, it is commonly assumed to have been eastern white cedar, to this day known as arborvitae (literally, "tree of life"). It also could have been white pine, eastern hemlock, balsam fir, or spruce, all of which contain vitamin C and can cure scurvy.

    "That Cartier was blind": Cook 1993, xxxvii.

    "their bloudy deede": Thomas Morton, quoted in Charles F. Adams Jr., ed., *The New English Canaan of Thomas Morton* (Boston: Prince Society, 1883), 131.

    "The Salvages": Ibid., 132.

101 *"Contrary wise":* Ibid., 132. Morton, a sometime lawyer, avid poet, hard-drinking Anglican, and Indian trader, was a thorn in the side of both the Pilgrims in Plymouth and the Puritans in Boston, who later burned his home in outrage at his wanton ways.

    "By this meanes": Ibid., 133–34.

102 *"some antient Plantations":* Thomas Dermer (1619), "To His Worshipfull Friend M. Samuel Purchas, Preacher of the Word, at the Church a Little Within Ludgate, London," in George Parker Winship, *Sailors Narratives of Voyages Along the New England Coast, 1524–1624* (Boston: Houghton Mifflin, 1905), 251.

    "good ground": Christopher Levett (1624), quoted in Charles H. Levermore, ed., *Forerunners*

and Competitors of the Pilgrims and Puritans (New York: New England Society of Brooklyn, 1912), 2:612.

102 *"Betsabés":* Pierre Biard (1611), in Thwaites 1896, 2:49.

"are astonished": Biard (1616), in Thwaites 1897, 3:105.

"a remnant remaines": Dermer, "To His Worshipfull Friend," 251.

103 *"Our ship had [not]":* Peter Lindeström, *Geographia Americae with an Account of the Delaware Indians Based on Surveys and Notes Made in 1654–1656,* trans. Amandas Johnson (Philadelphia: Swedish Colonial Society, 1925), 127–28.

"they were ten times": Adriaen van der Donck (1653), "Description of the New Netherlands," in *Collections of the New-York Historical Society,* ed. George Folsom, 2nd ser. (New York: H. Ludwig, 1841), 1:183.

"In my grandfather's time": Quoted in John W. Barber and Henry Howe, *Historical Collections of New Jersey, Past and Present* (New Haven, CT: John W. Barber, 1868), 23.

104 *"a sort of measles":* Paul Le Jeune (1634), in Thwaites 1897, 7:221.

"The people of the countries": Ibid.

106 *"attribute to the Faith":* Le Jeune (1656–57), in Thwaites 1899, 43:291.

"they assert": Ibid.

108 *"Generally covetous":* Smith, *Generall Historie,* 1:62.

məntu': For a discussion of the difficulty in translating the concept of *məntu'* (manitou) in Algonquian languages and the way supernatural associations clung to European objects well into the early twentieth century, see Frank G. Speck, *Naskapi: The Savage Hunters of Labrador* (Norman: University of Oklahoma Press, 1935).

109 *Martin's Hundred:* The remains of Martin's Hundred were discovered in 1971, and the excavation, which revealed in astonishing detail the life and death of a frontier community, is detailed in Hume 1982.

110 *"settled in such":* Smith, *Generall Historie,* 1:279.

"The poore weake Salvages": Ibid., 1:279–80.

"he held the peace": Ibid., 1:280.

"They came unarmed": Edward Waterhouse (1622), quoted in Edward D. Neill, *History of the Virginia Company of London* (Albany, NY: Joel Munsell, 1869), 318.

"Yea in some places": Ibid., 318–19.

111 *"Antonio the Negro":* Anthony Johnson's fascinating story is told in great detail in Breen and Innes 2005, *"Myne Owne Ground,"* the groundbreaking exploration of the rise and subsequent disappearance of free black communities on the Eastern Shore of Virginia.

"proper, civill": Smith, *Generall Historie,* 1:291.

both sides grew fatigued: Modern historians suspect that one invisible force working against the Indians may have been the accidental introduction by the colonists of malaria, which drains a person's—and even an entire culture's—energy.

112 *probably in his late nineties:* Nothing definitive is known about Opechancanough's birth, but English contemporaries guessed he was nearly one hundred at his death. See Fausz, "Opechancanough."

Chapter 4: *"Why Should You Be So Furious?"*

115 *"In consideration of Gods":* Francis Wyatt, quoted in Hume 1982.

"the archetype": Fausz, "Opechancanough."

116 *"ceded, transported":* "Affidavit of four men from the *Key of Calmar,* 1638," in *Narratives of Early Pennsylvania, West New Jersey and Delaware, 1630–1707,* ed. Albert Cook Myers (New York: Barnes & Noble, 1912), 88.

"paid and fully": Ibid.

117 *"What warrant have we":* John Winthrop (1629), "Generall Considerations for the Plantation
in New England, with an Answer to Several Objections," in *A Collection of Original Papers
Relative to the History of the Colony of Massachusetts* (Boston: Thomas & John Fleet, 1769), 30.
"That which is common": Ibid.
all such land deals: For a look at the relative honesty of Maine's Indian deeds, see Baker 1989.

118 *"Nothing in this Deed":* Warumbo deed granting lands to Richard Wharton, 1684, www.
mainememory.net/media/pdf/20270.pdf.
"made a generall alarm": Leonard Calvert to Richard Letchford, May 30, 1634, in *The Calvert
Papers,* no. 2, Fund Publication no. 34 (Baltimore: Maryland Historical Society, 1894), 20.

119 *the powerful Massawomeck:* The identity and homeland of this enigmatic tribe have been
debated since at least Thomas Jefferson's day (he thought they were the ancestors of the
Iroquois). For a detailed analysis, see Pendergast 1991.
Indian corn: The English used the word "corn" to describe a variety of cereal crops, including
barley, wheat, and oats. When I refer to "corn" in this book, I mean Indian corn, or maize.
held on to their identity: This overview of Piscataway history and resilience is drawn largely
from James H. Merrell, "Cultural Continuity Among the Piscataway Indians of Colonial
Maryland," *William and Mary Quarterly* 36 (October 1979): 548–70.

120 *the astringent Puritanism:* Pilgrims were "Puritans with a vengeance" (Philbrick 2006, 4), unwilling to remain within the existing Church of England. They had taken the then illegal
step of withdrawing from the church to Holland, from which they emigrated to New England.

121 *"there mayne end":* Winthrop, "Generall Considerations," 31.
"used too unfitt": Ibid.

122 "wall of separation": Roger Williams warned of "a gap in the hedge or wall of separation
between the garden of the church and the wilderness of the world" ("Mr. Cotton's Letter
Lately Printed, Examined and Answered, 1644," quoted in Perry Miller, *Roger Williams: His
Contribution to the American Tradition* [1953; repr., New York: Atheneum, 1970], 98).

123 *"the source and the mother":* Peter Stuyvesant (1660), quoted in *Documents Relative to the Colonial History of New York,* ed. B. Fernow (Albany, NY: Weed, Parson, 1883), 16:470.

124 *"Without wampum":* Ibid.

127 *"close together as they can":* Philip Vincent (1638), quoted in *History of the Pequot War,* ed.
Charles Orr (Cleveland: Helman-Taylor, 1897), 105.
the Mohegan into tributary: The exact relationship between the Mohegan and Pequot has
been debated for years. The issue is complicated by a lack of recognition for how complex
political identity could be among Native groups and villages. "We cannot understand Mohegan-Pequot relations by applying a model that presumes long-term 'tribal' stability and
continuity or that imagines hard-and-fast national boundaries dividing Pequots from Mohegans, Niantics, and Narragansetts. The loyalties and alliances of the Algonquian villages of
this region were rather fluid. The barriers between 'tribes' were far more permeable than those
that separated European states" (Cave 1996, 66).

128 *"clapt up":* Edward Winslow (1634), quoted in William Bradford, *Bradford's History of Plymouth Plantation, 1606–1646,* ed. William T. Davis (New York: Charles Scribner's Sons, 1920),
301–2.
killed a group of Narragansetts: Or Indians tributary to the Narragansett. Their identity is
unclear, but there was nothing subtle about Narragansett outrage.

129 *"might fight seven years":* John Underhill (1638), "Newes from America," in *History of the Pequot
War,* ed. Charles Orr (Cleveland: Helman-Taylor, 1897), 82.

131 *"a just ass":* John Stone, quoted in *Records of the Court of Assistants of the Colony of the Massachusetts Bay, 1630–1692,* ed. John Noble (Boston: County of Suffolk, 1904), 2:35. Here the insult
is politely altered to "just as*."
witnessed by other Natives: Accounts vary; the Indians involved may have been Pequots or

western Niantics, a tributary group living on the east side of the river, or both. The Pequot, however, consistently took responsibility for Stone's death, and claims by some historians, notably Francis Jennings (*The Invasion of America* [New York: W. W. Norton, 1976]), suggesting that others were responsible do not appear to hold water. The accounts of exactly what transpired that night are also contradictory. We have three conflicting Pequot narratives related secondhand by the Puritans. For these narratives, see John Mason, "A Brief History of the Pequot War," in *History of the Pequot War*, ed. Charles Orr (Cleveland: Helman-Taylor, 1897), 17; Underhill, "Newes from America," in *History of the Pequot War*, 57–58; William Bradford, in *Bradford's History*, 311–12; and John Winthrop, in *Winthrop's Journal*, ed. James K. Hosmer (New York: Charles Scribner's Sons, 1908), 139, and quoted in *Bradford's History*, 333.

"*Captain Stone, having drunk*": Underhill, "Newes from America," in *History of the Pequot War*, 57.

132 "*as friends to trade*": Winthrop, in *Winthrop's Journal*, 140.

133 "*If you make war*": Lion Gardener, "Leift Lion Gardener His Relation of the Pequot Warres," in *Collections of the Massachusetts Historical Society*, 3rd ser. (Cambridge, MA: E. W. Metcalf, 1833), 3:138. Although later editions of Gardener's account altered his name to "Gardiner," a form the family had adopted, the original spelling is used here.

"*Capt. Hunger*": Ibid.

"*do their utmost*": Ibid., 3:139.

"*indiscreet speaches*": Jonathan Brewster to John Winthrop Jr., June 18, 1636, in *Collections of the Massachusetts Historical Society*, ed. Robert C. Winthrop, 4th ser. (Boston: John Wilson & Son, 1865), 7:68.

"*whom I have found*": Ibid.

"*If they should not*": Henry Vane and John Winthrop to John Winthrop Jr., July 4, 1636, in *Collections of the Massachusetts Historical Society*, 3rd ser., 3:131.

134 *riding at anchor*: In some accounts, the ship was under sail but badly handled.

"*Being well acquainted*": John Winthrop, July 20, 1636, in *Winthrop's Journal*, 184.

135 "*one thousand fathoms*": Winthrop, August 25, 1636, in *Winthrop's Journal*, 186.

"*a strange mixture*": Edward Eggleston, *The Beginners of a Nation* (New York: D. Appleton, 1897), 201.

"*For, said I*": Gardener, "Leift Lion Gardener," 140.

136 "*Seeing you will go*": Ibid.

"*What cheere, Englishmen*": Underhill, "Newes from America," in *American History Told by Contemporaries*, ed. Albert B. Hart (New York: Macmillan, 1917), 1:439.

"*They not thinking*": Ibid.

"*What English man*": Ibid.

"*Seeing that they did*": Ibid., 1:59–60.

"*a pretty quantity*": Gardener, "Leift Lion Gardener," 142.

137 "*slew all they found*": Underhill, "Newes from America," in *American History Told by Contemporaries*, 60.

"*defending ourselves*": Gardener, "Leift Lion Gardener," 144.

"*said the arrows*": Ibid.

"*Have you fought*": Ibid., 132.

"*We said we knew*": Ibid.

"*We are Pequits*": Ibid.

138 "*having bene undertaken*": Massachusetts General Court (April 18, 1637), quoted in *History of the Pequot War*, xii.

"*were not fitted*": Gardener, "Leift Lion Gardener," 149.

139 "*commend our Condition*": Mason, "Brief History," 22.

"*very great Captains*": Ibid., 24.

"*upon peril of their Lives*": Ibid.

139 *"almost impregnable"*: Ibid., 26.

140 *"by which advantage"*: Roger Williams to John Winthrop, ca. August–October 1636, in *Collections of the Massachusetts Historical Society*, 3rd ser. (Boston: Charles C. Little & James Brown, 1846), 1:160.

"That it would be": Ibid., 1:161.

Shortly before daybreak: Mason ("Brief History," 26) suggests that the exhausted men overslept and dawn was already breaking when they awoke, but Underhill ("Newes from America," in *History of the Pequot War*, 78) says they awoke at one o'clock in the morning.

led by Uncas: There are two firsthand narratives of the attack—Mason's and Underhill's, with a third by Philip Vincent who, though he apparently drew on eyewitness accounts, was not at the melee himself. They differ in details, sometimes contradicting one another, as with the march time just noted and the participation of the Indian allies. Underhill's account, published a year after the war, mentions only in passing that they had three hundred Indians "without side our soldiers in a ring battalia" (Underhill, "Newes from America," in *History of the Pequot War*, 78), while Vincent credits the Narragansetts for covering the English retreat and says that the Mohegans "behaved themselves stoutly" (Vincent, quoted in *History of the Pequot War*, 106). Only Mason's "Brief History" gives Uncas a great deal of personal credit—but as scholars have pointed out, by the time Mason wrote his version of events years later, he considered Uncas a friend and ally and had been involved in several land deals with him.

"Owanux!": Mason, "Brief History," 27.

141 *"concluded to destroy"*: Ibid., 28.

"Seeing the fort": Underhill, "Newes from America," in *History of the Pequot War*, 80.

142 *"The fires of both"*: Ibid.

"Many were burnt": Ibid., 80–81.

"did swiftly over-run": Mason, "Brief History," 29.

"Some of them climbing": Ibid.

"God was above them": Ibid., 30. From Psalm 21 (Authorized [King James] Version), "Thou shalt make them as a fiery oven in the time of thine anger: the Lord shall swallow them up, and the fire shall devour them."

143 *"cried Mach it"*: Underhill, "Newes from America," in *History of the Pequot War*, 84.

"It was a fearfull": Bradford, in *Bradford's History*, 339.

"victory to the glory": Gardener, "Leift Lion Gardener," 137.

"It may be demanded": Underhill, "Newes from America," in *History of the Pequot War*, 81.

144 *blood-red hearts:* See William S. Simmons, "The Mystic Voice," in Hauptman and Wherry 1990, 149–52.

About three hundred: My discussion of Pequot slaves is largely informed by Fickes 2000. Some authors have referred to the Pequot captives as indentured servants, but English indenture contracts—while hardly a model of enlightened labor practices—set a specified period of servitude, were governed by a written contract, and required at least theoretical consent by the prospective servant. The Pequot women and children were, in all ways that mattered, slaves.

shipped to Bermuda: Although colonial records document only this single shipload of Pequot slaves sent to the islands after the 1637 war (which, according to those records, missed Bermuda and wound up in the Caribbean), the Mashantucket Pequot maintain that Pequots and members of other New England tribes were sold into slavery on Bermuda, where an annual "reconnection ceremony" is now held on St. David Island. See Carocci 2009.

"women and maid children": John Winthrop to Thomas Prence, July 28, 1637, quoted in *Bradford's History*, 429.

"Wee have heard": Hugh Peter to John Winthrop, 1637, in *Collections of the Massachusetts Historical Society*, 4th ser., 6:95.

145 *"Since the Most High":* Ibid., 6:225.

 "which they will": Ibid.

 "an incomparable way": Ibid.

 "shall no more be called": "Articles Between Ye Inglish in Connecticut and the Indian Sachems," in *Collections of the Rhode-Island Historical Society* (Providence: Marshall, Brown, 1835), 3:177. This is the only copy of the treaty that still exists.

 "suffer them to live": Ibid.

 "shall not presently": Ibid.

146 *"an Island of mine owne":* Lion Gardener, quoted in Ellsworth S. Grant, "To the Manor Born," *American Heritage,* October 1975, www.americanheritage.com/articles/magazine/ah/1975/6/1975_6_8.shtml.

Chapter 5: Between Two Fires

149 *"aged thirty years":* Robert Roules, quoted in Axtell 1974, 650.

 "What for you": Ibid.

150 *"If Englishmen shoot":* Ibid.

 Metacom, son of: There are various spellings of his name, including Metacomet. Although it may seem condescending to use an English name for the Wampanoag leader, it is unclear what the sachem actually called himself. In 1660, Metacom's older brother Wamsutta asked the court at Plymouth to give them both English names, and the court complied: Alexander and Philip. The names Philip used likely changed with the circumstances and audience. He is unlikely ever to have referred to himself as "King Philip"—that label was applied by the English, perhaps in equal parts sarcasm and recognition of his leadership—but he clearly considered himself an equal of the English king.

151 *"all the fishing ilandes":* Francis Card, quoted in Baxter 1900, 6:150.

 "[H]e doth make": Ibid., 6:150–51.

152 *"a dull and heavy-moulded":* William Hubbard, *A General History of New England,* 2nd ed. (Boston: Charles C. Little, 1848), 635.

 "either Skill or Courage": Ibid.

 "a greasy shirt": Roules, quoted in Axtell 1974, 650.

153 *"even to our clothes":* Ibid., 652.

 "that they might be": Ibid.

 "We found them": Ibid.

 "suffered neither constable": Ibid.

154 *"Christians in this Land":* Cotton Mather, *Magnalia Christi Americana* (Hartford: Silas Andrus & Son, 1820), 2:275.

 "the Immoralities": Gyles 1736, ii.

 "wer in everi thing": John Easton, "A Relacion of the Indyan Warr, by Mr. Easton, of Roade Island, 1675," in Lincoln 1913, 10.

155 *"Some of our captives":* Samuel Penhallow, "The History of the Wars of New-England with the Eastern Indians" (1726; repr., Cincinnati: Dodge & Harpel, 1859), 25.

 "gave satisfaction": Ibid.

158 *Pokanoket sachem:* The Pokanoket, which by the 1670s lived near Mount Hope, Rhode Island, were a group of affiliated villages within the larger Wampanoag culture. As Jill Lepore points out in *The Name of War* (New York: Alfred A. Knopf, 1998), Indian society in the late seventeenth century was in immense flux, and it is difficult to know with any certainty who was part of which group. Like her, I generally use "Wampanoag" to refer to the Pokanoket, Pocasset, Sakonnet, and other closely related groups in southeastern Massachusetts.

 "saied they had bine": Easton, "Relacion of the Indyan Warr," 10.

159 *a jury of twelve:* Supplementing white juries with Indian jurors, especially Christian Indians, was not uncommon, especially in capital cases. For one thing, Native participation ensured the presence of interpreters.

hanged on June 8: One of the three actually survived the hanging when the rope broke.

"The next day": Easton, "Relacion of the Indyan Warr," 12.

"So the war begun": Ibid.

"In this time": Daniel Witherell to John Winthrop Jr., June 30, 1675, in *Collections of the Massachusetts Historical Society,* 3rd ser. (Boston: Charles C. Little & James Brown, 1846), 10:119.

160 *Nipmucs in Massachusetts:* The names, I realize, can be confusing. By the mid-seventeenth century, the Nipmuc occupied what is now central Massachusetts, while the Massachusett were a closely related Wampanoag division originally occupying the coast south of Massachusetts Bay.

"Yet after all": "The Examination and Relation of James Quannapaquait," January 24, 1675, in Salisbury 1997, 120.

"leaving & loosing": Ibid.

161 *"If they came":* Ibid.

"alleaging that Philip": Ibid., 115.

163 *"not deliver up":* William J. Miller, *King Philip and the Wampanoags of Rhode Island,* 2nd ed. (Providence, RI: Sidney S. Sides, 1885), 139.

164 *"young and old":* James Oliver (December 26, 1675), quoted in George M. Bodge, ed., *Soldiers in King Philip's War,* 3rd ed. (Boston: Rockwell & Churchill Press, 1906), 174.

165 *"They within":* "A Farther Brief and True Narration of the Late Wars Risen in New-England," reprinted in *A Farther Brief and True Narration of the Great Swamp Fight in the Narragansett Country, December 19, 1675* (Providence, RI: Society of Colonial Wars, 1912), 10.

"We no sooner": Ibid.

"We have slain": Ibid.

"300 fighting men": Oliver, quoted in Bodge, *Soldiers in King Philip's War,* 174–75.

166 *"a Renagado English Man":* Nathaniel Saltonstall (1676), "The Present State of New-England with Regard to the Indian War," in Lincoln 1913, 67.

"a sad wretch": Oliver, quoted in Bodge, *Soldiers in King Philip's War,* 175.

"shot 20 times": Ibid. In *The Name of War,* Lepore notes that Tift's story and the accusations against him are a morass of inconsistencies, ranging from Oliver's contention that he was an active fighter, to Nathaniel Saltonstall's contention that Tift didn't fire a shot, to suggestions that he'd married a Wampanoag woman. Lepore argues that Tift's gruesome fate had as much to do with his gender—a male turncoat was especially dangerous—and English fears about colonists going native as with his actions.

"Counsel of War": Saltonstall, "Present State," 67.

"destroyed with exquisite": Nathaniel Saltonstall (1676), "A New and Further Narrative of the State of New-England," in Lincoln 1913, 97.

"It is computed": Ibid.

"all young Serpents": Hubbard 1803, 216.

"13 Squawes": John Hull (1676), quoted in Bodge, *Soldiers in King Philip's War,* 480.

167 *"Why shall wee":* "The Examination and Relation of James Quannapaquait," 125.

"Let us live": Ibid.

"Know by this paper": Quoted in Daniel Gookin, "An Historical Account of Doings and Sufferings of the Christian Indians in New England," in *Transactions and Collections of the American Antiquarian Society* (Cambridge, MA: Harvard University Press, 1836), 2:494.

"You must consider": Ibid.

"Always brutal": Lepore, *The Name of War,* xiii.

"a landscape of ashes": Ibid.

168 *Menameset:* The village was also known as Menemesseg or Wenimesset.

168 *"Thou hast been":* "The Examination and Relation of James Quannapaquait," 122.
     "None shall wrong thee": Ibid.
169 *"Hee will not beeleve":* Ibid., 117.
     Kattananit, however, chose: Kattananit left Menameset the night before the planned attack, having made arrangements with other pro-English Indians in the village to meet him secretly some weeks later and deliver his children. But some English authorities, not trusting Kattananit's motives, prevented the rendezvous, and he didn't get his children back until after the war ended.
     "If God please": "The Examination and Relation of James Quannapaquait," 117.
170 *"Some in our house":* Mary Rowlandson (1682), quoted in Salisbury 1997, 69. Only four pages of the original edition of Rowlandson's book survive. This annotated version uses the "second Addition," one of four published that year, and is less modernized than later editions.
     "Lord, let me dy": Ibid.
     "[stood] amazed": Ibid.
     "but when it came": Ibid., 70.
171 *"I shall dy":* Ibid., 73.
172 *Wetamo (or Weetamoo):* Although the sachem's name is usually spelled Weetamoo, Strong 1999 argues that Wetamo is probably closer to the original Algonquian, which linguist Frank Speck (1928) translated as "lodge-keeper." Rowlandson's own attempt to render Wetamo's name as "Wettimore" would seem to support this spelling.
     "full of wormes": Rowlandson, quoted in Salisbury 1997, 106.
     "I could not but": Ibid., 82.
     "a bait": Ibid.
173 *"I was utterly":* Ibid., 83.
     "For the Indian Sagamores": Quoted in Nourse 1884, 110.
     "if you have any": Ibid.
     "Though they were": Rowlandson, quoted in Salisbury 1997, 97.
174 *"for these sundry":* John Easton (1676), "Record of a Court Martial Held at Newport, R.I. in August, 1676," in *Narrative of the Causes Which Led to Philip's Indian War, of 1675 and 1676,* ed. Franklin B. Hough (Albany, NY: J. Munsell, 1858), 175–76.
     "to show such gentlemen": Thomas Church, *A History of King Philip's War,* ed. Samuel G. Drake (Boston: Howe & Norton, 1825), 101. This account was first published in 1716 by the son of Captain Benjamin Church, who was Alderman's commander at the time of Philip's death. Noting that Philip's head went for the same bounty as those of other enemy Indians, the younger Church groused, "Methinks this was scanty reward and poor encouragement."
175 *"made Publick":* The cover of the second edition of Rowlandson's book contained the inscription, "Written by Her own Hand for Her private Use, and now made Publick at the earnest Desire of some Friends, and for the Benefit of the Afflicted."
     first American woman: The first American woman author was poet Anne Bradstreet, whose *The Tenth Muse Lately Sprung Up in America, By a Gentlewoman of Those Parts* was published in 1650 in England.
     "we were not ready": Rowlandson, quoted in Salisbury 1997, 80.
     "but be acknowledged": Ibid., 23.
     "that Hazardous service": Nourse 1884, 114.
     "in sattisfaction": Ibid.
     "that abominable Indian": Increase Mather, *Early History of New England,* ed. Samuel G. Drake (Boston: Samuel G. Drake, 1864), 257.
176 *"James the printer":* Rowlandson, quoted in Salisbury 1997, 103.
     "A Revolter": "A True Account of the most Considerable Occurences that have hapned in the Warre between the English and the Indians in New-England" (1676), in *The Old Indian Chronicle,* ed. Samuel G. Drake (Boston: Antiquarian Institute, 1836), 130.

177 *"Honoured Mother":* Thaddeus Clark to his mother, June 14, 1676, Maine Historical Society, coll. 420, box 7/32, www.mainememory.net/media/pdf/22345.pdf.

retaking the remaining vessels: Having lost their boats to the Abenakis, some Marbleheaders were even more incensed to see them retaken by colonial officials from New York, which controlled parts of the Maine coast. "Some of those owners have said they had rather the Indyans had kept their Ketches, then that they should come into the hands of New-Yorke Governm[ent]" (Anthony Brockholls [January 7, 1677], quoted in Franklin B. Hough, *Papers Relating to Pemaquid and Adjacent Areas in the Present State of Maine, Known as Cornwall County, When Under the Colony of New-York* [Albany, NY: Weed, Parsons, 1856], 12).

179 *"that none of said Indians":* George Wadleigh, *Notable Events in the History of Dover, New Hampshire* (Dover, NH: Tufts College Press, 1913), 81.

cochecho or qouchecho: Beginning in the late nineteenth century, the spelling changed to Cocheco, apparently due to a clerical error (Thompson 1892, 44).

William Hathorne: Sometimes spelled Hawthorne.

"seize all Indians": Chandler E. Potter, *The History of Manchester, Formerly Derryfield, in New Hampshire* (Manchester, NH: C. E. Potter, 1856), 77.

"Sham battles": Mock battles were also sometimes used for political or religious purposes, such as one especially elaborate drama staged by the Spanish in Mexico in 1539. Reenacting the siege of Jerusalem, complete with Saint James on a white stallion helping to rout the Moors, this mock battle was a way of reinforcing the legitimacy of Catholic Spanish rule over the conquered Aztecs.

180 *"taken like so many":* Drake 1910, 17.

Chapter 6: *"Our Enimies Are Exceedeing Cruell"*

183 *"My Moose-skin Coat":* Gyles 1736, 16.
184 *"wonderfully reviv'd":* Ibid.
     "but again my Spirits": Ibid.
     "The Indians cry'd out": Ibid.
     "which were as void": Ibid.
     "The Indians said": Ibid., 17.
     "whole like a Shoe": Ibid.
185 *"I follow'd them":* Ibid.
     "the Common Abuses": John Pike, quoted in Cotton Mather, "Decennium Luctuosum, or The Remarkables of a Long War with Indian Savages" (1688), in Lincoln 1913, 186.
     "at which [the Indians]": Ibid.
187 *"plant their pumpkins":* George Wadleigh, *Notable Events in the History of Dover, New Hampshire* (Dover, NH: Tufts College Press, 1913), 91.
     "dull weather": Gyles 1736, 6n.
188 *"Brother Waldron":* Ibid.
189 *"I cross out":* R. H. Howard and Henry E. Crocker, eds., *A History of New England* (Boston: Crocker, 1881), 2:217.
     "Our Enimies are": Quoted in Baxter 1897, 5:2.
190 *"one watch coate":* Quoted in Baxter 1907, 9:29.
     "Tow hwendred musquitt": Ibid.
     "Briches": Ibid., 9:30.
     "the air of that place": Gyles 1736, i.
     "considerable Difficulties": Ibid., ii.
191 *several hundred Abenakis:* Estimates on the size of the war party vary from about one hundred (Parkman 1911) to three hundred to four hundred in seventy canoes, according to English captives and contemporary accounts (Paltsits 1905).

191 *led by Madockawando:* As with many of the Indians—both individuals and groups—living on the Maritime Peninsula in the seventeenth century, Madockawando's precise ethnicity is debatable. Though usually described as a Penobscot, he is also referred to by French sources as a Maliseet (Bourque 1989). Moxus is usually described as a Canibas from the Kennebec River. That tribal name is now considered synonymous with Penobscot and Kennebec.

193 *"The Baron of Saint Castiens": Lahontan's New Voyages to North-America,* ed. Ruben Thwaites (Chicago: A. C. McClurg, 1905), 1:328.

"The Yelling": Gyles 1736, 2.

"glittering in his Hand": Ibid.

194 *"offered me no abuse":* Ibid.

"strange Indians": Ibid., 3. Moxus, who lived along the mid-coast of Maine and had many encounters with the English settlers, may have been genuinely upset that Maliseets in the war party had wounded Thomas Gyles, although John Gyles gives no hint of a special relationship between Moxus and his father. Human nature being what it is, Moxus may simply have been passing the responsibility to the Maliseets, the "strange Indians" in the war party, for what would have been a bloody attack regardless. It's unlikely, after all, that Moxus and other local Wapánahkis would have participated in the raid but not attacked the English personally.

"My Father replied": Ibid.

"He parted with": Ibid.

"She asked me": Ibid., 4.

195 *"I . . . dare not":* Ibid., 5.

"I had rather follow": Ibid.

Mrs. Williams was killed: Age and gender played a remarkably consistent role in determining which captives were likely to survive after a raid. Of the 112 English settlers taken from Deerfield, for example, 20 died on the march north, notes John Demos in *The Unredeemed Captive* (1994), his highly readable account of the attack and its aftermath. Three out of 4 infants taken were killed, while 31 of the 35 children ages three to twelve survived. All 21 teenagers made it to Canada, while 10 of the 26 women died (including one killed after suffering a miscarriage). Only 4 of the 26 adult men died, 1 by murder, 1 from wounds suffered in the attack, and 2 from starvation. "The moral of all this seems clear. If you are living at Deerfield in 1704, and if capture is your fate, it's better by far to be a grown man than a woman, and best of all to be a teenager" (Demos 1994, 39).

"scattered sheep": John Williams, *The Redeemed Captive Returning to Zion* (1795; repr., Springfield, MA: H. R. Hunting, 1908), 61.

197 *her Mohawk name:* Williams's Mohawk name is spelled in various ways in Kahnawake mission records, including 8aongot and Gonoangote; see Demos 1994.

she remained firmly: Williams was married to a Mohawk by 1713, when she was sixteen. Her husband, François Xavier Arosen, was probably a decade older. A New York trader who met Eunice that year, in hopes of negotiating her release, reported that she had a fierce desire to remain in Kahnawake. In the trader's opinion, the marriage was a love match. In fact, the couple remained together until his death in 1765. She died twenty years later, at age eighty-nine. That year, one of their grandsons began regular visits to Deerfield, and in 1800 two of his sons—Eunice's great-grandchildren—started studies at a nearby school.

"At Home I had": Gyles 1736, 5.

"where the other Squaws": Ibid.

"laid down a Pledge": Ibid.

"A Captive among": Ibid.

198 *"By and by":* Ibid., 6.

wolastoqiyik, *"the people of wolastokuk":* These and other Maliseet words are from Francis and Leavitt 2008. In Maliseet-Passamaquoddy, the name for the cultural group as a whole is *waponahkiyik,* and the name for their homeland is *waponahkik.* I have retained the Penobscot form used earlier in the book to avoid confusion.

198 *Meductic, the chief village:* Like many place names of Native origin, there are multiple spellings of the name of this fort, including Medoctec. Gyles rendered it as "Medoctack-Fort." Although it was the primary Maliseet settlement in the 1680s, it was abandoned less than a century later.

"*I comforted myself*": Gyles 1736, 6.

199 "*seiz'd by each Hand*": Ibid., 7.

"*'till one would think*": Ibid.

"*'till the Blood*": Ibid.

"*and if he cry out*": Ibid.

"*got his head*": John Smith, *The Generall Historie of Virginia, New-England & the Summer Isles* (1624; repr., Glasgow: James MacLehose & Sons, 1907), 1:101.

"*I look'd on one*": Gyles 1736, 7.

200 "*I met with no Abuse*": Ibid., 8.

"*It is not really accurate*": Haviland and Power 1994, 162.

201 "*shaking their Hands*": Gyles 1736, 10.

"*GOD wonderfully provides*": Ibid., 9.

202 "*This was occasioned*": Ibid., 12.

"*a stout, ill-natur'd*": Ibid., 19.

"*I told him*": Ibid.

203 "*I . . . follow'd him*": Ibid.

"*And I don't remember*": Ibid.

"*expecting every Minute*": Ibid., 18.

"*I went to him*": Ibid.

204 "*a most ambitious*": Ibid., 14.

"*puffing & blowing*": Ibid.

"*At every turn*": Ibid., 15.

"*the heat of the Weather*": Ibid.

"*not brought over*": Ibid., 21.

"*Were it not*": Ibid., 20.

205 "*He said, that I*": Ibid., 36.

"*Malicious persons*": Ibid., 32.

"*I . . . went into the Woods*": Ibid., 33.

"*Monsieur Decbouffour*": Ibid.

206 "*I had not liv'd*": Ibid., 34.

"*Little English*": Ibid., 37.

"*'Little English,' we have*": Ibid.

207 "*that he would do*": Ibid., 39.

"*I made him*": Ibid., 34.

208 "*a gross Mistake*": Ibid., 10n.

"*part horror story*": Vaughan and Clark 1981, 94.

the story of Hannah Duston: The family's last name is spelled several ways in historical documents, including Durtson, Dunstan, Dustin, and Dustan. "Duston" is how the name appears in Haverhill town records from the time, although the only extant signature by Thomas is "Dustan."

209 "*go back into the woods*": George Wingate Chase, *The History of Haverhill, Massachusetts* (Haverhill, MA: G. W. Chase, 1861), 86.

"*his cruel and excessive*": Ibid., 122.

"*that wicked house*": Quoted in *Records and Files of the Quarterly Courts of Essex County, Massachusetts, 1683–1686,* ed. George F. Dow (Salem, MA: Essex Institute, 1975), 9:603.

209 *heavy winter clothes:* In part because concealing a winter pregnancy was fairly easy, April and May were the most common months for neonaticide—the murder of a newborn—by desperate mothers bearing children no one knew about. During the moralistic hysteria whipped up around the Salem witch trials, however, it appears that innocent mothers of stillborn infants may have been prosecuted for murder, since even today it can be difficult to determine whether a newborn was killed or died naturally.

210 *"I was always":* Cotton Mather (1693), *Warnings from the Dead, or Solemn Admonitions unto All People,* quoted in Melvin Yazawa, ed., *The Diary and Life of Samuel Sewall* (Boston: Bedford Books, 1998), 39.

"Disobedience to my Parents": Ibid.

"one of the greatest": Cotton Mather, "Diary of Cotton Mather, 1681–1708," in *Collections of the Massachusetts Historical Society,* 7th ser. (Boston: Plimpton Press, 1911), 7:165.

"greedily bought up": Ibid.

On March 15: The date has been given variously as March 5 (most notably by Mather), 10, 15, and 16. Mirick 1832 consulted original town records, which he said listed March 15 as the date.

211 *Jonathan Haynes and his:* As was often the case, the captors split the hostages among several bands. Haynes and his sixteen-year-old son, Thomas, were taken to Maine, where they escaped. The other three children were sold to the French in Canada. Mary, age nineteen, was ransomed for a hundred pounds of tobacco, but her two brothers Jonathan and Joseph, ages twelve and seven when captured, remained in Canada for the rest of their lives, marrying into French families and becoming successful farmers. Their father, having once escaped Indian raiders, was not as lucky the second time. He was tomahawked in the winter of 1698 during an ambush in which another of his sons was captured.

"Tawney": Cotton Mather (1692), "A Brand Pluck'd out of the Burning," in *Narratives of the Witchcraft Cases, 1648–1706,* ed. Lincoln Burr (New York: Charles Scribner's Sons, 1914), 16:261.

"Indian Sagamores": Ibid., 282.

213 *he couldn't risk:* According to some accounts, including Cotton Mather's retelling based on his interview of Hannah, and Mirick's 1832 history of Haverhill, Thomas fired on their pursuers. Family tradition contends that he held off the attackers by bluff alone, which, given the limitations of reloading a musket, seems likely.

214 *about forty-five miles:* Mather's account states that Duston and the others were marched "a long travel of an hundred and fifty miles, more or less, within a few days ensuing" (*Magnalia Christi Americana,* with an introduction and occasional notes by Thomas Robbins [1702; repr., Hartford: Silas Andrus & Son, 1853], 2:635). Doubtless it felt that far, and most retellings since then have repeated the distance, but a straight-line measurement from Haverhill to Penacook, New Hampshire, is forty-five miles.

the end of March: Mather's account places these events on April 30, while most other accounts say March 30, which, given the distance covered, seems more likely. Because of the difference between the Julian and Gregorian calendars, the modern date would be around April 8, when there was a new moon.

mol8demak: Laurent 1884 rendered this as "Morôdemak," meaning "deep river" (p. 215).

nikn tekw ok: Like most Indian place names, this is descriptive and not necessarily specific to one location. Laurent 1884 rendered it as *nikattegw* or *nikôntegw,* meaning "first branch, or outrunning stream or river . . . a channel" (p. 216).

iglizmôniskwak: Spelling per Day 1994, 236.

215 *"Not one of them":* Mary Rowlandson (1682), quoted in Salisbury 1997, 107.

rape of Indian women: Perhaps because of different underlying cultural norms, sexual assaults against white female captives on the Great Plains were more common in the eighteenth and nineteenth centuries, though by no means the rule. There was no one, universal "Indian" behavior, just as there was no universal "white" or "European" behavior.

215 *"sore and heavy blows"*: Abigail Willy [Willey] (1683), quoted in *Collections of the New-Hampshire Historical Society*, ed. Nathaniel Bouton (Concord, NH: McFarland & Jenks, 1866), 8:147.

216 *"pray'd the English way"*: Quoted in Sewall 1927, 453. Sewall spoke with Duston, as did Mather, during her visit to Boston shortly after her escape.
"If your God shall": Mather, *Magnalia Christi Americana*, 2:635.
"Strike 'em dere": This quote, passed down through Duston family tradition, does not appear in either Mather's or Sewall's account.
"heartened the nurse": Mather, *Magnalia Christi Americana*, 2:635.

217 *"[lay] Hold"*: Mather, "Decennium Luctuosum," 209.
"he heard him Cry": Ibid.
an additional twenty acres: Based on "the old and generally received tradition" in Haverhill that the land "was *bought with scalp money*" (Chase, *History of Haverhill*, 195).

219 *"[cut] off the scalps"*: Mather, *Magnalia Christi Americana*, 2:636.
"raging tigress": Nathaniel Hawthorne (1836), "The Duston Family," quoted in Lucy Maddox, *Removals* (New York: Oxford University Press, 1991), 117.
"a bloody old hag": Ibid.
"nothing ever seen": Ibid.
"These tired women": Henry David Thoreau, *A Week on the Concord and Merrimack Rivers*, ed. Odell Shapard (New York: Charles Scribner's Sons, 1921), 239.
"shifted into a dangerous": Bruchac 2006.
"mercy-killed": Ibid.

220 *"What, was Lizzy"*: Shawn Regan, "Hannah Duston: Heroine or Villainess?" *Haverhill (MA) Eagle-Tribune*, August 18, 2006.
"a psychotic murderer": Constantine A. Valhouli, "Duston Is an Appropriate Symbol for Haverhill," *Haverhill (MA) Eagle-Tribune*, August 25, 2006.
"More than being": Quoted in Brad Perriello, "Duston Tribute Appalls Abenaki," *Haverhill (MA) Eagle-Tribune*, August 27, 2006.

221 *"Must we be haunted"*: Bruchac 2006.

Chapter 7: "Oppressions, Grievances & Provocacons"

223 *"Apalatchia"*: James Moore (April 16, 1704), "An Account of What the Army from Thence Had Done, Under the Command of Colonel Moore," *Boston News-Letter*, April 24–May 1, 1704, 2. The letter appeared in the second issue of what is generally credited with being the first successful newspaper in the English colonies.

224 *"I now have"*: Ibid.
"the entire population": José de Zúñiga y la Cerda to Philip V (1704), quoted in Mark F. Boyd, Hale G. Smith, and John W. Griffin, *Here They Once Stood* (Gainesville: University Press of Florida, 1951), 50.

225 *up to 51,000 Indians:* Estimates range from 24,000 to 51,000 Indians enslaved by the Carolinians, but historians admit that even the upper range is a conservative figure and the real total may have been much higher. Because the Carolinians were carrying on much of the slave trade under the bureaucratic radar—selling them off the books to buyers in the Caribbean, Middle Colonies, and New England to avoid paying taxes—records were probably intentionally misleading.

226 *"gain [by] all"*: Commons House of Assembly, September 17, 1703, in *Journals of the Commons House of Assembly*, quoted in Verner W. Crane, *The Southern Frontier, 1670–1732* (Ann Arbor: University of Michigan, 1956), 79.

227 holahtas: The term *holahta* seems to have been used for a chief (possibly a paramount chief) only among the Apalachee and Timucua, while the Yamasee, Guale, and Tama to the north used *mico*. See Hann 1992 for a detailed discussion. Moore, for his part, referred to Don Patri-

cio Hinachuba as "the cassik" ("An Account," 2), *cacique* being a Spanish term of Caribbean
origin applied to American Indian leaders.

227 *"the crown of my king":* Don Patricio Hinachuba to José de Zúñiga y la Cerda, May 29, 1705,
translated in Mark F. Boyd, "Further Consideration of the Apalachee Missions," *Americas* 9
(April 1953): 474.

228 *"having already almost":* Daniel Defoe (1705), "Party-Tyranny, or An Occasional Bill in Min-
iature, as Practiced in Carolina," in *Narratives of Early Carolina, 1650–1708,* ed. Alexander S.
Salley Jr. (New York: Charles Scribner's Sons, 1911), 240.

"a dark kingdom": Peter Silver, "The Older South?" *Reviews in American History* 31 (2003): 193.

229 *"the vilest race":* Gideon Johnson, quoted in William Stevens Perry, *The History of the American
Episcopal Church, 1587–1883* (Boston: James R. Osgood, 1885), 1:379.

230 *"Deareskinnes dressed":* Thomas Hariot, *A Briefe and True Report of the New Found Land of
Virginia,* introduction by Luther S. Livingston (1588; facs. ed., New York: Dodd, Mead, 1903),
B2v.

"such infinite Herds": Thomas Ashe (1682), "Carolina, or A Description of the Present State
of That Country," in *Narratives of Early Carolina, 1650–1708,* ed. Alexander S. Salley Jr. (New
York: Charles Scribner's Sons, 1911), 150.

232 *haunting, blood-soaked:* One example is recounted in Richard White's excellent *The Middle
Ground: Indians, Empires, and Republics in the Great Lakes Region, 1650–1815* (Cambridge:
Cambridge University Press, 1991). Seneca warriors raiding deep into what today would be
Indiana came upon a Miami village while its fighting men were away. The Senecas killed or
captured virtually everyone. Each night when they camped, the Senecas—who, like a number
of northeastern tribes, practiced ritualized cannibalism—killed and cooked a child. "And
every morning, they took a small child, thrust a stick through its head and sat it up on the
path with its face toward the Miami town they had left. Behind the Senecas came the pursuing Miamis, and at every Seneca campsite, brokenhearted parents recognized their child" (pp.
4–5). According to Miami tradition, the bereaved pursuers ambushed the Senecas just shy of
their own town somewhere in western New York, killing all but six of them.

Yamasis, or Yamasee: The origin of the name is unclear. English spelling varied widely, and
even today scholars use both Yamasee and Yamassee. I have followed Worth, "Yamasee,"
2004.

234 *"We have been intirely":* Thomas Nairne to Edward Marston, August 20, 1705, quoted in Frank
J. Klingberg, "Early Attempts at Indian Education in South Carolina: A Documentary,"
South Carolina Historical Magazine 61 (January 1960): 2.

"are now obliged": Thomas Nairne to Charles Spencer, July 10, 1708, in *Nairne's Muskhogean
Journals,* ed. Alexander Moore (Jackson: University Press of Mississippi, 1988), 75.

235 *"assur'd me that":* John Lawson, *History of North Carolina* (1831; repr., Charlotte, NC: Observer
Printing House, 1903), xiii.

"the meaner sort": Ibid., xi.

five-hundred- to six-hundred-mile arc: Lawson put the distance at a thousand miles, which is
an understandable exaggeration for a man laboriously following waterways.

"which are so numerous": Lawson, *History of North Carolina,* 22.

236 *"Tuskeraro":* Ibid., 31.

"Tho' this be called": John Barnwell, February 4, 1712, in "The Tuscarora Expedition: Letters of
John Barnwell," *South Carolina Historical and Genealogical Magazine* 9 (January 1908): 32.

"a natural Failing": Lawson, *History of North Carolina,* 116.

"split the Pitch-Pine": Ibid.

238 *"Inside of 18 months":* Christoph von Graffenried's Account of the Founding of New Bern, ed. Vincent H. Todd (Raleigh, NC: Edwards & Broughton Printing, 1920), 228.

"The land is good": Ibid., 312.

"This country is praised": Ibid.

239 *leaders of the Conestoga:* The Conestoga, named for their town along the lower Susquehanna River, were the remnants of the Susquehannock, who had been badly diminished in seventeenth-century wars with the Iroquois.

"elder women": J. N. B. Hewitt, "Tuscarora," in *Handbook of American Indians North of Mexico,* ed. F. W. Hodge (Washington, DC: Bureau of American Ethnology, 1910), 2:843.

"past carriage": Ibid., 844.

"I am credibly": Alexander Spotswood to Edward Hyde, January 1711, in *The Official Letters of Alexander Spotswood,* ed. R. A. Brock (Richmond: Virginia Historical Society, 1882), 1:48.

"to fall upon": Ibid.

"a Train of ill": Ibid.

240 *"a just reproach":* Ibid.

"saying there was": Christoph von Graffenried, "Copy of the account written Mr. Edward Hyde, Governor in North Carolina, the 23d of October, 1711, with reference to my miraculous deliverance from the savages," in *Christoph von Graffenried's Account of the Founding of New Bern,* ed. Vincent H. Todd (Raleigh, NC: Edwards & Broughton Printing, 1920), 263.

"fine and apparently": Ibid.

"this was of no consequence": Ibid.

"There came . . . a general": Ibid., 266.

241 *"This spoiled everything":* Ibid.

"I turned toward": Ibid., 267.

242 *"dressed like a Christian":* Ibid.

"in the most dignified": Ibid.

"made a short speech": Ibid., 269.

"Surveyor General Lawson": Ibid.

"The threat had": Ibid., 270.

243 *"cannot find a member":* Alexander Spotswood to Lord Dartmouth, December 28, 1711, in *Official Letters of Alexander Spotswood,* 1:137.

"for carrying on": Ibid., 1:135.

"out of a humor": John Page to John Harleston, December 1, 1708, quoted in A. S. Salley, "Barnwell of South Carolina," *South Carolina Historical and Genealogical Magazine* 2 (January 1901): 47.

"my brave Yamassees": Barnwell, February 4, 1712, 34.

244 *"small forts":* Ibid., 32.

"within the Fort": Ibid.

"did not yield": Ibid.

"got all the slaves": Ibid., 33.

"to stimulate themselves": Graffenried, quoted in Hewitt, "Tuscarora," 2:846.

the most ambitious: Barnwell reported that an escaped Virginia slave named Harry had been the architect of the fort, but this may reflect a refusal (as among the Puritans when confronting the Narragansett Great Swamp fort; see chapter 5) to believe that Indians could design such a "civilized" fortification themselves.

"strong as well": Barnwell, March 12, 1712, in "Tuscarora Expedition," 43.

"large reeds & canes": Ibid.

245 *"without a dram":* Ibid., 50.

"that I ordered": Ibid.

"contrived several sorts": Ibid., 50–51.

"This siege": Ibid., 51.

"hardy boys": Ibid., 54.

246 *"quit all pretentions":* Ibid., 52.

"never to build": Ibid.

246 *"I want words"*: Thomas Pollock (December 23, 1712), quoted in Joseph W. Barnwell, "The Second Tuscarora Expedition," *South Carolina Historical and Genealogical Magazine* 10 (January 1909): 36.

"the ignorance and obstinacy": Ibid.

"The destructions his indians": Thomas Pollock (June 15, 1713), quoted in Barnwell, "Second Tuscarora Expedition," 36.

a fort even more: By far the best exploration of Tuscarora innovation is military historian Wayne E. Lee's "Fortify, Fight, or Flee: Tuscarora and Cherokee Defensive Warfare and Military Culture Adaptation," *Journal of Military History* 68 (July 2004): 713–40, from which much of this discussion is drawn. Recent excavations at Neoheroka have confirmed Moore's descriptions concerning the size and sophistication of the fort.

247 *"Of ye Enemies"*: James Moore to Thomas Pollock, March 27, 1713, in Barnwell, "Second Tuscarora Expedition," 39.

248 *enjoyed a windfall*: It's not known what the Yamasee and other Indian allies received for the seven hundred Tuscarora captives—the wholesale price, so to speak—but estimates of the price they fetched on the slave market vary from £7,000 to £24,000, a fortune in those days (Olexer 2005).

the value of slaves: According to French sources, English traders would pay the equivalent of up to £37 for an Indian slave, although this is almost certainly exaggerated; £5 to £10 may be closer to the mark, about what the French paid when they had the resources. See Gallay 2002, 311–14, for a discussion of comparable slave valuations.

"not now so much": Thomas Nairne (1710), "A Letter from South Carolina," quoted in Haan 1981, 347.

249 *"sum of their Indians"*: David Crawley to William Byrd, July 30, 1715, in *Calendar of State Papers: Colonial*, ed. William N. Sainsbury (London: Public Records Office, 1964), 28:248.

"when out among": Ibid., 28:247–48.

"cut off the said": James Douglas (May 6, 1714), quoted in *Journals of the Commissioners of the Indian Trade, September 20, 1710–August 29, 1718*, ed. William L. McDowell Jr. (Columbia: South Carolina Department of Archives and History, 1955), 1:56.

"to prevent their falling": Ibid.

"seemed pleased with": Ibid.

250 *"that the white men"*: Quoted in Ramsey 2008, 228, who details how he uncovered this account—long rumored but lost for centuries—which explains Wright's role and confirms that the Yamasee fully expected to be attacked and enslaved. Dictated in the early days of the war by the Yamasee chief known as the Huspaw King to an English boy, it was written "in gunpowder ink" and addressed to Governor Charles Craven. It lay unnoticed since the 1700s in the archives of the British Admiralty until Ramsey stumbled upon it.

"vex'd the great": Ibid.

252 *"Oppressions, Grievances"*: Francis Le Jau, May 10, 1715, in *The Carolina Chronicles of Dr. Francis LeJau, 1706–1717*, ed. Frank J. Klingberg (Berkeley: University of California Press, 1956), 153.

"In my humble": Ibid.

armed black slaves: African slaves made up a significant proportion of the Carolina militia during the Yamasee War and nearly a third of the standing army formed in the wake of the militia's spotty performance. As Oatis 2004 has pointed out, the slaves proved especially adept at frontier warfare and played important roles as scouts and interpreters. Except for one slave freed by an act of the Commons House, however, "most of the several hundred blacks who served on behalf of South Carolina were unceremoniously returned to their plantations by the war's end" (p. 170).

254 *"How to hold both"*: Letter to Joseph Boone, April 25, 1717, in *Colonial State Papers*, National Archives (UK), catalogue CO 5/1265, no. 69; calendar ref. item 542, vol. 29 (1716–1717), 290–91.

"in such a temper": John Barnwell to Robert Johnson, April 20, 1719, in *Colonial State Papers*,

National Archives (UK), catalogue CO 5/1265, nos. 127, i., ii., calendar ref. item 164, vol. 31 (1719–1720), 80–81.

255 *"Our Southward"*: Ibid.
"I fear much": Ibid.
"the Golden Islands": Robert Montgomery and John Barnwell, *The Most Delightful Golden Islands* (1720; repr., Atlanta: Cherokee Publishing, 1969), 1.

Chapter 8: "One Head One Mouth, and One Heart"

260 *Tarachiawagon:* Tarachiawagon (spelled a variety of ways, including Tharonhiawá:kon and, in Mohawk, *tharǫhyawâ·kǫ*) is the Good Twin in Iroquois cosmology, who battled and eventually triumphed over his evil brother, Tawiskaron. The name is perhaps more accurately translated as "sky grasper." See Fenton 1978 and Richter 1992.
"king of the Indian traders": George Patterson Donehoo, *A History of the Cumberland Valley in Pennsylvania* (Harrisburg, PA: Susquehanna History Association, 1930), 1:96.
"the Buck": Wainwright 1959, 177.

261 *"So You may see":* George Croghan to William Murray, July 12, 1765, in *Collections of the Illinois State Historical Library,* ed. Clarence W. Alvord (Springfield: Illinois State Historical Library, 1916), 11:58.
"Kantucqui": The name, referring to both the region and a river, first appears in a 1753 deposition in Pennsylvania, in which it is also spelled Cantucky (F. W. Hodge, ed., *American Anthropologist* [Lancaster, PA: American Anthropological Association, 1908], 10:340).
"the Half-King": Tanaghrisson and Scarouady (who was similarly appointed "half-king" by the Iroquois League to supervise the Shawnee) appear in colonial records beginning in 1748. See Conrad Weiser to Provincial Council, September 29, 1748, in *Minutes of the Provincial Council of Pennsylvania,* 5:358 (hereafter cited as *MPCP*).
Onondaga: With the establishment of the Iroquois League, the Onondaga—whose homeland sat in the middle of the original Five Nations, south of Oneida Lake—became the "fire keepers" for the confederacy, and their villages were the ceremonial heart of the league and the setting for most of its council deliberations. See Harold Blau, Jack Campisi, and Elisabeth Tooker, "Onondaga," in Trigger 1978, 491–99.
Madame Montour: There is little about Madame Montour's early life that can be pinned down with any certainty, although claims that she was the daughter of the French governor were a fallacy. "Madame" was a title; she may have been born Elizabeth Couc in 1665 or Catherine Montour in 1684, and whether she was of pure French or métis descent is anyone's guess. Likewise, there are conflicting accounts of her captivity among the Iroquois—whether she was taken as a child or as a young married woman, and whether by English-backed Iroquois while she was living in Canada or by French-allied Iroquois while living in New York, although the former seems more likely. By 1704, however, she was in French Detroit, creating scandals with French officers, and by 1709 she was working for the New York government, having taken over as interpreter for her murdered brother Louis Couc Montour. She may have had one or more marriages prior to her union with Andrew's father, the Oneida war chief Carondowana, who died while raiding the Catawba in 1729. Madame Montour was involved in the Five Nations' negotiations before and during the Tuscarora War, but around 1727 she moved to Pennsylvania, where Carondowana was the Iroquois council's overseer among the Shawnee. She remained there with Andrew after her husband's death in 1729. For recent examinations of Madame Montour's life, see Hunter, "Tanaghrisson," and Royot 2007.

262 *"countenance [was] decidedly":* Count Nicolaus Ludwig graf von Zinzendorf (1742), quoted in William C. Reichel, ed., *Memorials of the Moravian Church* (Philadelphia: J. B. Lippincott, 1870), 1:95–96.
"of brass and other": Ibid., 1:96.

262 *"set a Horn":* Scarouady (October 4, 1753), in *MPCP,* 5:683.

"one of their Counsellors": Ibid.

"a French woman": Conrad Weiser, "Narrative of a Journey, Made in the Year 1737, by Conrad Weiser," trans. Hiester H. Muhlenberg, in *Collections of the Historical Society of Pennsylvania,* ed. John Pennington and Henry C. Baird (Philadelphia: Merrihew & Thompson, 1853), 1:8.

263 *"of all the Traders":* Horatio Sharpe to Robert Dinwiddie, December 26, 1754, in *Correspondence of Governor Horatio Sharpe,* ed. William H. Browne (Baltimore: Maryland Historical Society, 1888), 1:151.

"He abused me": Conrad Weiser to Richard Peters, September 13, 1754, in *Pennsylvania Archives,* 2nd ser., 7:246 (hereafter cited as *PA*).

"faithful, knowing, & prudent": Conrad Weiser (June 23, 1748), in *MPCP,* 5:290.

"a French man": Richard Peters to Proprietaries, November 6, 1753, in *Penn Manuscripts, Official Correspondence* (Philadelphia: Historical Society of Pennsylvania), 6:115.

"at a lost": Conrad Weiser to Richard Peters, August 4, 1754, in *Penn Manuscripts,* 6:12.

264 *loosely related groups:* See Becker 1989 for a contrary view, suggesting that the Lenape/Delaware as usually defined were three linguistically and culturally distinct groups.

Unlike the Susquehannock: See Sugrue 1992.

266 *"must live in love":* William Penn (1683), quoted in Jean R. Soderlund, ed., *William Penn and the Founding of Pennsylvania, 1680–1684* (Philadelphia: University of Pennsylvania Press and Historical Society of Pennsylvania, 1983), 307.

"noe man shall": "Indian sections in certain conditions or concessions agreed upon between William Penn and his Pennsylvania land purchasers" (July 11, 1681), in *William Penn's Own Account of the Lenni Lenape or Delaware Indians,* ed. Albert Cook Myers, rev. ed. (Moorestown, NJ: Middle Atlantic Press, 1970), 55.

"unique blend": Jennings 1984, 241.

"sachemakers": William Penn, quoted in Soderlund, *William Penn,* 156.

"two handfull": Ibid., 161.

"so much Wampum": Deed (June 22, 1683), quoted in *William Penn's Own Account,* 86.

267 *"4 hansfull Bells":* Ibid., 88.

"bought lands": William Penn, quoted in Soderlund, *William Penn,* 292.

"as long as": "Letter of Pennsylvania Indians to the King of England Commending Their Friend William Penn, 1701," in *William Penn's Own Account of the Lenni Lenape or Delaware Indians,* ed. Albert Cook Myers, rev. ed. (Moorestown, NJ: Middle Atlantic Press, 1970), 66.

268 *a level of diplomatic:* The league was and remains a remarkable achievement, but it has been increasingly credited—on fairly thin evidence—with having served as the direct inspiration for the federal structure established by the American colonies' Articles of Confederation, an argument made in works such as Donald A. Grinde, *The Iroquois and the Founding of the American Nation* (San Francisco: Indian Historian Press, 1977), and Bruce E. Johansen, *Forgotten Founders* (Ipswich, MA: Gambit, 1982). The Onondaga sachem Canasatego, at a treaty in Lancaster in 1744, praised to colonial officials the "Union and Amity" of the Iroquois League and told them, "We heartily recommend Union and a good Agreement between you Brethren" (quoted in Payne 1996, 609). But while Benjamin Franklin and others among the founders were aware of the league's history and structure and may well have been influenced by it, much of the argument for such influence boils down to a single, decidedly sarcastic remark by Franklin. Despairing of colonial unity, Franklin in 1750 wrote, "It would be a very strange Thing, if six Nations of ignorant Savages should be capable of forming a Scheme for such an Union . . . and yet that a like Union should be impracticable for ten or a Dozen English Colonies" (Franklin to James Parker, March 20, 1751, in *The Papers of Benjamin Franklin,* ed. Leonard W. Labaree [New Haven, CT: Yale University Press, 1961], 4:118–19).

271 *British embassy warned:* With the 1707 Acts of Union, England and Scotland were joined into the United Kingdom of Great Britain.
"You may have half ": James Dayrolle (June 14, 1709), quoted in Wallace 1945, 6.

272 *"I suffered much":* Conrad Weiser, "Copy of a Family Register in the Handwriting of Conrad Weiser," trans. Hiester H. Muhlenberg, in *Collections of the Historical Society of Pennsylvania,* ed. John Pennington and Henry C. Baird (Philadelphia: Merrihew & Thompson, 1853), 1:2.
"were so barbarous": Ibid.
"learned the greater part": Ibid., 1:3.
"there were always": Ibid.

273 *"could not believe":* Sassoonan (June 5, 1728), in *MPCP,* 3:322. Like many Natives, Sassoonan was known by multiple names, depending on the audience. He often appears in colonial records as Olumapies.

274 *"without the Consent":* James Logan (June 5, 1728), in *MPCP,* 3:340.
Ann Eva Feck: Ann's name was sometimes recorded as Anne or Anna.

275 *"Shikellima, of the five Nations":* *MPCP,* 3:357. Many colonial scribes were slow to adopt the term "Six Nations" following the Tuscarora's entry into the league in the early 1720s.
"forty Shillings": Ibid., 3:425.
"a trusty good Man": Ibid., 3:410.
"whose Services had been": Ibid., 3:337.

276 *"Also in the day":* Numbers 10:10 (*The Holy Bible, Quatercentenary Edition,* ed. Gordon Campbell [1611; facs. ed., Oxford: Oxford University Press, 2010], 33).

277 *"at a previous period":* Conrad Weiser (1737), quoted in "Narrative of a Journey, Made in 1737, by Conrad Weiser," trans. Hiester H. Muhlenberg, in *Collections of the Historical Society of Pennsylvania,* ed. John Pennington and Henry C. Baird (Philadelphia: Merrihew & Thompson, 1853), 1:16.
"but not without": Ibid., 7.

280 *"a French woman by birth":* Ibid., 8.
Sheshequin Path: Because of flooding, Weiser's party was forced to detour off the main path for the first dozen miles or so, where they encountered the worst ice.
"We came to a thick": Weiser, quoted in "Narrative of a Journey," 9.
"frightful": Ibid.
"The passage through": Ibid.
"terrible mountains": Ibid.

281 *"stood still in astonishment":* Ibid., 10.
"Indian shoe strings": Ibid., 16.
"on which we sustained": Ibid.
"now their children": Ibid., 17.
"You kill [the game]": Ibid.
"through a dreadful": Ibid., 18.
"I stepped aside": Ibid.
"the sensible reasoning": Ibid.

282 *"let justice have":* William Penn to Gulielma Penn [his wife] and children, August 4, 1682, in Soderlund, *William Penn,* 169.
The Penns' justification: For a damning examination of the Walking Purchase and the attempts in decades thereafter to cover up the essential fraud, see Jennings 1984, especially Appendix B, "Documents of the Walking Purchase." It is ironic that Jennings suspected that one of the infamous walkers, Solomon Jennings, might have been his own ancestor.

284 *"all the best":* Lappawinzo (1737), quoted in William J. Buck, *The Indian Walk* (Philadelphia: Edwin S. Stuart, 1886), 95.

285 *"The other Day":* Canasatego (July 12, 1742), in *MPCP,* 4:578. How precisely the written record of Indian speeches like this one, given at treaty councils and other formal occasions, matches

what was actually said is a question that has vexed historians for centuries. The orator's words had to pass through many linguistic layers to reach the printed page—in this case from spoken Onondaga (presumably) to spoken English, interpreted by a man whose first language was German; then to a scribe feverishly scribbling his version of the translated speech in English, and whose notes were often later assembled, edited, and reworked by a provincial official such as Logan, who submitted them for publication; and finally subjected to *further* polishing by someone like Benjamin Franklin, who reprinted volumes of treaty speeches over the years and who often added rhetorical flourishes of his own.

Each step was rife with opportunities for mistakes, misunderstandings, heavy-handed editorial grooming, and intentional tampering with a speaker's words. Sloth, drunkenness, and exhaustion at the time of the translation and transcription all played roles, as did the opportunity to gore a particular political ox. The Reverend Richard Peters, provincial secretary of Pennsylvania during the 1740s and 1750s, seems to have frequently altered the speeches of Indian orators to provide a more colony-friendly reading. Yet is also clear that some scribes and interpreters would go to great lengths to ensure they were accurately capturing the letter and spirit of an oration. On one occasion in 1756, when Conrad Weiser and another interpreter differed in their understanding of a Seneca leader's speech, they met with him privately to iron out the kinks before giving a public translation (Hagedorn 1988).

Imperfect as the council minutes and speech transcriptions are, "in no other source did ethnocentric Euro-Americans preserve with less distortion a memoir of Indian thoughts, concerns, and interpretations of events" (Richter 1982, 47–48), and they may be our best opportunity to hear the eloquence of an oral society speaking for itself. For an exploration of one especially well-documented example of varying accounts flowing from a single man's speeches, see Merrell 2006.

285 *"We see with":* Canasatego (July 12, 1742), in *MPCP,* 4:579.
 "You ought to be": Ibid.

288 *"has never been deemed":* Robert Hunter Morris to Horatio Sharpe, January 7, 1754, in *PA,* 2nd ser., 2:114.
 "amusing, provided it": Albert T. Volwiler, "George Croghan and the Westward Movement, 1741–1782," *Pennsylvania Magazine of History and Biography* 46 (1922): 273–311.
 "Youl Excuse": George Croghan to Richard Peters, September 26, 1758, in *PA,* 1st ser., 3:545.

289 *at whose nexus:* For a detailed examination of Logan's Indian trade dealings, see Jennings 1966.

290 *"The French man offering":* Conrad Weiser (June 17, 1747), in *MPCP,* 5:86–87.

291 *Canada, New England:* Though largely sparing the Middle Colonies, the war took a fearsome toll farther north, with major battles for the French fortress of Louisbourg (in the Maritimes) and frontier attacks that emptied much of New York north of Albany. The war ended with a peace treaty in 1748 that restored all boundaries and possessions to their prewar status, a situation no one expected to last. It didn't.
 "I have had a dream": All quotations are from Wallace 1945, 152. Another version of this exchange appears in C. Z. Weiser, *The Life of (John) Conrad Weiser,* 2nd ed. (Reading, PA: Daniel Miller, 1899), 106.
 "some Honest Trader": Conrad Weiser to Richard Peters, July 20, 1747, in *PA,* 1st ser., 1:762.
 "Twightwees": The Miamis living in the Ohio country are usually referred to in colonial records as Twightwees, possibly derived from their own name for themselves, *twaatwãã;* their name in Lenape, *tuwéhtuwe;* or similar Iroquoian forms (Callender 1978, 689).

292 *"It is manifest":* Benjamin Franklin, "A treaty held by commissioners, members of the Council of the Province of Pennsylvania, at the town of Lancaster" (July 22, 1748), quoted in Susan Kalter, ed., *Benjamin Franklin, Pennsylvania, and the First Nations: The Treaties of 1736–62* (Urbana-Champaign: University of Illinois Press, 2006), 157.
 "loathsome idolatry": Conrad Weiser (September 3, 1743), "Letter of Conrad Weiser to the Leaders at Ephrata," in C. Z. Weiser, *The Life of (John) Conrad Weiser,* 2nd ed.

292 *"spiritual and physical"*: Ibid.
    "fulsome self-praise": Ibid.
    "I take leave": Ibid.

293 *"Brethren, you came"*: Conrad Weiser, "Copy of a Journal of the Proceedings of Conrad Weiser,
    in His Journey to Ohio," trans. Hiester H. Muhlenberg, in *Collections of the Historical Society of
    Pennsylvania*, ed. John Pennington and Henry C. Baird (Philadelphia: Merrihew & Thomp-
    son, 1853), 1:27.
    "We give you": Ibid.

294 *"We therefore remove"*: Ibid., 1:28.
    "back inhabitants": See *MPCP*, 2:43, for one of the first uses of the term in the Pennsylvania
    records (sometimes written as "back Christian Inhabitants").
    "Brethren, some of you": Weiser, "Copy of a Journal," 1:30.
    "a civil and brotherly present": Ibid.

295 *"shall serve to strengthen"*: Ibid., 1:31.
    "You will have peace": Ibid., 1:32.
    "Let us keep": Ibid.

296 *"You can't be insensible"*: James Hamilton to George Clinton, September 20, 1750, in *Documents
    Relative to the Colonial History of the State of New-York*, ed. E. B. O'Callaghan (Albany, NY:
    Weed, Parsons, 1855), 6:594
    "not the least danger": Conrad Weiser, quoted in Wallace 1945, 270.
    "The land on both": Ibid.
    "a midling good": Ibid., 271.
    "a good start": Ibid.
    "one thinks it": Ibid.
    "the 6 nations": Ibid.

297 *"ye Scum"*: Richard Peters to Thomas Penn, October 24, 1748, quoted in Wallace 1945, 270.
    "may be made": Ibid.

Chapter 9: The Long Peace Ends

299 de boeufs illinois: Joseph-Pierre de Bonnécamps, "Account of the voyage on the Beautiful
    river made in 1749, under the direction of Monsieur de Celoron, by father Bonnecamps," in
    Thwaites 1900, 69:178.
    "somber and dismal": Ibid., 69:169.
    serpents à sonnettes: Ibid., 69:166.

300 *"a monument"*: Pierre-Joseph Céloron de Blainville, "Céloron's Journal," ed. A. A. Lambing, in
    Expedition of Celoron, ed. C. B. Galbreath (Columbus, OH: F. J. Heer, 1921), 18.
    between the upper Allegheny: Bonnécamps's "Account of the voyage" contains many references
    to what was then a remarkable wilderness, from the virgin forest along the upper Ohio, where
    twenty-nine of the party were able to sit, shoulder to shoulder, inside a giant, hollow syca-
    more tree, to the "vast prairies, where the herbage was sometimes of extraordinary height" (p.
    189) along the Great Miami River—the edge of the midwestern tallgrass prairie. The priest
    was sorely disappointed to learn he'd come within a few miles of Big Bone Lick in Ken-
    tucky, "the famous salt-springs where are the skeletons of immense animals" (ibid.). George
    Croghan twice visited the lick in the 1760s, collecting a six-foot mastodon tusk and teeth,
    which were described in a paper for the Royal Society—the frontier trader's only scientific
    publication (Collinson and Croghan 1767).
    "a cup of the milk": Céloron, "Céloron's Journal," 29.
    "Each village": Bonnécamps, "Account of the voyage," 185.

300 *working for George Croghan:* See the testimony of Joseph Fortiner, "An Extract of the Examination of Four English Traders," in *The Olden Time,* ed. Neville B. Craig (Cincinnati: Robert Clarke, 1876), 2:184.

300 *"They had some":* Bonnécamps, "Account of the voyage," 171.

300 *"lodged in miserable":* Ibid.

301 *"a bad village":* Céloron, "Céloron's Journal," 34.

"who saw us": Bonnécamps, "Account of the voyage," 177.

302 *"give Encouragement":* John Lawson (1710), quoted in *History of North Carolina* (1831; repr., Charlotte, NC: Observer Printing House, 1903), 142.

303 *"the white and the brown":* Conrad Weiser to Richard Peters, July 20, 1747, in *PA,* 1st ser., 1:762.
"Asia [is] chiefly": Benjamin Franklin, "Observations concerning the increase of mankind, peopling of countries, etc.," in *The Writings of Benjamin Franklin,* ed. Albert Henry Smyth (New York: Macmillan, 1905), 3:72. Although Franklin's essay appeared in full at the end of the original 1755 tract *Observations on the Late and Present Conduct of the French* by William Clarke, the next edition four years later replaced the final paragraph containing Franklin's musings on race with another concerning trade.
"purely white People": Ibid.
"Spaniards, Italians, French": Ibid.
"by excluding all": Ibid., 3:73.
"the lovely White": Ibid.
Ironically, "red": See Shoemaker 1997. While Shoemaker traced the origins of the word "red" for American Indians back to Native societies in the Southeast, she was unable to determine whether the term emerged entirely from Indian self-identification or whether it was a reflection of European self-description as "white," since, as noted, the colors red and white were potent duality symbols in southeastern Indian belief.
"desire always to be": George Chicken, October 28, 1725, in "Colonel Chicken's Journal to the Cherokees, 1725," in *Travels in the American Colonies,* ed. Newton D. Mereness (New York: Macmillan, 1916), 169.

304 *"white nothings":* James Adair, *The History of the American Indians,* ed. Kathryn E. Holland Braund (Tuscaloosa: University of Alabama Press, 2005), 138.
shuwánakw: See Silver 2008, 16.
"half one, half t'other": Cotton Mather, *Magnalia Christi Americana* (Hartford: Silas Andrus & Son, 1820), 2:517.

305 *something uniquely his own:* This interpretation of Montour's goals, as expressed through his land dealings, and their eventual betrayal at the Albany Congress of 1754, reflects the work of James H. Merrell, especially "'The Cast of His Countenance': Reading Andrew Montour" (1997).
"Andrew Montour has": Conrad Weiser to Richard Peters, August 4, 1748, in *PA,* 1st ser., 2:12.
"to Accommodate": Jonathan Wright, Tobias Hendricks, and Samuel Blunston to Peter Chartier, November 19, 1731, in *PA,* 1st ser., 1:299.
"Remain free": Ibid.
"earnestly and repeatedly": April 24, 1752, *MPCP,* 5:566.

306 *"many Persons":* Ibid., 5:567.

307 *"the lower sort":* Richard Peters to Thomas Penn, July 5, 1749, quoted in Wallace 1945, 278.
"but Mr. Weiser": Ibid., 279.
"the Proprietaries": Ibid.
"all the lands": Deed (July 2, 1744), quoted in Jennings 1984, 361.

308 *"Virginia resumes":* Robert Dinwiddie to Lords Commissioners for Trade, January 1755, in *Official Records of Robert Dinwiddie,* ed. Robert A. Brock (Richmond: Virginia Historical Society, 1883), 1:381.

308 *"Damn it":* Thomas Cresap, quoted in "Deposition of George Aston, 1736," in *PA,* 1st ser., 1:510. For an overview of the border conflict, see Dutrizac 1991.

"the heedless greed": Jennings 1988, 13.

"beautiful natural Meadows": Christopher Gist, February 17, 1751, in "Christopher Gist's First and Second Journals, September 11, 1750–March 29, 1752," in *George Mercer Papers Relating to the Ohio Company of Virginia,* ed. Lois Mulkearn (Pittsburgh: University of Pittsburgh Press, 1954), 18.

"good level farming": Ibid., 33.

310 *"polishing and strengthening":* "The Treaty of Logg's Town, 1752," *Virginia Magazine of History and Biography* 13 (October 1905): 145.

"An Indian who spoke": Christopher Gist, March 12, 1752, in "Christopher Gist's First and Second Journals," 39.

"I was at a Loss": Ibid.

"a very numerous Army": Conrad Weiser, journal, August 13, 1753, in *MPCP,* 5:644–45.

"Raise about 2000 men": Conrad Weiser to John Taylor? (recipient uncertain), n.d., quoted in Wallace 1945, 349, with minor editing.

311 *"Therefore now":* Report of Richard Peters, Isaac Norris, and Benjamin Franklin, September 22, 1753, in *MPCP,* 5:668.

"very contemptuous manner": Ibid., 5:669.

"You are to require": Earl of Holderness to Robert Dinwiddie, August 28, 1753, quoted in *The Diaries of George Washington,* ed. Donald Jackson and Dorothy Twohig (Charlottesville: University Press of Virginia, 1976), 1:126.

"We do hereby": Ibid.

313 *"four others as Servitors,":* George Washington, "Journey to the French Commandant 31 October 1753–16 January 1754," in *Diaries of George Washington,* 1:130.

"extream good & bad": Ibid., 1:133.

"equally well situated": Ibid., 1:132–33.

"dos'd themselves": Ibid., 1:144.

"with the greatest": Ibid.

"an elderly Gentleman": Ibid., 1:148.

314 *"Sir, in Obedience":* Ibid., 1:127n.

"ploting every Scheme": Ibid., 1:151.

"As to the Summons": Jacques Legardeur de Saint-Pierre to Robert Dinwiddie, December 15, 1753, in *Executive Journals of the Council of Colonial Virginia,* ed. Wilmer L. Hall (Richmond: Commonwealth of Virginia, 1945), 5:459.

"I was unwilling": Christopher Gist, quoted in *Diaries of George Washington,* 1:157n.

"The Cold incres'd": George Washington, December 26, 1753, in *Diaries of George Washington,* 1:155. The term "match coat" or "matchcoat" appears frequently in colonial records and not always in reference to a garment. Through the mid-seventeenth century, "match coat" often meant a length of rough-woven cloth, such as duffel or stroud, usually about six feet long—a common item included in Indian land purchases. By the late seventeenth and early eighteenth centuries, however, the term could as easily refer to a tailored outer garment made of the same coarse fabric, as in Penn's land deeds with the Lenape. See Becker 2005.

315 *"much fatigued":* Christopher Gist, quoted in *Diaries of George Washington,* 1:157n.

"a slender man": Robert C. Alberts, *The Most Extraordinary Adventures of Major Robert Stobo* (Boston: Houghton Mifflin, 1965), 1.

"If ye government": George Croghan to Richard Peters, February 3, 1754, in *PA,* 1st ser., 2:118.

316 *"Can't talk":* George Croghan to James Hamilton, February 3, 1754, in *PA,* 1st ser., 2:119.

"had enlisted about": George Croghan to James Hamilton, March 23, 1754, in *MPCP,* 6:21.

"I am convinced": Ibid., 6:22.

316 *"demanding a Reinforcement":* George Washington, April 19, 1754, in *Diaries of George Washington,* 1:177.

"Mr. Ward": George Washington to James Hamilton, April 1754, in *MPCP,* 6:28.

317 *"Whatever may be":* Robert Dinwiddie to James Hamilton, April 27, 1754, in *MPCP,* 6:31.

"The whole of ye Ohio": George Croghan to James Hamilton, May 14, 1754, in *PA,* 1st ser., 2:144.

"hearty Concurrence": Ibid.

"is only atacking": Ibid.

"I ashure": Ibid.

"I am a fread": Ibid.

318 *"A smart Action":* Adam Stephen, "Account of the Engagement That Happen'd at a Place Call'd *The Flats,*" Pennsylvania Gazette, September 12, 1754, 2.

319 *"Therefore keeping up":* Ibid.

Jumonville and several other: There are multiple, contradictory accounts of what happened during the skirmish, perhaps reflecting the fog of war as much as any intentional deceit. Washington's journal says he ordered his men to fire when the French "discovered us" (*Diaries of George Washington,* 1:195). suggesting that Jumonville's men shot first, while Stephen's "Account of the Engagement" makes no mention of how the battle began. Neither says anything about Tanaghrisson executing Jumonville. A French survivor gave the most damning account, saying Washington's troops fired two unprovoked volleys, waited for Jumonville to begin reading his message, and fired again. The witness escaped before the end of the battle, however, and made no mention of Jumonville's death (Sieur de Contrecoeur to Marquis Duquesne, June 2, 1754, quoted in *The Olden Time,* ed. Neville B. Craig [Pittsburgh, Wright & Charlton, 1848], 2:190). A young French officer captured at the fight said that when the British attacked, "neither we nor our own Men took to our arms," and through an interpreter the French invited Washington to "come to our Cabbin, that we might confer" (quoted in *Official Records of Robert Dinwiddie,* 1:225). An English private, not present but reporting what his comrades told him, said that a Frenchman fired the first shot, and he blamed Tanaghrisson for killing a wounded Frenchman, presumably Jumonville (Baker-Crothers and Hudnut 1942). Finally, Denis Kaninguen, believed to be a Catholic Iroquois who was with Tanaghrisson, specifically told the French that Tanaghrisson hatcheted the fallen Jumonville to death and dipped his hands in Jumonville's brains (Denis Kaninguen [June 30, 1754], translated in Fred Anderson, *Crucible of War* [New York: Alfred A. Knopf, 2000], 57).

"Thou are not dead": Kaninguen (June 30, 1754), in Anderson, *Crucible of War,* 57.

"I heard the bullets": George Washington to John Augustine Washington, May 31, 1754, quoted in *Diaries of George Washington,* 1:195n59.

"He would not": Quoted ibid., 1:197n.

"a charming field": George Washington to Robert Dinwiddie, May 27, 1754, in *Official Records of Robert Dinwiddie,* 1:175.

320 *"an absolute Falsehood":* George Washington, May 27, 1754, in *Diaries of George Washington,* 1:198.

"with the utmost": Robert Dinwiddie to George Muse, June 2, 1754, in *Official Records of Robert Dinwiddie,* 1:187.

"would be of singular": George Washington to Robert Dinwiddie, June 3, 1754, in *Official Records of Robert Dinwiddie,* 1:200n.

321 *"We have been":* George Washington to Robert Dinwiddie, June 12, 1754, in *Writings of George Washington from the Original Manuscript Sources, 1745–1799,* ed. John C. Fitzpatrick (Washington, DC: U.S. Government Printing Office, 1931), 1:76.

"Intentions toward us": George Washington, June 18, 1754, in *Diaries of George Washington,* 1:202.

"a good-natured man": Conrad Weiser to James Hamilton, September 3, 1754, "Journal of the

Proceedings of Conrad Weiser in His Way to and at Aughwick, by Order of His Honour Governor Hamilton, in the Year 1754, in August and September," in *MPCP,* 6:151.

321 *"took upon him[self]":* Ibid.
"He lay at one Place": Ibid., 6:152.

323 *"At Night the ffrench":* John Shaw (September 16, 1754), "Affidavit of John Shaw," in "A Private Soldier's Account of Washington's First Battles in the West: A Study in Criticism," *Journal of Southern History* 8 (February 1942): 25.
"to be Reinforced": Ibid.

324 "la assasain": Quoted in Culm Villiers, "Washington's Capitulation at Fort Necessity, 1754," *Virginia Magazine of History and Biography* 6 (January 1899): 268.
"a plausible Pretence": George Washington, May 27, 1754, in *Diaries of George Washington,* 1:197.
"Instead of coming": Ibid., 1:198.

325 *"could hardly be":* Shaw, "Affidavit," 26.
"parole of Honour": Villiers, "Washington's Capitulation," 270.
"As the English have": Ibid.
"The Misfortune": Robert Dinwiddie to James Innes, July 20, 1754, in *Official Records of Robert Dinwiddie,* 1:232.

326 *"for each Man":* House of Burgesses, minutes, July 18, 1754, in *Letters to Washington,* ed. Stanislaus Murray Hamilton (Boston: Houghton Mifflin, 1898), 1:28.

327 *"I counted above":* Conrad Weiser to James Hamilton, September 13, 1754, in *MPCP,* 6:149.
"There is such": George Croghan to James Hamilton, August 30, 1754, in *PA,* 1st ser., 2:161.
"I here enclose": Ibid.
"When an Officer": Humphrey Bland, *A Treatise of Military Discipline,* 9th ed. (London: Baldwin, Richardson, Longman, Crowder, 1762), 132.

329 *"being hurried on":* Bland, *Treatise,* 145.
"is greatly wanted": Robert Stobo to George Washington, July 28, 1754, in *MPCP,* 6:162.
"Strike this Fall": Ibid., 6:162–63.
Moses the Song: Moses is also sometimes identified as Scarouady's son-in-law (Steele 2005).

330 *"When We engaged":* Stobo to Washington, 6:163.
"If they should know": Ibid.

331 *"the Covenant Chain":* Hendrick (June 16, 1753), in *Documents Relative to the Colonial History of the State of New-York,* ed. E. B. O'Callaghan (Albany, NY: Weed, Parsons, 1855), 6:788.
"Tho' I can scarce": James Hamilton to Roger Wolcott, March 4, 1754, in *MPCP,* 5:768.
"Enemies to their": Ibid.
"raising a Civil War": Ibid.
"I beseech your Honour": Ibid., 5:768–69.

332 *"almost starved":* Conrad Weiser to Richard Peters, March 15, 1754, quoted in Wallace 1945, 353.
"I am old": Ibid.
"all thought there was": Benjamin Franklin, *Autobiography of Benjamin Franklin,* ed. Frank Woodworth Pine (New York: Henry Holt, 1916), 243.
"for the whole Province": Weiser to Peters, March 15, 1754, 355.
"as far Westward": John Penn and Richard Peters to James Hamilton, July 9, 1754, in *MPCP,* 6:115.

333 *"a Devil [who] has stole":* Conochquiesie (July 3, 1755), in *Documents Relative to the Colonial History of the State of New-York,* ed. E. B. O'Callaghan (Albany, NY: Weed, Parsons, 1855), 6:984.
"Indians slyly": Ibid.
"He went in": Daniel Claus to James Hamilton, September 1754, in *PA,* 1st ser., 2:175.
"by many false": Ibid.
"Lidius treated them": Ibid.
"for our hunting Ground": Hendrick (July 6, 1754), in *MPCP,* 6:119.

333 *"Orders not to suffer":* Ibid.
    "We will never part": Ibid., 6:116.
    "that reeked": Fred Anderson, *Crucible of War* (New York: Alfred A. Knopf, 2000), 79.
    "I may perhaps": Weiser to Peters, March 15, 1754, 353.
334 *"What We are now":* Hendrick (July 5, 1754), in *MPCP,* 6:115–16.
335 *"Andrew Montour had already":* Conrad Weiser to Richard Peters, May 2, 1754, quoted in Wallace 1945, 355.
    "He is vexed": Conrad Weiser to Richard Peters, September 13, 1754, in *PA,* 2nd ser., 7:246.
    "He abused me": Ibid.
336 *"The Indians here":* John Harris Jr. to James Hamilton, August 30, 1754, in *PA,* 1st ser., 2:178.
    "What the french": Weiser to Taylor, 349.

Chapter 10: War Chief, Peace Chief

338 *"To acquire it":* *Memoirs of Major Robert Stobo of the Virginia Regiment,* ed. Neville B. Craig (Pittsburgh: John S. Davidson, 1854), 20.
    "he still preserved": Ibid., 22.
    "thoughtless stunt": Robert Stobo, quoted in Robert C. Alberts, *The Most Extraordinary Adventures of Major Robert Stobo* (Boston: Houghton Mifflin, 1965), 141.
339 *two regiments of regular:* Descriptions of the soldiers, uniforms, and hierarchy of Braddock's forces are drawn from Franklin T. Nichols, "The Organization of Braddock's Army," *William and Mary Quarterly* 4 (April 1947): 125–47.
340 *"men from sixty":* Robert Orme, 1755, in "Captain Orme's Journal," in *The History of an Expedition Against Fort Du Quesne,* ed. Winthrop Sargent (Philadelphia: J. B. Lippincott, 1856), 286.
341 *"I do hereby Certifie":* Edward Braddock (June 9, 1755), quoted in *PA,* 1st ser., 2:348.
    "the Officers": June 2, 1755, *MPCP,* 6:397.
    "My Lodgings": Charlotte Brown, June 5, 1755, in "The Journal of Charlotte Brown, Matron of the General Hospital with the English Forces in America, 1754–1756," in *Colonial Captivities, Marches and Journeys,* ed. Isabel M. Calder (Port Washington, NY: Kennikat Press, 1967), 180.
    "At 2 in the Morning": Brown, June 12, 1754, in ibid., 182
    "a Rattle snake Colonels named": Ibid.
    "a Rattle Snake Colonel": Anonymous, May 8, 1755, in "Journal of the Proceedings of the Seamen," in *The History of an Expedition Against Fort Du Quesne,* ed. Winthrop Sargent (Philadelphia: J. B. Lippincott, 1856), 372.
    "the most desolate Place": Brown, June 13, 1755, in "Journal of Charlotte Brown, 182–83.
342 *"I wish for nothing":* George Washington to Robert Orme, March 15, 1755, in *George Washington Papers at the Library of Congress, 1741–1799,* 2nd ser., letterbook 1, image 30, http://lcweb2.loc.gov/cgi-bin/ampage?collId=mgw2&fileName=gwpage001.db&recNum=29.
    "None of these": Orme, n.d, in "Captain Orme's Journal," 313.
    "[I] am sorry": Robert Dinwiddie to Edward Braddock, May 9, 1755, in *Official Records of Robert Dinwiddie,* ed. Robert A. Brock (Richmond: Virginia Historical Society, 1884), 2:34.
    "stormed like a Lyon": George Croghan to Robert Hunter Morris, April 16, 1755, in *MPCP,* 6:368.
343 *"by Fire and Sword":* Ibid., 6:368–69.
    "He shall take": Edward Shippen IV to Edward Shippen III, March 19, 1755, quoted in Wallace 1945, 381.
    "To which Genl. Braddock": Shingas (November 3, 1755), quoted in "The Captivity of Charles Stuart, 1755–57," *Mississippi Valley Historical Review* 13 (June 1926): 63.
344 *"On which Genl. Braddock":* Ibid.
    "if they might not": Ibid.
    "Mr. [Richard] Peters related": June 2, 1755, *MPCP,* 6:397.

344 *"The Truth is"*: William Shirley Jr. to Robert Hunter Morris, June 7, 1755, in *PA*, 1st ser., 2:346–47.

347 *"die as an old Roman"*: Charles Swaine to Richard Peters, August 5, 1755, in Howard N. Eavenson, *Map Maker and Indian Trader* (Pittsburgh: University of Pittsburgh Press, 1949), 190.
"I luckily escaped": George Washington to Mary Ball Washington, July 18, 1755, in *George Washington Papers at the Library of Congress, 1741–1799*, 2nd ser., letterbook 1, image 86, http://lcweb2.loc.gov/cgi-bin/ampage?collId=mgw2&fileName=gwpage001.db&recNum=0.
"chiefly Regular Soldiers": Ibid.
"It is not possible": Brown, July 11, 1755, in "Journal of Charlotte Brown," 184.
"General Braddock cou'd": Thomas Dunbar to William Shirley, August 21, 1755, in *MPCP*, 6:593.

348 *"Mismanagement"*: Ibid.
"It was the pride": Scarouady (August 22, 1755), in *MPCP*, 6:589.
"But let us unite": Ibid.
"We are now": "Petition of the Inhabitants of Cumberland County," July 15, 1754, in *MPCP*, 6:131.

349 *"He desires me"*: George Croghan to Charles Swaine, October 9, 1755, in *MPCP*, 6:642.
"The Enemy came down": "Petition from Inhabitants of the West Side of Sasquehannah," October 20, 1755, in *MPCP*, 6:647.
"If the white people": Conrad Weiser to Robert Hunter Morris, October 25, 1755, in *MPCP*, 6:649–50.

350 *"I pray, good Sir"*: Ibid., 6:650.
"There is two-thirds": John Potter to Richard Peters, November 3, 1755, in *MPCP*, 6:674.

351 *"First our Fingers"*: Charles Stuart, quoted in "Captivity of Charles Stuart," 61–62.

352 *"Frequently Called"*: Shingas, quoted in "Captivity of Charles Stuart," 62.
"[Shingas] said": John Heckewelder, *History, Manners, and Customs of the Indian Nations Who Once Inhabited Pennsylvania*, ed. William C. Reichel (Philadelphia: Historical Society of Pennsylvania, 1876), 270n.
"to carry the war": Pierre de Rigaud de Vaudreuil to Jean-Baptiste de Machault d'Arnouville, June 8, 1756, in *PA*, 2nd ser., 6:352.

353 *"If Onontio finds"*: Scarouady (November 8, 1755), in *MPCP*, 6:683.
"I must deal plainly": Ibid., 6:686.
"Instead of . . . providing": Robert Hunter Morris to Pennsylvania Assembly, November 8, 1755, in *MPCP*, 6:684.

354 *"[They] called me"*: Conrad Weiser to Robert Hunter Morris, November 19, 1755, in *PA*, 1st ser., 2:505.
"a lusty, rawbon'd": Anonymous (July 1756), quoted in *PA*, 1st ser., 2:725.

355 *"he can drink"*: Ibid.
"Oh, we're all": Conversation quoted in Hal C. Sipe, *The Indian Chiefs of Pennsylvania* (1929; repr., Lewisburg, PA: Wennawoods Publishing, 1994), 370.
"immediately to declare": Joseph Fox et al. to Robert Hunter Morris, April 10, 1756, in *MPCP*, 7:78.

356 *"the following Rewards"*: Ibid.
"every Male Indian": Ibid.

357 *"to let the Indians know"*: Benjamin Franklin, *Autobiography of Benjamin Franklin*, ed. Frank Woodworth Pine (New York: Henry Holt, 1916), 279.
"Have we, the Governor": William Denny (November 12, 1756), in *MPCP*, 7:320.
"venal, lazy and inept": Nicholas B. Wainwright, "Governor William Denny of Pennsylvania," *Pennsylvania Magazine of History and Biography* 81 (April 1957): 172.

358 *"idle, fatuous, greedy"*: Ralph L. Ketcham, "Conscience, War, and Politics in Pennsylvania, 1755–1757," *William and Mary Quarterly* 20 (July 1963): 430.

NOTES 427

358 *"the Joy which appear'd":* "Easton Treaty Texts, July and November 1756," 47, supplementary material, for James H. Merrell, "'I desire all that I have said . . . may be taken down aright': Revisiting Teedyuscung's 1756 Treaty Council Speeches," *William and Mary Quarterly* 63 (October 2006): 777–826.

"The Times are not": Teedyuscung (November 13, 1756), in *MPCP,* 7:321. This rendering of Teedyuscung's speech is based on the polished and edited version later published by Benjamin Franklin; for other versions captured by a variety of witnesses, see "Easton Treaty Texts, July and November 1756."

"I take away": Ibid.

"I have not far": Ibid., 7:324.

"A bargain is": Ibid.

"his Children forge": Ibid., 7:325. Emphasis added.

359 *"following my conscience":* Robert Stobo, quoted in Alberts, *Most Extraordinary Adventures,* 154. Stobo's remarks are based on the published French-language transcripts of his interrogation and trial ("Procès de Robert Stobo et de Jacob Wambram pour crime de haute trahison," in *Rapport de l'Archiviste de la Province de Quebec pour 1922–23* [Montreal: Province of Quebec, Ls-A. Proulx, 1923]), translated and rendered into conversational English by Alberts.

"I believed myself": Ibid., 167.

360 *"very little foundation":* August 8, 1755, *MPCP,* 6:534.

"Filius Gallicae": Francis Jennings makes a convincing case for the ever busy Reverend William Smith as the source of the "Filius Gallicae" letters, as part of Smith's anti-Quaker screeds. See Jennings 1988.

Montour found his life: Relatively little is documented regarding Montour's home life, much of it contained in a single entry in the minutes of the Pennsylvania council from April 20, 1756: "The Indians had a long Conference with the Governor. They put Andrew Montour's children under his care, as well as the three that are here, to be independent of the Mother, as a Boy of twelve years old, that he had by a former Wife, a Delaware, a Grand Daughter of Allomipis [Olumapies, or Sassoonan]. They added, that he had a girl among the Delawares called Kayodaghscroony, or Madelina, and desired she might be distinguished, enquired after, and sent for, which was promised" (*MPCP,* 7:95–96).

Sarah Ainse, Montour's second wife, is usually described as an Oneida, and after their divorce she took their newborn, a boy named Nicholas, to live with the Oneida in New York. However, she sometimes described herself as a Shawnee, and Montour said she was related to a Conoy chief. Montour's debt has been blamed on her spendthrift ways, but on her own, Sarah became a very successful trader, working the Great Lakes as far west as Michilimackinac. Nicholas, in turn, became active in the North West Company, trading furs. He married well and became a man of property, installing a horse track on his estate. He died in 1808 at age fifty-two, but his gritty mother outlived him by fifteen years, surviving into her nineties (Béland, "Montour, Nicholas"; Clarke, "Ainse [Hands], Sarah").

361 *"It is now Come":* Conrad Weiser to Richard Peters, October 4, 1757, in *PA,* 1st ser., 3:283.

hundreds of Pennsylvanians: Ward 1995 argues that the number of dead and captured settlers has been significantly underestimated by historians and places the total dead on the Pennsylvania and Virginia frontier (including land that is now part of Maryland) at fifteen hundred, with more than a thousand—perhaps many more—taken hostage. The rate of dead or captured in frontier counties was 3 percent of the population. "Such figures are not incomparable to the Revolutionary War or even the Civil War . . . The conflict also resulted in more collateral damage than any other war in the region, including perhaps the Civil War" (p. 316).

363 *"that he ate":* Harvey Hostetler and William Franklin Hochstetler, *Descendants of Jacob Hochstetler, the Immigrant of 1736* (Elgin, IL: Brethren Publishing, 1912), 39.

"a common jail": Memoirs of Major Robert Stobo, 26.

363 *"other services":* Alberts, *Most Extraordinary Adventures,* 174.

    "I touched Mr. Stobo": Quoted ibid., 176.

365 *"Mr. Weiser informed":* Pennsylvania Provincial Council, minutes, July 28, 1756, in *PA,* 1st ser., 2:727.

    "the King and his wild": Pennsylvania Provincial Council, minutes, July 26, 1756, in *PA,* 1st ser., 2:724.

    "We would have": Teedyuscung via George Croghan to William Denny, July 30, 1757, in *MPCP,* 7:682.

    Wallangundowngen: Per Post et al. 1999.

366 *"If I died":* Christian Frederick Post, July 22, 1758, in "The Journal of Christian Frederick Post, from Philadelphia to the Ohio," in *Early Western Travels, 1748–1846,* ed. Reuben Gold Thwaites (Cleveland: Arthur H. Clark, 1904), 1:188. As Thwaites notes, Post had limited education, and English was not his native language; the material quoted here was heavily edited and corrected, since the original is all but indecipherable.

    "The Quakers submit": Conrad Weiser, quoted in Wallace 1945, 524.

    "to which [the Indians]": Post, August 2, 1758, in "Journal of Christian Frederick Post," 1:190.

    "not to stir": Post, August 25, 1758, in ibid., 204.

    "if I did not think": Post, August 28, 1758, in ibid., 211–12.

367 *"observed that the Shirts":* Pennsylvania Provincial Council, minutes, July 26, 1756, in *PA,* 1st ser., 2:725.

    "Captain Henry Montour": October 8, 1758, *MPCP,* 8:176.

369 *"Brethren: With this String":* Ibid.

    "Mr. Weiser interpreted": Ibid., 8:177.

    "When you have settled": Pisquetomen (October 13, 1758), in *MPCP,* 8:189.

370 *"Mr. Weiser was ordered":* October 13, 1758, *MPCP,* 8:190.

    "at a private Conference": Ibid.

    "I tell you we": Sagughsuniunt (October 15, 1758), in *MPCP,* 8:191.

    "We believe[d]": William Denny (October 16, 1758), in *MPCP,* 8:193.

371 *"I sit here as":* Teedyuscung (October 20, 1758), in *MPCP,* 8:203.

    "If you take": William Denny (October 20, 1758), in *MPCP,* 8:207.

    "Honble. Sir": George Washington to Francis Fauquier, November 25, 1758, in *George Washington Papers at the Library of Congress, 1741–1799,* 2nd ser., letterbook 5, images 160 and 163, http://lcweb2.loc.gov/cgi-bin/ampage?collId=mgw2&fileName=gwpage005.db&recNum=160.

372 *"His looks grew pale":* Memoirs of Major Robert Stobo, 32.

    "a lady fair": Ibid.

    "close confinement": "General Whitmore's Interrogations of Robert Stobo," quoted in Alberts, *Most Extraordinary Adventures,* 351–52.

    "about Ten a Clock": "A journal of Lieut. Simon Stevens, from the time of his being taken, near Fort William-Henry," quoted in George M. Kahrl, "Captain Robert Stobo (concluded)," *Virginia Magazine of History and Biography* 49 (July 1941): 254.

373 *"I took hold":* Ibid., 255.

    "I am Monsieur": Memoirs of Major Robert Stobo, 50.

    "a lofty frigate": Ibid., 52.

    "The Tide being": "Journal of Lieut. Simon Stevens," 255.

374 *"I hope the Reader":* Ibid., 256.

    "where we all": Ibid., 257.

    "Monsieur Stobo's name": Memoirs of Major Robert Stobo, 68.

    "Ill went the victuals": Ibid., 70.

375 *"by a fatality":* Thomas Pownall to Jeffery Amherst, October 3/4, 1759, quoted in Alberts, *Most Extraordinary Adventures,* 263.

Chapter 11: Endings

376 *"to your particular":* Jeffery Amherst to Francis Fauquier, October 25, 1759, quoted in Robert C. Alberts, *The Most Extraordinary Adventures of Major Robert Stobo* (Boston: Houghton Mifflin, 1965), 271.

"as I on my own": Ibid.

"for his steady": House of Burgesses, November 19, 1759, in *Journals of the House of Burgesses of Virginia, 1758–1761,* ed. H. R. McIlwaine (Richmond: Virginia State Library, 1908), 150.

377 *"he never afterwards":* "Petition of Robert Stobo's Sisters," March 24, 1774, in Alberts, *Most Extraordinary Adventures,* 356.

"In the name of God": "Conrad Weiser's Will," November 24, 1759, in C. Z. Weiser, *The Life of (John) Conrad Weiser* (Reading, PA: Daniel Miller, 1876), 100.

"beloved wife": Ibid., 101

378 *"beyond the Kittochtany":* Ibid., 102.

"Aug. 13, 1762": Jacob Hochstetler to James Hamilton, August 13, 1762, in *PA,* 1st ser., 4:99. As was typical of the freewheeling spelling of the day, Hochstetler's name was spelled two different ways by whoever wrote the petition for the illiterate farmer, who merely made an X—his mark—on the paper.

379 *"On the proprietaries'":* James Hamilton (August 18, 1762), quoted in *PA,* 4th ser., 3:157.

380 *"Brother," the sachem said:* Tamaqua (November 28, 1758), quoted in "Second Journal of Christian Frederick Post," *The Olden Time* 1 (April 1846): 167.

381 *"against all the Indians":* William Trent, June 23, 1763, in "William Trent's Journal at Fort Pitt, 1763," A. T. Volwiler, ed., *Mississippi Valley Historical Review* 11 (December 1924): 400.

"Out of our regard": Trent, June 24, 1763, in ibid.

"Sundries got to Replace": Trent, quoted in Elizabeth A. Fenn, "Biological Warfare in Eighteenth-Century North America: Beyond Jeffery Amherst," *Journal of American History* 86 (March 2000): 1554.

"often visited them": Harvey Hostetler and William Franklin Hochstetler, *Descendants of Jacob Hochstetler, the Immigrant of 1736* (Elgin, IL: Brethren Publishing, 1912), 45.

383 *"Allein, und doch":* The usual translation, "Alone, yet not alone am I / though in this solitude so drear," strays far from the actual German verse. Quoted in Sipe 1998, 215.

"My name is": Hostetler and Hochstetler, *Descendants of Jacob Hochstetler,* 37.

384 *"A thick Scull":* George Croghan to William Murray, July 12, 1765, in *Collections of the Illinois State Historical Library,* ed. Clarence W. Alvord (Springfield: Illinois State Historical Library, 1916), 11:58.

385 *"I shall send Montour":* William Johnson to George Croghan, March 1766, in *Papers of Sir William Johnson,* ed. Alexander C. Flick (Albany: University of New York, 1927), 5:120.

"Captain Montour": Isaac Hamilton to Thomas Gage, January 22, 1772, in *Correspondence of General Thomas Gage,* 2 vols. (New Haven, CT: Yale University Press, 1931–1933), quoted in Lewin 1966, 186.

386 *"any lands":* King George III, proclamation, October 7, 1763, Guilder Lehrman Institute of American History, GLC05214, http://gilderlehrman.pastperfect-online.com/33267cgi/mweb. exe?request=record;id=244F978B-2CBF-4295-9B03-924022753327;type=301.

"the several Nations": Ibid.

"I can never look": George Washington to William Crawford, September 21, 1767, in *George Washington Papers at the Library of Congress, 1741–1799,* 5th ser., account book 2, image 14, http:// memory.loc.gov/cgi-bin/ampage?collId=mgw5&fileName=gwpage004.db&recNum=13.

388 *"Having committed an act":* Alberts, *Most Extraordinary Adventures,* 318.

Bibliography

Academy of Natural Sciences. "Harlan's Musk Ox (*Bootherium bombifrons*)." http://www.ansp.org/museum/jefferson/otherFossils/bootherium.php.

Adams, J. M., and H. Faure, eds. "Review and Atlas of Palaeovegetation: Preliminary Land Ecosystem Maps of the World Since the Last Glacial Maximum." Quaternary Environments Network, Oak Ridge National Laboratory, TN, 1997. http://www.esd.ornl.gov/projects/qen/adams1.html.

Adams, Simon, Alan Bryson, and Mitchell Leimon. "Walsingham, Sir Francis (c. 1532–1590)." In *Oxford Dictionary of National Biography,* edited by H. C. G. Matthew and Brian Harrison. Oxford: Oxford University Press, 2004. http://www.oxforddnb.com/view/article/28624.

Adney, Edwin Tappan, and Howard I. Chapelle. *The Bark Canoes and Skin Boats of North America.* Washington, DC: Smithsonian Institution, 1964.

Alaska Office of History and Archaeology. "Broken Mammoth Archaeological Project." July 9, 2008. http://www.dnr.state.ak.us/parks/oha/mammoth/mammoth1.htm.

Alderfer, E. G. *The Ephrata Commune.* Pittsburgh: University of Pittsburgh Press, 1985.

Allen, John L. "From Cabot to Cartier: The Early Exploration of Eastern North America, 1497–1543." *Annals of the Association of American Geographers* 82 (September 1992): 500–21.

Anderson, David G., and Robert C. Mainfort Jr. *The Woodland Southeast.* Tuscaloosa: University of Alabama Press, 2002.

Anderson, David G., and Kenneth E. Sassaman. "Early and Middle Holocene Periods, 9500 to 3750 B.C." In *Handbook of North American Indians,* edited by Raymond D. Fogelson, 14:87–100. Washington, DC: Smithsonian Institution, 2004.

Anderson, Fred. *Crucible of War.* New York: Alfred A. Knopf, 2000.

Andrews, Kenneth R. *Trade, Plunder and Settlement: Maritime Enterprise and the Genesis of the British Empire, 1480–1630.* Cambridge: Cambridge University Press, 1984.

Archer, John. "Puritan Town Planning in New Haven." *Journal of the Society of Architectural Historians* 34 (May 1975): 140–49.

Arner, R. D. "The Story of Hannah Duston: Cotton Mather to Thoreau." *American Transcendental Quarterly* 18 (1973): 19–23.

Axtell, James. "The Power of Print in the Eastern Woodlands." *William and Mary Quarterly* 44 (April 1987): 300–309.

———. "The Vengeful Women of Marblehead: Robert Roules's Deposition of 1677." *William and Mary Quarterly* 31 (October 1974): 647–52.

Axtell, James, and William C. Sturtevant, "The Unkindest Cut, or Who Invented Scalping." *William and Mary Quarterly* 37 (July 1980): 451–72.

Aydelotte, Frank. "Elizabethan Seamen in Mexico and Ports of the Spanish Main." *American Historical Review* 48 (October 1942): 1–19.

Baichtal, Jim. "A Summary of the Ongoing Paleontological and Associated Research on the Tongass National Forest." http://www.fs.fed.us/geology/paleo_research_overview_ak.pdf.

Baker, Emerson W. "'A Scratch with a Bear's Paw': Anglo-Indian Land Deeds in Early Maine." *Ethnohistory* 36 (Summer 1989): 235–56.

Baker, Emerson W., Edwin A. Churchill, Richard S. D'Abate, et al., eds. *American Beginnings*. Lincoln: University of Nebraska Press, 1994.

Baker-Crothers, Hayes, and Ruth Allison Hudnut. "A Private Soldier's Account of Washington's First Battles in the West: A Study in Criticism." *Journal of Southern History* 8 (February 1942): 23–62.

Bakker, Peter. "'The Language of the Coast Tribes Is Half Basque': A Basque-American Indian Pidgin in Use Between Europeans and Native Americans in North America, ca. 1540–ca. 1640." *Anthropological Linguistics* 31 (Fall–Winter 1989): 117–47.

——— . "Two Basque Loanwords in Micmac." *International Journal of American Linguistics* 55 (April 1989): 258–61.

Baudry, René. "Louis-Pierre Thury." In *Dictionary of Canadian Biography Online*. http://www.biographi.ca/009004-119.01-e.php?&id_nbr=562&&PHPSESSID=m4b9emgtqumu6euaac7oka2ha5&PHPSESSID=m4b9emgtqumu6euaac7oka2ha5.

Bauxar, J. Joseph. "Yuchi Ethnoarchaeology: Parts II–V." *Ethnohistory* 4 (Autumn 1957): 369–464.

Baxter, James P. *Documentary History of the State of Maine*. Vols. 5, 6, 9. Portland, ME: Thurston Print, 1897, 1900, 1907.

Becker, Marshall J. "Lenape Population at the Time of European Contact: Estimating Native Numbers in the Lower Delaware Valley." *Proceedings of the American Philosophical Society* 133 (1989): 112–22.

——— . "Matchcoats: Cultural Conservatism and Change in One Aspect of Native American Clothing." *Ethnohistory* 52 (Fall 2005): 727–87.

Béland, François. "Montour, Nicholas." In *Dictionary of Canadian Biography Online*. http://www.biographi.ca/009004-119.01-e.php?&id_nbr=2563&interval=25&&PHPSESSID=8e1vu3cuadb43naijj90ibt297,2000.

Bogen, David S. "Mathias de Sousa: Maryland's First Colonist of African Descent." *Maryland Historical Magazine* 96 (Spring 2001): 68–85.

Bonatto, Sandro L., and Francisco M. Salzano. "A Single and Early Migration for the Peopling of the Americas Supported by Mitochondrial DNA Sequence Data." *Proceedings of the National Academy of Sciences* 94 (March 1997): 1866–71.

Bourque, Bruce J. "Ethnicity on the Maritime Peninsula, 1600–1759." *Ethnohistory* 36 (Summer 1989): 257–84.

——— . *Twelve Thousand Years*. Lincoln: University of Nebraska Press, 2001.

Bourque, Bruce J., and Ruth Holmes Whitehead. "Tarrentines and the Introduction of European Trade Goods in the Gulf of Maine." *Ethnohistory* 32 (Autumn 1985): 327–41.

——— . "Trade and Alliances in the Contact Period." In *American Beginnings*, edited by Emerson W. Baker, Edwin A. Churchill, Richard S. D'Abate, et al. Lincoln: University of Nebraska Press, 1994.

Bowne, Eric E. "'Carying Away Their Corne and Children': The Effects of Westo Slave Raids on the Indians of the Lower South." In *The Mississippi Shatter Zone*, edited by Robbie Ethridge and Sheri M. Shuck-Hall, 104–14. Lincoln: University of Nebraska Press, 2009.

Boyce, Douglas W. "Iroquoian Tribes of the Virginia–North Carolina Coastal Plain." In *Handbook of North American Indians*, edited by Bruce G. Trigger, 15:282–89. Washington, DC: Smithsonian Institution, 1978.

Brack, H. G. *Norumbega Reconsidered*. Davistown, ME: Davistown Museum, 2008.

Bradley, Bruce, and Dennis Stanford. "The North Atlantic Ice-Edge Corridor: A Possible Palaeo-lithic Route to the New World." *World Archaeology* 36 (December 2004): 459–78.

Bragdon, Kathleen J. *Native People of Southern New England, 1500–1650*. Norman: University of Oklahoma Press, 1999.

Breen, T. H., and Stephen Innes. *"Myne Owne Ground."* New York: Oxford University Press, 2005.

Brinsfield, John W. "Daniel Defoe: Writer, Statesman, and Advocate of Religious Liberty in South Carolina." *South Carolina Historical Magazine* 76 (July 1975): 107–11.

Bromber, Robert. "The Liar and the Bard: David Ingram, William Shakespeare and *The Tempest*." *SEDERI: Yearbook of the Spanish and Portuguese Society for English Renaissance Studies*, no. 12 (2002): 123–34. http://dialnet.unirioja.es/servlet/oaiart?codigo=1977019.

Brown, James A. "Exchange and Interaction Until 1500." In *Handbook of North American Indians*, edited by Raymond D. Fogelson, 14:677–85. Washington, DC: Smithsonian Institution, 2004.

Brown, Jennifer H. S. "Métis, Halfbreeds and Other Real People: Challenging Cultures and Categories." *History Teacher* 27 (November 1993): 19–26.

Brown, Nancy Marie. *The Far Traveler*. New York: Harcourt, 2007.

Brown, Philip M. "Early Indian Trade in the Development of South Carolina: Politics, Economics, and Social Mobility During the Proprietary Period, 1670–1719." *South Carolina Historical Magazine* 76 (July 1975): 118–28.

Bruchac, Margaret. "Reclaiming the Word 'Squaw' in the Name of the Ancestors." November 1999 and March 29, 2001. http://www.nativeweb.org/pages/legal/squaw.html.

———. "Reconsidering Hannah Duston and the Abenaki." *North Andover (MA) Eagle-Tribune*, August 28, 2006.

Burton, William, and Richard Lowenthal. "The First of the Mohegans." *American Ethnologist* 1 (November 1974): 589–99.

Cahokia Mounds State Historic Site and Cahokia Mounds Museum Society. "Cahokia Mounds State Historic Site." 2008. http://www.cahokiamounds.com/.

Callender, Charles. "Miami." In *Handbook of North American Indians*, edited by Bruce G. Trigger, 15:681–89. Washington, DC: Smithsonian Institution, 1978.

Carocci, Max. "Written Out of History: Contemporary Native American Narratives of Enslavement." *Anthropology Today* 25 (June 2009): 18–22.

Cave, Alfred A. *The Pequot War*. Amherst: University of Massachusetts Press, 1996.

———. "Who Killed John Stone? A Note on the Origins of the Pequot War." *William and Mary Quarterly* 49 (July 1992): 509–21.

Ceci, Lynn. "The First Fiscal Crisis in New York." *Economic Development and Cultural Change* 28 (July 1980): 839–47.

———. "The Value of Wampum Among the New York Iroquois: A Case Study in Artifact Analysis." *Journal of Anthropological Research* 38 (Spring 1982): 97–107.

Chamberlain, Kathleen Egan. "Shickellamy." In *American National Biography Online*. http://www.anb.org/articles/20/20-00935.html.

Champagne, Duane. "The Delaware Revitalization Movement of the Early 1760s: A Suggested Reinterpretation." *American Indian Quarterly* 12 (Spring 1988): 107–26.

Chang, Kenneth. "New Evidence of Meteor Bombardment." *New York Times*, January 2, 2009.

Chapman, Joseph A., and George A. Feldhammer. *Wild Mammals of North America*. Baltimore: Johns Hopkins University Press, 1982.

Chatters, James C. "Kennewick Man: Encounter with an Ancestor." Smithsonian Institution, 2004. http://www.mnh.si.edu/arctic/html/kennewick_man.html.

Chu, Jonathan M. "Endecott, John." In *American National Biography Online*. http://www.anb.org. silk.library.umass.edu:2048/articles/01/01-00264.html.

Clark, Charles E. "Gorges, Sir Ferdinando (1568–1647)." In *Oxford Dictionary of National Biography*, edited by H. C. G. Matthew and Brian Harrison. Oxford: Oxford University Press, 2004. http://www.oxforddnb.com/view/article/11098.

Clarke, John. "Ainse (Hands), Sarah (Montour; Maxell; Willson [Wilson])." In *Dictionary of Canadian Biography Online.* http://www.biographi.ca/009004-119.01-e.php?&id_nbr=2729&interval=25&&PHPSESSID=bp5ccmrlt5090pu2dptb9mub51,2000.

"The Cochecho Massacre." Dover (NH) Public Library, 1989. http://www.dover.lib.nh.us/Dover-History/cocheco.htm.

Collinson, Peter, and George Croghan. "An Account of Some Very Large Fossil Teeth, Found in North America, and Described by Peter Collinson." *Philosophical Transactions* 57 (1767): 464–67.

Conkey, Laura E., Ethe Boissevain, and Ives Goddard. "Indians of Southern New England and Long Island: Late Period." In *Handbook of North American Indians,* edited by Bruce G. Trigger, 15:177–89. Washington, DC: Smithsonian Institution, 1978.

Conkling, Philip W. *Islands in Time.* Camden, ME: Down East Books, 1999.

———. "Time Capsules: The Ecology of Mid-Coast Maine in 1605." In *One Land—Two Worlds,* edited by Mac Deford and Philip Conkling. Rockland, ME: Island Institute, 2005.

Connole, Dennis A. *Indians of the Nipmuck Country in Southern New England, 1630–1750.* Jefferson, NC: McFarland, 2000.

Cook, Ramsay, ed. *The Voyages of Jacques Cartier.* Toronto: University of Toronto Press, 1993.

Côté, Pierre-L. "Duquesne de Menneville, Ange, Marquis Duquesne." In *Dictionary of Canadian Biography Online.* http://www.biographi.ca/009004-119.01-e.php?&id_nbr=1872&interval=25&&PHPSESSID=j8sa4ji6bgqf4tsarudc5pd550.

Covington, James W. "Apalachee Indians, 1704–1763." *Florida Historical Quarterly* 50 (April 1972): 366–84.

Cowasuck Band of the Pennacook Abenaki People. "N'dakina—Our Homelands and People." http://www.cowasuck.org/history/ndakina.cfm.

Crompton, Samuel Willard. "Gardiner, Lion." In *American National Biography Online.* http://www.anb.org.silk.library.umass.edu:2048/articles/01/01-00312.html.

Cummings, Hubertis M. "William Penn of Worminghurst Makes His First Sales of Lands in Pennsylvania." *Pennsylvania History* 30 (July 1963): 264–71.

D'Abate, Richard. "A Nation Above All Others: Empire and Race in the Waymouth Celebration of 1905." In *One Land—Two Worlds,* edited by Mac Deford and Philip Conkling. Rockland, ME: Island Institute, 2005.

Dana, Carol, Peter Paul, Roberto Mendoza, et al. *The Wabanakis of Maine and the Maritimes.* Philadelphia: American Friends Service Committee, 1989.

Day, Gordon M. "The Indian as an Ecological Factor in the Northeastern Forest." *Ecology* 34 (April 1953): 329–46.

———. *Western Abenaki Dictionary.* Hull, QC: Canadian Museum of Civilization, 1994.

Demallie, Raymond J. "Tutelo and Neighboring Groups." In *Handbook of North American Indians,* edited by Raymond D. Fogelson, 14:286–300. Washington, DC: Smithsonian Institution, 2004.

Demos, John. *The Unredeemed Captive.* New York: Alfred A. Knopf, 1994.

Denevan, William M., ed. *The Native Population of the Americas in 1492.* Madison: University of Wisconsin Press, 1992.

Denevan, William M. "The Pristine Myth: The Landscape of the Americas in 1492." *Annals of the Association of American Geographers* 82 (1992): 369–85.

Dillehay, T. D., C. Ramírez, M. Pino, M. B. Collins, J. Rossen, and J. D. Pino-Navarro. "Monte Verde: Seaweed, Food, Medicine, and the Peopling of South America." *Science* 320 (May 9, 2008): 784–86.

Dixon, David. *Never Come to Peace Again.* Norman: University of Oklahoma Press, 2005.

Dobyns, Henry F. "Disease Transfer at Contact." *Annual Review of Anthropology* 22 (1993): 273–91.

Donaghue, Margaret. "A Four-Hundred-Year-Old Mystery." *Farmington First: The University of Maine Farmington Alumni Magazine,* June 2005, 10–11.

Donehoo, George P. *A History of Indian Villages and Place Names in Pennsylvania.* 1928. Facs. ed. Lewisburg, PA: Wennawoods Publishing, 1998.

Donnelly, Jeffrey P., Neal W. Driscoll, Elazar Uchupi, et al. "Catastrophic Meltwater Discharge Down the Hudson Valley: A Potential Trigger for the Intra-Allerød Cold Period." *Geology* 33 (February 2005): 89–92.

Drake, Francis. *The Town of Roxbury: Its Memorable Persons and Places.* Boston: Municipal Printing Office, 1905.

Drake, James D. *King Philip's War.* Amherst: University of Massachusetts Press, 1999.

Drake, Samuel Adams. *The Border Wars of New England.* New York: Charles Scribner's Sons, 1910.

Dufour, Ronald P. "Mason, John." In *American National Biography Online.* http://www.anb.org.silk. library.umass.edu:2048/articles/01/01-00578.html.

———. "Oldham, John." In *American National Biography Online.* http://www.anb.org.silk.library. umass.edu:2048/articles/01/01-00686.html.

———. "Underhill, John." In *American National Biography Online.* http://www.anb.org.silk.library. umass.edu:2048/articles/01/01-00906.html.

Duncan, David Ewing. *Hernando de Soto.* Norman: University of Oklahoma Press, 1997.

Durzan, Don J. "Arginine, Scurvy and Cartier's 'Tree of Life.'" *Journal of Ethnobotany and Ethnomedicine* (February 2, 2009). http://www.ethnobiomed.com/content/5/1/5.

Dutrizac, Charles D. "Local Identity and Authority in a Disputed Hinterland: The Pennsylvania-Maryland Border in the 1730s." *Pennsylvania Magazine of History and Biography* 115 (January 1991): 35–61.

Ebenesersdóttir, Sigríður Sunna, Ásgeir Sigurðsson, Frederico Sánchez-Quinto et al. "A New Subclade of mtDNA Haplogene C1 Found in Icelanders: Evidence of Pre-Columbian Contact?" *American Journal of Physical Anthropology,* (January 2011) 92–99.

Eccles, W. J. "Coulon de Villiers, Louis." In *Dictionary of Canadian Biography Online.* http://www. biographi.ca/009004-119.01-e.php?&id_nbr=1284&interval=25&&PHPSESSID=7hrlqio3 oqlp1gcafohd403fd5.

———. "Coulon de Villiers de Jumonville, Joseph." In *Dictionary of Canadian Biography Online.* http://www.biographi.ca/009004-119.01-e.php?&id_nbr=1283&interval=25&&PHPSESS ID=7hrlqio3oqlp1gcafohd403fd5.

Erickson, Vincent O. "Maliseet-Passamaquoddy." In *Handbook of North American Indians,* edited by Bruce G. Trigger, 15:123–36. Washington, DC: Smithsonian Institution, 1978.

Fausz, J. Frederick. "An 'Abundance of Blood Shed on Both Sides': England's First Indian War, 1609–1614." *Virginia Magazine of History and Biography* 98 (January 1990): 3–56.

———. "Opechancanough." In *American National Biography Online.* http://www.anb.org.silk.library.umass.edu:2048/articles/20/20-00744.html.

———. "Opechancanough: Indian Resistance Leader." In *Struggle and Survival in Colonial America,* edited by David G. Sweet and Gary B. Nash. Berkeley: University of California Press, 1981.

Feest, Christian F. "More on Castoreum and Traps in Eastern North America." *American Anthropologist* 77 (September 1975): 603.

———. "North Carolina Algonquians." In *Handbook of North American Indians,* edited by Bruce G. Trigger, 15:271–81. Washington, DC: Smithsonian Institution, 1978.

———. "Virginia Algonquians," In *Handbook of North American Indians,* edited by Bruce G. Trigger, 15:253–70. Washington, DC: Smithsonian Institution, 1978.

Felt, Joseph B. *The Annals of Salem, from Its First Settlement.* Salem, MA: W & S. B. Ives, 1827.

Fenton, William N. "Northern Iroquoian Cultural Patterns." In *Handbook of North American Indians,* edited by Bruce G. Trigger, 15:296–321. Washington, DC: Smithsonian Institution, 1978.

Fickes, Michael L. "'They Could Not Endure That Yoke': The Captivity of Pequot Women and Children After the War of 1637." *New England Quarterly* 73 (March 2000): 58–81.

Firestone, R. B., A. West, J. P. Kennett, et al. "Evidence for an Extraterrestrial Impact 12,900 Years Ago That Contributed to the Megafaunal Extinctions and the Younger Dryas Cooling." *Proceedings of the National Academy of Sciences* 104 (October 9, 2007): 16016–21.

Fischer, David Hackett. *Champlain's Dream*. New York: Simon & Schuster, 2008.

Flannery, Tim. *The Eternal Frontier*. New York: Atlantic Monthly Press, 2001.

Foster, David R., and Glenn Motzkin. "Interpreting and Conserving the Openland Habitats of Coastal New England: Insights from Landscape History." *Forest Ecology and Management* 185 (2003): 127–50.

Francis, David A., and Robert M. Leavitt. *A Passamaquoddy-Maliseet Dictionary/Peskotomuhkati Wolastoqewi Latuwewakon*. Fredericton, NB: Goose Lane Editions, 2008. http://www.lib.unb.ca/Texts/Maliseet/dictionary/.

Franklin, Benjamin. "Rules by Which a Great Empire May Be Reduced to a Small One." In *The Writings of Benjamin Franklin*, edited by Albert Henry Smyth. Vol. 6. New York: Macmillan, 1907.

Franks, Kenny A. "Tanacharison." In *American National Biography Online*. http://www.anb.org/articles/20/20-01693.html.

Gallay, Alan. "Barnwell, John (c. 1671–1724)." In *Oxford Dictionary of National Biography*, edited by H. C. G. Matthew and Brian Harrison. Oxford: Oxford University Press, 2004. http://www.oxforddnb.com/view/article/68433.

——— . *The Indian Slave Trade*. New Haven, CT: Yale University Press, 2001.

Gallivan, Martin D. "Powhatan's Werowocomoco: Constructing Place, Polity, and Personhood in the Chesapeake, C.E. 1200–C.E. 1609." *American Anthropologist* 109 (March 2007): 85–100.

Garraty, John A. "Tisquantum." In *American National Biography Online*. http://www.anb.org.silk.library.umass.edu:2048/articles/20/20-00975.html.

Gibbon, Guy, ed. *Archaeology of Prehistoric Native America*. New York: Garland Publishing, 1998.

Goddard, Ives. "Delaware." In *Handbook of North American Indians*, edited by Bruce G. Trigger, 15:213–39. Washington, DC: Smithsonian Institution, 1978.

——— . "Eastern Algonquian Languages." In *Handbook of North American Indians*, edited by Bruce G. Trigger, 15:70–77. Washington, DC: Smithsonian Institution, 1978.

——— . "A Further Note on Pidgin English." *International Journal of American Linguistics* 44 (January 1978): 73.

——— . "Some Early Examples of American Indian Pidgin English from New England." *International Journal of American Linguistics* 43 (January 1977): 37–41.

——— . "A True History of the Word 'Squaw.'" http://anthropology.si.edu/goddard/squaw_1.pdf.

Goebel, Ted. "Pleistocene Human Colonization of Siberia and the Peopling of the Americas: An Ecological Approach." *Evolutionary Anthropology* 8 (December 1999): 208–27.

Gradie, Charlotte M. "Spanish Jesuits in Virginia: The Mission That Failed." *Virginia Magazine of History and Biography* 96 (April 1988): 131–56.

Gragson, Ted L., and Paul V. Bolstad. "A Local Analysis of Early-Eighteenth Century Cherokee Settlement." *Social Science History* 31 (Fall 2007): 435–68.

Grant, William L. *Voyages of Samuel de Champlain, 1604–1618*. New York: Charles Scribner's Sons, 1907.

Griffin, James B. "Late Prehistory of the Ohio Valley." In *Handbook of North American Indians*, edited by Bruce G. Trigger, 15:547–59. Washington, DC: Smithsonian Institution, 1978.

Gugliotta, Guy. "The Great Human Migration." *Smithsonian*, July 2008, 56–64.

Gyles, John. *Memoirs of Odd Adventures, Strange Deliverances, &c. in the Captivity of John Gyles, Esq; Commander of the Garrison on St. George's River. Written by Himself*. Boston: S. Knefland & T. Green, 1736.

Haan, Richard L. "The 'Trade Do's Not Flourish as Formerly': The Ecological Origins of the Yamassee War of 1715." *Ethnohistory* 28 (Autumn 1981): 341–58.

Hacker, Barton C. "Women and Military Institutions in Early Modern Europe: A Reconnaissance." *Signs* 4 (Summer 1981): 643–71.

Haensch, Stephanie, Raphaella Bianucci, and Michel Signoli et al. "Distinct Clones of *Yersinia pestis* Caused the Black Death." *PLoS Pathogens* 6 (October 2010): e1001134. doi:10.1371/journal.ppat.1001134.

Hagedorn, Nancy L. "'A Friend to Go Between Them': The Interpreter as Cultural Broker During Anglo-Iroquois Councils, 1740–1770." *Ethnohistory* 35 (Winter 1988): 60–80.

Haile, Edward Wright, ed. *Jamestown Narratives*. Champlain, VA: RoundHouse, 1998.

Hakluyt, Richard. *Principall Navigations, Voiages and Discoveries of the English Nation.* London: George Bishop & Ralph Newberrie, 1589.

Hall, Michael G. *The Last American Puritan: The Life of Increase Mather.* Middletown, CT, and Hanover, NH: Wesleyan University Press/University Press of New England, 1988.

Hambleton, Else. *Daughters of Eve.* New York: Routledge, 2004.

Hamilton, William B. *Place Names of Atlantic Canada.* Toronto: University of Toronto Press, 1996.

Hann, John H. *Indians of Central and South Florida, 1513–1763.* Gainesville: University Press of Florida, 2003.

——. "Political Leadership Among the Natives of Spanish Florida." *Florida Historical Quarterly* 71 (October 1992): 188–208.

——. "St. Augustine's Fallout from the Yamasee War." *Florida Historical Quarterly* 68 (October 1989): 180–200.

——. "Summary Guide to Spanish Florida Missions and *Visitas* with Churches in the Sixteenth and Seventeenth Centuries." *Americas* 46 (April 1990): 417–513.

Hanna, Charles A. *The Wilderness Trail.* Vol. 1. New York: G. P. Putnam's Sons, 1911.

"Hannah Duston Bobblehead Sparks Controversy." *Haverhill (MA) Eagle-Tribune,* July 29, 2008. http://www.eagletribune.com/newhampshire/x1876442987/Hannah-Duston-bobblehead-sparks-controversy/print.

Hannay, James. *Nine Years a Captive, or John Gyles' Experience Among the Malicite Indians.* St. John, NB: Daily Telegraph Steam Job Press, 1875.

Harington, C. R. "North American Short-Faced Bear." Canadian Museum of Nature, March 1996. http://www.beringia.com/research/bear.html.

Harper, K. N., Paolo S. Ocampo, Bret M. Steiner, et al. "On the Origin of the Treponematoses: A Phylogenetic Approach." *PLoS Neglected Tropical Diseases* 2 (January 16, 2008): e148. http://www.plosntds.org/article/info:doi%2F10.1371%2Fjournal.pntd.0000148.

Hauptman, Laurence M., and Ronald G. Knapp. "Dutch-Aboriginal Interaction in New Netherlands and Formosa: An Historical Geography of Empire." *Proceedings of the American Philosophical Society* 121 (April 29, 1977): 166–82.

Hauptman, Laurence M., and James D. Wherry. *The Pequots of Southern New England.* Norman: University of Oklahoma Press, 1990.

Haviland, William A., and Marjory W. Power. *The Original Vermonters.* Lebanon, NH: University Press of New England, 1994.

Heard, J. Norman. *Handbook of the American Frontier.* Vol. 1. Metuchen, NJ: Scarecrow Press, 1987.

Heaton, Timothy H. "Ice Age Paleontology of Southeast Alaska." http://www.usd.edu/esci/alaska/.

Heckewelder, John, and Peter S. Du Ponceau. "Names which the Lenni Lenape or Delaware indians, who once inhabited this country, had given to rivers, streams, places &c." *Transactions of the American Philosophical Society* 4 (1834): 351–396.

Henige, David. "Primary Source by Primary Source: On the Role of Epidemics in New World Depopulation." *Ethnohistory* 33 (Summer 1986): 293–312.

Hildeburn, Charles R. "Sir John St. Clair, Baronet, Quarter-Master General in America, 1755–1767." *Pennsylvania Magazine of History and Biography* 9 (April 1885): 1–14.

Hodges, Glenn. "Cahokia: America's Forgotten City." *National Geographic,* January 2011, 126–45.

Hoffman, Paul E. "The Chicora Legend and Franco-Spanish Rivalry in La Florida." *Florida Historical Quarterly* 62 (April 1984): 419–38.

Holliday, V. T., and D. J. Meltzer. "The 12.9-ka ET Impact Hypothesis and North American Paleoindians." *Current Anthropology* 51 (October 2010): 575–607.

Holmes, Peter. "Stucley, Thomas (c. 1520–1578)." In *Oxford Dictionary of National Biography*, edited by H. C. G. Matthew and Brian Harrison. Oxford: Oxford University Press, 2004. http://www.oxforddnb.com/view/article/26741.

Hubbard, William. *A Narrative of the Indian Wars in New-England.* Stockbridge, MA: Willard, 1803.

Hudson, Charles H. *Black Drink.* Athens: University of Georgia Press, 2004.

Hughes, Susan S. "Getting to the Point: Evolutionary Change in Prehistoric Weaponry." *Journal of Archaeological Method and Theory* 5 (December 1998): 345–408.

Hume, Ivor Noël. *Martin's Hundred.* New York: Alfred A. Knopf, 1982.

Hunter, William A. "History of the Ohio Valley." In *Handbook of North American Indians,* edited by Bruce G. Trigger, 15:588–93. Washington, DC: Smithsonian Institution, 1978.

———. "Tanaghrisson." In *Dictionary of Canadian Biography Online.* http://www.biographi.ca/009004-119.01-e.php?&id_nbr=1674.

Ibbetson, David. "Popham, Sir John (c. 1531–1607)." In *Oxford Dictionary of National Biography,* edited by H. C. G. Matthew and Brian Harrison. Oxford: Oxford University Press, 2004. http://www.oxforddnb.com/view/article/22543.

Jacobs, Wilbur R. "Wampum: The Protocol of Indian Diplomacy." *William and Mary Quarterly* 6 (October 1949): 596–604.

Jefferies, Richard W. "Regional Cultures, 700 B.C.–A.D. 1000." In *Handbook of North American Indians,* edited by Raymond D. Fogelson, 14:115–27. Washington, DC: Smithsonian Institution, 2004.

Jennings, Francis. *The Ambiguous Iroquois Empire.* New York: W. W. Norton, 1984.

———. "The Delaware Interregnum." *Pennsylvania Magazine of History and Biography* 89 (April 1965): 174–98.

———. *Empire of Fortune.* New York: W. W. Norton, 1988.

———. "Glory, Death, and Transfiguration: The Susquehannock Indians in the Seventeenth Century." *Proceedings of the American Philosophical Society* 112 (February 15, 1968): 15–53.

———. "The Indian Trade of the Susquehanna Valley." *Proceedings of the American Philosophical Society* 110 (December 1966): 406–24.

———. "Miantonomo." In *American National Biography Online.* http://www.anb.org.silk.library.umass.edu:2048/articles/20/20-00671.html.

———. "Sassacus." In *American National Biography Online.* http://www.anb.org.silk.library.umass.edu:2048/articles/20/20-00899.html.

Johnston, A. J. B. *Mathieu Da Costa and Early Canada: Possibilities and Probabilities.* Halifax, NS: Parks Canada, 2009. http://canadachannel.ca/HCO/index/php/Mathieu_Da_Costa_and_Early_Canada,_by_A._J._B._Johnston.

"Jonathan Haynes." The Morris Clan. http://www.themorrisclan.com/GENEALOGY/HAYNES%20Jonathan%20F3744.html.

Jones, David S. "Virgin Soils Revisited." *William and Mary Quarterly* 60 (October 2003): 703–42.

Justice, Noel D. *Stone Age Spear and Arrow Points of the Midcontinental and Eastern United States.* Bloomington: Indiana University Press, 1988.

Karr, Ronald Dale. "'Why Should You Be So Furious?': The Violence of the Pequot War." *Journal of American History* 85 (December 1988): 876–909.

Kawashima, Yasuhide. "Forest Diplomats: The Role of Interpreters in Indian-White Relations on the Early American Frontier." *American Indian Quarterly* 13 (Winter 1989): 1–14.

Kelso, William A. *Jamestown: The Buried Truth.* Charlottesville: University of Virginia Press, 2006.

Kelton, Paul. "Avoiding the Smallpox Spirits: Colonial Epidemics and Southeastern Indian Survival." *Ethnohistory* 51 (Winter 2004): 45–71.

Kemp, Brian M., Ripan S. Malhi, John McDonough, et al. "Genetic Analysis of Early Holocene Skeletal Remains from Alaska and Its Implications for the Settlement of the Americas." *American Journal of Physical Anthropology* 132 (April 2007): 605–21.

Kennett, Douglas J., James P. Kennett, Allen West, Chris Mercer, et al. "Nanodiamonds in the Younger Dryas Boundary Sediment Layer." *Science* 323 (January 2, 2009): 94.

Kennett, Douglas J., James P. Kennett, Allen West, G. James West, et al. "Shock-Synthesized Hexagonal Diamonds in Younger Dryas Boundary Sediments." *Proceedings of the National Academy of Sciences* 106 (August 4, 2009): 12623–28.

Kenney, Alice P. "Neglected Heritage: Hudson River Valley Dutch Material Culture." *Winterthur Portfolio* 20 (Spring 1985): 49–70.

Kerr, R. S. "Planetary Impacts: Did the Mammoth Slayer Leave a Diamond Calling Card?" *Science* 323 (January 2, 2009): 26.

King, Anna. "Asatru Leader Visits Kennewick Museum." *Tri-City (WA) Herald,* August 4, 2006.

Kitchen, Andrew, Michael M. Miyamoto, and Connie J. Mulligan. "A Three-Stage Colonization Model for the Peopling of the Americas." *PLoS ONE* 3 (February 13, 2008): e1596. http://plosntds.org/article/info:doi:10.1371/journal.pone.0001596.

Knight, Janice. "Telling It Slant: The Testimony of Mercy Short." *Early American Literature* 36 (2002): 39–69.

Kupperman, Karen Ordahl. "John Smith." In *American National Biography Online.* http://www.anb.org.silk.library.umass.edu:2048/articles/20/20-00960.html.

Kurlansky, Mark. *The Basque History of the World.* New York: Walker, 1999.

———. *Cod.* New York: Walker, 1997.

Lambert, Paul F. "Scarouady." In *American National Biography Online.* http://www.anb.org/articles/20/20-00903.html.

———. "Shingas." In *American National Biography Online.* http://www.anb.org/articles/20/20-00936.html.

Landy, David. "Tuscarora Among the Iroquois." In *Handbook of North American Indians,* edited by Bruce G. Trigger, 15:518–24. Washington, DC: Smithsonian Institution, 1978.

Larsen, Clark Spencer. "Telltale Bones." *Archaeology* 45 (March–April 1992): 43–46.

Laurent, Joseph. *New Familiar Abenakis and English Dialogues.* Montreal: Leger Brousseau, 1884.

Lee, Mike. "Tribes, Asatru Pay Respects to Old Bones Before Move to Seattle Museum." *Tri-City (WA) Herald,* October 30, 1998.

Lee, Wayne E. "Early American Ways of War: A New Reconnaissance, 1600–1815." *Historical Journal* 44 (March 2001): 269–89.

———. "Peace Chiefs and Blood Revenge: Patterns of Restraint in Native American Warfare, 1500–1800." *Journal of Military History* 71 (July, 2007): 701–41.

Leland, Charles G. *The Algonquin Legends of New England.* Boston: Houghton Mifflin, 1884.

Lemonick, Michael D., and Andrea Dorfman. "Who Were the First Americans?" *Time,* March 5, 2006. http://www.time.com/time/magazine/article/0,9171,1169905,00.html.

Lewin, Howard. "Frontier Diplomat: Andrew Montour." *Pennsylvania History* 33 (April 1966): 153–86

Lincoln, Charles H., ed. *Narratives of the Indian Wars, 1675–1699.* New York: Charles Scribner's Sons, 1913.

Lindsay, Jay. "Haverhill Residents Consider Indian Scalper as Symbol of Rebirth." Associated Press, August 18, 2006.

Livi-Bacci, Massimo. "Return to Hispaniola: Reassessing a Demographic Catastrophe." *Hispanic American Historical Review* 83 (February 2003): 3–51.

Magnusson, Magnus, and Hermann Pálsson, eds. *The Vinland Sagas.* London: Penguin, 1965.

Malone, Patrick M. "Changing Military Technology Among the Indians of Southern New England, 1600–1677." *American Quarterly* 25 (March 1973): 48–63.

Mann, Charles C. "America, Found and Lost." *National Geographic,* May 2007, 32–55.

———. *1491.* New York: Alfred A. Knopf, 2005.

Marquardt, William H. "Calusa." In *Handbook of North American Indians,* edited by Raymond D. Fogelson, 14:204–12. Washington, DC: Smithsonian Institution, 2004.

Marshall, O. H. "De Céloron's Expedition to the Ohio in 1749." *Magazine of American History* 2 (March 1878): 129–50.

Mashantucket Pequot Museum and Research Center. "From Nameag to Noank." http://www.pequotmuseum.org/SocietyCulture/AftermathofthePequotWar/AftermathofthePequotWar.htm.

———. "The Mashantucket Land Grant." http://www.pequotmuseum.org/ExhibitGalleries/LifeontheReservationI/TheMashantucketLandGrant.htm.

———. "Mashantucket Pequot Tribal Nation History." http://www.pequotmuseum.org/TribalHistory/TribalHistoryOverview/TribalHistoryOverview.htm.

Matter, Robert A. "Economic Basis of the Seventeenth-Century Florida Missions." *Florida Historical Quarterly* 52 (July 1973): 18–38.

———. "Missions in the Defense of Spanish Florida, 1566–1710." *Florida Historical Quarterly* 54 (July 1975): 18–38.

McBride, Kevin A. "The Legacy of Robin Cassacinamon: Mashantucket Pequot Leadership in the Historic Period." In *Northeastern Indian Lives, 1632–1816,* edited by Robert S. Grumet and Anthony F. C. Wallace. Amherst: University of Massachusetts Press, 1996.

Mensforth, Robert R. "Human Trophy Taking in Eastern North America During the Archaic Period." In *The Taking and Displaying of Human Body Parts as Trophies by Amerindians,* edited by Richard J. Chacon and David H. Dye. New York: Springer-Verlag, 2007.

Merrell, James H. "'The Cast of His Countenance': Reading Andrew Montour." In *Through a Glass Darkly,* edited by Ronald Hoffman, Mechal Sobel, and Fredrika J. Teute, 13–39. Chapel Hill: Omohundro Institute of Early American History and Culture/North Carolina Press, 1997.

———. "'I desire all that I have said . . . may be taken down aright': Revisiting Teedyuscung's 1756 Treaty Council Speeches." *William and Mary Quarterly* 63 (October 2006): 777–826.

———. *Into the American Woods.* New York: W. W. Norton, 1999.

Merwick, Donna. *The Shame and Sorrow.* Philadelphia: University of Pennsylvania Press, 2006.

Milanich, Jerald T. "Prehistory of the Lower Atlantic Coast After 500 B.C." In *Handbook of North American Indians,* edited by Raymond D. Fogelson, 14:229–37. Washington, DC: Smithsonian Institution, 2004.

———. "Timucua." In *Handbook of North American Indians,* edited by Raymond D. Fogelson, 14:219–28. Washington, DC: Smithsonian Institution, 2004.

Miller, Jay. "Teedyuscung." In *American National Biography Online.* http://www.anb.org/articles/20/20-01006.html.

Minutes of the Provincial Council of Pennsylvania. Vols. 3–8. Harrisburg, PA: Theo. Fenn, 1840, 1851, 1852.

Mirick, Benjamin L. *History of Haverhill, Massachusetts.* Haverhill, MA: A. W. Thayer, 1832.

Mithun, Marianne. *The Languages of Native North America.* Cambridge: Cambridge University Press, 1999.

Modugno, Joseph R. "Introduction to 'The Duston Family.'" Hawthorne in Salem. http://www.hawthorneinsalem.org/page/11763/.

Mook, Maurice A. "Virginia Ethnology from an Early Relation." *William and Mary Quarterly* 23 (April 1943): 101–29.

Moore, Alexander. "Moore, James." In *American National Biography Online.* http://www.anb.org/articles/01/01-00618.html.

Morey, David C. "A Race for the American Coast." In *One Land—Two Worlds*, edited by Mac Deford and Philip Conkling. Rockland, ME: Island Institute, 2005.

———, ed. *The Voyage of the Archangell*. Gardiner, ME: Tilbury House, 2005.

Morgan, Basil. "Hawkins, Sir John (1532–1595)." In *Oxford Dictionary of National Biography*, edited by H. C. G. Matthew and Brian Harrison. Oxford: Oxford University Press, 2004. http://www.oxforddnb.com/view/article/12672.

National Research Council of the National Academies. Committee on Atlantic Salmon in Maine. *Atlantic Salmon in Maine*. Washington, DC: National Academies Press, 2004.

"News from New England Concerning the Indians." In *Documentary History of the State of Maine*, edited by James P. Baxter. Vol. 5. Portland, ME: Thurston Print, 1897.

Nicholar, Joseph. *The Life and Traditions of the Red Man*. Edited by Annette Kolodny. Durham, NC: Duke University Press, 2007.

Norton, Mary Beth. *In the Devil's Snare*. New York: Alfred A. Knopf, 2002.

Nourse, Henry Stedman. *Early Records of Lancaster, Massachusetts, 1643–1725*. Lancaster, MA: W. J. Coulter, 1884.

Nyholm, Christine. "Obstinate Nuns, Industrious Wives, and Independent Widows: Women and Power in New France, 1689–1730." *UCI Undergraduate Research Journal* 7 (2004): 61–72.

Oatis, Steven J. *A Colonial Complex*. Lincoln: University of Nebraska Press, 2004.

Oberg, Michael L. "'We Are All the Sachems from East to West': A New Look at Miantonomi's Campaign of Resistance." *New England Quarterly* 77 (September 2004): 478–99.

O'Leary, Jamie. "Basque Whaling in Red Bay, Labrador." Newfoundland and Labrador Heritage, 1997. http://www.heritage.nf.ca/exploration/basque.html.

Olexer, Barbara. *The Enslavement of the American Indian in Colonial Times*. Columbia, MD: Joyous Publishing, 2005.

O'Shea, John M., and Guy A. Meadows. "Evidence for Early Hunters Beneath the Great Lakes." *Proceedings of the National Academy of Sciences* 106 (June 23, 2009): 10120–23.

Paltsits, Victor Hugo. *Depredation at Pemaquid in August, 1689*. Portland, ME: Lefavor-Tower, 1905.

Panum, Peter L. *Observations Made During the Epidemic of Measles on the Faroe Islands in the Year 1846*. Translated by A. S. Hatcher. New York: American Public Health Association, 1940.

Paquay, François S., Steven Goderis, Greg Ravizza, et al. "Absence of Geochemical Evidence for an Impact Event at the Bølling-Allerød/Younger Dryas Transition." *Proceedings of the National Academy of Sciences* (December 15, 2009). doi:10.1073/pnas.0908874106.

Park, Robert W. "The Dorset-Thule Succession in Arctic North America: Assessing Claims for Culture Contact." *American Antiquity* 58 (1993): 203–34.

Parkman, Francis. *Count Frontenac and New France Under Louis XIV*. Boston: Little, Brown, 1911.

Parsons, John. "Bjarni Herjolfsson." In *The Canadian Encyclopedia*. http://www.thecanadianency-clopedia.com/index.cfm?PgNm=TCE&Params=A1ARTA0000790.

Payne, Samuel B. Jr. "The Iroquois League, the Articles of Confederation, and the Constitution." *William and Mary Quarterly* 53 (July 1996): 605–20.

Pearce-Duvet, Jessica M. C. "The Origin of Human Pathogens: Evaluating the Role of Agriculture and Domestic Animals in the Evolution of Human Disease." *Biological Reviews* 81 (2006): 369–82.

Pendergast, James F. "The Massawomeck: Raiders and Traders into the Chesapeake Bay in the Seventeenth Century." *Transactions of the American Philosophical Society* 81, pt. 2 (1991): i–101.

Pennsylvania Archives. Edited by Samuel Hazard. 1st ser., vols. 1–4. Philadelphia: Joseph Severns, 1852.

Pennsylvania Archives. Edited by John B. Linn and William H. Engle. 2nd ser., vol. 2. Reprint. Harrisburg, PA: E. K. Meyers, 1890.

Pennsylvania Archives. Edited by John B. Linn and William H. Engle. 2nd ser., vol. 7. Harrisburg, PA: Lane S. Hart, 1898.

Pennsylvania Archives. Edited by George Edward Reed. 4th ser., vol. 3. Harrisburg, PA: William Stanley Ray, 1900.

Perry, Marilyn Elizabeth. "Uncas." In *American National Biography Online.* http://www.anb.org. silk.library.umass.edu:2048/articles/20/20-01035.html.

Pestana, Carla Gardina. "Mason, John (1601–1672)." In *Oxford Dictionary of National Biography,* edited by H. C. G. Matthew and Brian Harrison. Oxford: Oxford University Press, 2004. http://www.oxforddnb.com/view/article/18281.

Philbrick, Nathaniel. *Mayflower.* New York: Viking, 2006.

Pielou, E. C. *After the Ice Age.* Chicago: University of Chicago Press, 1991.

Piscataway Conoy Tribe. "History." http://www.piscatawayconoy.com/content/learn/history.html.

Post, Christian Frederick, John Hays, Robert S. Grumet, and James A. Rementer. "Journey on the Forbidden Path: Chronicles of a Diplomatic Mission to the Allegheny Country, 1760." *Transactions of the American Philosophical Society* 89 (1999): i–156.

Pothier, Bernard. "Le Moyne d'Iberville et d'Ardilliéres, Pierre." In *Canadian Biography Online.* http://www.biographi.ca/009004-119.01-e.php?BioId=35062.

Price, David A. *Love and Hate in Jamestown.* New York: Alfred A. Knopf, 2003.

Pringle, Heather. "Crow Creek's Revenge." *Science* 279 (March 27, 1998): 2039.

Prins, Harald E. L. "Children of Gluskap: Wabanaki Indians on the Eve of European Invasion." In *American Beginnings,* edited by Emerson W. Baker, Edwin A. Churchill, Richard S. D'Abate, et al. Lincoln: University of Nebraska Press, 1994.

——— . "To the Land of the Mistigoches: American Indians Traveling to Europe in the Age of Exploration." *American Indian Culture and Research* 17 (1993): 175–95.

Quinn, David Beers. *England and the Discovery of America.* New York: Alfred A. Knopf, 1974.

——— . "Ingram, David." In *Dictionary of Canadian Biography Online.* http://www.biographi.ca/009004-119.01-e.php?&id_nbr=350&interval=15&&PHPSESSID=mepksm94i59g6hsa81fos6duc2.

——— . "Some Spanish Reactions to Elizabethan Colonial Enterprises." *Transactions of the Royal Historical Society,* 5th ser., 1 (1951): 1–23.

Quinn, David B., and Allison M. Quinn, eds. *The English New England Voyages.* London: Hakluyt Society, 1983.

Quint, A. H. "The Journal of the Rev. John Pike." In *Proceedings of the Massachusetts Historical Society, 1875–76.* Boston: Massachusetts Historical Society, 1876.

Ramsey, William L. "A Coat for 'Indian Cuffy': Mapping the Boundary Between Freedom and Slavery in Colonial South Carolina. *South Carolina Historical Magazine* 103 (January 2002): 48–66.

——— . "'Something Cloudy in Their Looks': The Origins of the Yamasee War Reconsidered." *Journal of American History* 90 (June 2003): 44–75.

——— . *The Yamasee War.* Lincoln: University of Nebraska Press, 2008.

Ransome, David R. "Rosier, James (1573–1609)." In *Oxford Dictionary of National Biography,* edited by H. C. G. Matthew and Brian Harrison. Oxford: Oxford University Press, 2004. http://www.oxforddnb.com/view/article/24109.

——— . "Waymouth, George (fl. 1587–1611)." In *Oxford Dictionary of National Biography,* edited by H. C. G. Matthew and Brian Harrison. Oxford: Oxford University Press, 2004. http://www.oxforddnb.com/view/article/29155.

Raoult, Didier, David L. Reed, Katharina Dittmar, et al. "Molecular Identification of Lice from Pre-Columbian Mummies." *Journal of Infectious Diseases* 197 (February 15, 2008): 535–43.

Ray, C. E., B. N. Cooper, and W. S. Benninghoff. "Fossil Mammals and Pollen in a Late Pleistocene Deposit at Saltville, Virginia." *Journal of Paleontology* 41 (May 1967): 608–22.

Richter, Daniel K. *Facing East from Indian Country.* Cambridge, MA: Harvard University Press, 2001.

Richter, Daniel K. *The Ordeal of the Longhouse: Peoples of the Iroquois League in the Era of European Colonization.* Chapel Hill: University of North Carolina Press, 1992.

———. "Rediscovered Links in the Covenant Chain: Previously Unpublished Transcripts of New York Indian Treaty Minutes, 1677–1691." *Proceedings of the American Antiquarian Society* 92 (April 1982): 45–85.

Rioux, Jean de la Croix. "Sagard, Gabriel." In *Dictionary of Canadian Biography Online.* http://www.biographi.ca/009004-119.01-e.php?&id_nbr=251&interval=25&&PHPSESSID=fkc ku402n3vfd5re77frcckki6.

Robinson, Paul A. "Lost Opportunities: Miantonomi and the English in Seventeenth Century Narragansett Country." In *Northeastern Indian Lives, 1632–1816,* edited by Robert S. Grumet. Amherst: University of Massachusetts Press, 1996.

Rolde, Neil. "The World Around Waymouth." In *One Land—Two Worlds,* edited by Mac Deford and Philip Conkling. Rockland, ME: Island Institute, 2005.

Roper, L. H. "Moore, James, Jr." In *American National Biography Online.* http://www.anb.org/articles/01/01-01315.html.

Roth, Randolph A. "Child Murder in New England." *Social Science History* 25 (Spring 2001): 101–47.

Rountree, Helen C. *Pocahontas's People: The Powhatan Indians of Virginia Through Four Centuries.* Norman: University of Oklahoma Press, 1996.

Rountree, Helen C., and Thomas E. Davidson. *Eastern Shore Indians.* Charlottesville: University of Virginia Press, 1997.

Royot, Daniel. *Divided Loyalties in a Doomed Empire.* Newark: University of Delaware Press, 2007.

Rudes, Blair A., Thomas J. Blumer, and J. Alan May. "Catawba and Neighboring Groups." In *Handbook of North American Indians,* edited by Raymond D. Fogelson, 14:301–18. Washington, DC: Smithsonian Institution, 2004.

Rushforth, Brett. "'A Little Flesh We Offer You': The Origins of Indian Slavery in New France." *William and Mary Quarterly* 60 (October 2003): 777–808.

Russell, Emily W. B. "Indian-Set Fires in the Forests of the Northeastern United States." *Ecology* 64 (February 1983): 78–88.

Sachse, Julius Friedrich. *The German Sectarians of Pennsylvania, 1708–1742.* Vol. 1. Philadelphia: P. C. Stockhausen, 1899.

Salagnac, Georges Cerbelaud. "Abbadie de Saint-Castin, Bernard-Anselme d', Baron de Saint-Castin." In *Dictionary of Canadian Biography Online.* http://www.biographi.ca/009004-119.01-e.php?&id_nbr=593.

———. "Abbadie de Saint-Castin, Jean-Vincent d', Baron de Saint-Castin." In *Dictionary of Canadian Biography Online.* http://www.biographi.ca/009004-119.01-e.php?&id_nbr=594&&PHPSESSID=8ipm4cqrdt7mbst4m1k9h63034.

Salisbury, Neal. "Embracing Ambiguity: Native Peoples and Christianity in Seventeenth-Century North America." *Ethnohistory* 50 (Spring 2003): 247–59.

———, ed. *The Sovereignty and Goodness of God, by Mary Rowlandson, with Related Documents.* Boston: Bedford/St. Martin's, 1997.

Salwen, Bert. "Indians of Southern New England and Long Island: Early Period." In *Handbook of North American Indians,* edited by Bruce G. Trigger, 15:160–76. Washington, DC: Smithsonian Institution, 1978.

Sassaman, Kenneth E., and David G. Anderson. "Late Holocene Period, 3750 to 650 B.C." In *Handbook of North American Indians,* edited by Raymond D. Fogelson, 14:101–14. Washington, DC: Smithsonian Institution, 2004.

Saunt, Claudio. "History Until 1776." In *Handbook of North American Indians,* edited by Raymond D. Fogelson, 14:128–38. Washington, DC: Smithsonian Institution, 2004.

Schafer, Dave. "Ancient Tri-Citian's Bones Raise Controversy." *Tri-City (WA) Herald,* September 9, 1996.

———. "Skull Likely Early White Settler." *Tri-City (WA) Herald,* July 30, 1996.

Schroeder, M. A., and L. A. Robb. "Greater Prairie-Chicken." In *The Birds of North America*, edited by A. Poole, P. Stettenheim, and F. Gill, no. 36. Philadelphia and Washington DC: Academy of Natural Sciences and American Ornithologists' Union, 1993.

Sewall, Samuel. *Samuel Sewall's Diary*. Edited by Mark van Doren. New York: Macy-Masius, 1927.

Shoemaker, Nancy. "How the Indians Got to Be Red." *American Historical Review* 102 (June 1997): 625–44.

Siebert, Frank T. "Mammoth or 'Stiff-Legged Bear.'" *American Anthropologist* 4 (October–December 1937): 721–25.

Silver, Peter. *Our Savage Neighbors*. New York: W. W. Norton, 2008.

Silverman, David J. *Faith and Boundaries*. New York: Cambridge University Press, 2005.

Simmons, Matt. "June 1605: When the Creation of America Began." In *One Land—Two Worlds*, edited by Mac Deford and Philip Conkling. Rockland, ME: Island Institute, 2005.

Sipe, Hal C. *The Indian Wars of Pennsylvania*. 2nd ed. 1929. Reprint. Lewisburg, PA: Wennawoods Publishing, 1998.

Smith, H. M., Jr. "Fort Necessity." *Virginia Magazine of History and Biography* 41 (July 1943): 204–14.

Smith, Marvin T. "Aboriginal Depopulation in the Post-Contact Southeast." In *The Forgotten Centuries: Indians and Europeans in the American South*, edited by Charles Hudson and Carmen C. Tesser. Athens: University of Georgia Press, 1994.

Snow, Dean R. "Eastern Abenaki." In *Handbook of North American Indians*, edited by Bruce G. Trigger, 15:137–47. Washington, DC: Smithsonian Institution, 1978.

——. "The Ethnohistoric Baseline of the Eastern Abenaki." *Ethnohistory* 23 (Summer 1976): 291–306.

Snow, Dean R., and Kim M. Lanphear. "European Contact and Indian Depopulation in the Northeast: The Timing of the First Epidemics." *Ethnohistory* 35 (Winter 1988): 15–33.

Snyderman, George S. "The Functions of Wampum." *Proceedings of the American Philosophical Society* 98 (December 23, 1954): 469–94.

Socotomah, Donald. "A Wabanaki Perspective on the Voyage of George Waymouth." In *One Land—Two Worlds*, edited by Mac Deford and Philip Conkling. Rockland, ME: Island Institute, 2005.

Sohn, Emily. "Dangerous History: The Genetic Secrets of a Savvy Killer." *Science News* 171 (May 26, 2007): 330–32.

Speck, Frank G. *Penobscot Man*. Philadelphia: University of Pennsylvania Press, 1940.

——. *Penobscot Man*. Orono: University of Maine Press, 1998.

——. "Territorial Subdivisions and Boundaries of the Wampanoag, Massachusett, and Nauset Indians." *Indian Notes and Monographs* 44 (1928): 67.

Spiess, Arthur E., and Bruce D. Spiess. "New England Pandemic of 1616–1622: Cause and Archaeological Implication." *Man in the Northeast* 34 (Fall 1987): 71–83.

Stang, John. "Skull Found on Shore of Columbia." *Tri-City (WA) Herald*, July 29, 1996.

Starbuck, David R. *The Archeology of New Hampshire*. Durham: University of New Hampshire Press, 2006.

Starna, William A. "The Biological Encounter: Disease and the Ideological Realm." *American Indian Quarterly* 16 (Autumn 1992): 511–19.

Steele, Ian K. "Hostage-Taking 1754: Virginians vs. Canadians." *Journal of the Canadian Historical Society* 16 (2005): 49–73.

——. "Shawnee Origins of Their Seven Years' War." *Ethnohistory* 53 (Fall 2006): 657–87.

Strong, Pauline Turner. *Captive Selves, Captivating Others*. New York: Westview Press, 2000.

Sugrue, Thomas J. "The Peopling and Depeopling of Early Pennsylvania: Indians and Colonists, 1680–1720." *Pennsylvania Magazine of History and Biography* 116 (January 1992): 3–31.

Tauber, Alfred I. *Henry David Thoreau and the Moral Agency of Knowing*. Berkeley: University of California Press, 2003.

Thompson, Mary P. *Landmarks in Ancient Dover, New Hampshire.* Concord, NH: Republican Press Association, 1892.

Thornton, Russell. "Aboriginal North American Population and Rates of Decline, ca. A.D. 1500–1900." *Current Anthropology* 38 (April 1997): 310–15.

———. "Demographic History." In *Handbook of North American Indians,* edited by Raymond D. Fogelson, 14:48–52. Washington, DC: Smithsonian Institution, 2004.

Thornton, Russell, Tim Miller, and Jonathan Warren. "American Indian Recovery Following Smallpox Epidemics." *American Anthropologist* 93 (March 1991): 28–45.

Thwaites, Reuben Gold, ed. *The Jesuit Relations and Allied Documents.* Vols. 2, 3, 7, 20, 43, 64, 69. Cleveland: Burrows Brothers, 1896, 1897, 1898, 1899, 1900.

Tooker, Elisabeth. "The League of the Iroquois: Its History, Politics, and Ritual." In *Handbook of North American Indians,* edited by Bruce G. Trigger, 15:418–41. Washington, DC: Smithsonian Institution, 1978.

Tooker, William W. "Some More About Virginia Names." *American Anthropologist* 7 (July 1905): 524–28.

Trigger, Bruce G., and James F. Pendergast. "Saint Lawrence Iroquoians." In *Handbook of North American Indians,* edited by Bruce G. Trigger, 15:357–61. Washington, DC: Smithsonian Institution, 1978.

Turgeon, Laurier. "French Fishers, Fur Traders, and Amerindians During the Sixteenth Century: History and Archaeology." *William and Mary Quarterly* 55 (October 1998): 585–610.

Ulrich, Laurel Thatcher. *Good Wives.* New York: Vintage, 1991.

University of Iowa Hospitals and Clinics Medical Museum. "Yaupon: *Ilex vomitoria.*" http://www.uihealthcare.com/depts/medmuseum/galleryexhibits/naturespharmacy/ilexvomitplant/ilexvomitoria.html.

Vaughan, Alden T. "Pequots and Puritans: The Causes of the War of 1637." *William and Mary Quarterly* 21 (April 1964): 256–69.

———. *Transatlantic Encounters.* New York: Cambridge University Press, 2006.

———. "Trinculo's Indian: American Natives in Shakespeare's England." In *"The Tempest" and Its Travels,* edited by Peter Hulme and William H. Sherman. Philadelphia: University of Pennsylvania Press, 2000.

Vaughan, Alden T., and Edward W. Clark. *Puritans Among the Indians.* Cambridge, MA: Harvard University Press, 1981.

Verney, Jack. *The Good Regiment.* Montreal: McGill-Queen's University Press, 1991.

Vickers, Daniel. *Farmers and Fishermen.* Chapel Hill: University of North Carolina Press, 1994.

Vigneras, L. A. "New Light on the 1497 Cabot Voyage to America." *Hispanic American Historical Review* 36 (November 1956): 503–6.

"Virginia Troops in the French and Indian Wars." *Virginia Magazine of History and Biography* 1 (January 1894): 278–87.

Volwiler, Albert T. *George Croghan and the Westward Movement, 1741–1782.* 1926. Reprint. Lewisburg, PA: Wennawoods Publishing, 2000.

Wainwright, Nicholas B. *George Croghan: Wilderness Diplomat.* Chapel Hill: University of North Carolina Press, 1959.

———. "An Indian Trade Failure: The Story of the Hockley, Trent and Croghan Company, 1748–1752." *Pennsylvania Magazine of History and Biography* 72 (October 1948): 343–75.

Walker, Phillip L. "A Bioarchaeological Perspective on the History of Violence." *Annual Review of Anthropology* 30 (2001): 573–96.

Walker, Willard B. "Creek Confederacy Before Removal." In *Handbook of North American Indians,* edited by Raymond D. Fogelson, 14:373–92. Washington, DC: Smithsonian Institution, 2004.

Wallace, Birgitta. "Norse Voyages." In *The Canadian Encyclopedia.* http://www.thecanadianencyclopedia.com/index.cfm?PgNm=TCE&Params=A1ARTA0005791.

Wallace, Paul A. W. *Conrad Weiser: Friend of Colonist and Mohawk*. Philadelphia: University of Pennsylvania Press, 1945.

———. *Indian Paths of Pennsylvania*. Harrisburg: Pennsylvania Historical and Museum Commission, 1965.

———. *Indians in Pennsylvania*. Harrisburg: Pennsylvania Historical and Museum Commission, 1968.

Wang, Sijia, Cecil M. Lewis Jr., Mattias Jakobsson, et al. "Genetic Variation and Population Structure in Native Americans." *PLoS Genetics* 3 (November 23, 2007): e185. http://www.plosntds .org/article/info:doi:10.1371/journal.pgen.0030185.

Ward, Matthew C. "Fighting the 'Old Women': Indian Strategy on the Virginia and Pennsylvania Frontier, 1754–1758." *Virginia Magazine of History and Biography* 103 (July 1995): 297–320.

Warrick, Gary. "European Infectious Disease and Depopulation of the Wendat-Tionontate (Huron-Petun)." *World Archaeology* 35 (October 2003): 258–75.

Waters, Michael R., Steven L. Forman, Thomas A. Jennings, et al. "The Buttermilk Creek Complex and the Origins of Clovis at the Debra L. Friedkin Site, Texas." *Science* 331 (March 25, 2011): 1599–1603.

Webber, Mabel L. "The First Governor Moore and His Children." *South Carolina Historical and Genealogical Magazine* 37 (January 1936): 1–6.

Weber, David J. *The Spanish Frontier in North America*. New Haven, CT: Yale University Press, 1992.

Weir, Robert M. *Colonial South Carolina*. Millwood, NY: KTO Press, 1983.

Wheeler, George A., and Henry W. Wheeler. *History of Brunswick, Topsham and Harpswell, Maine*. Boston: Alfred Mudge & Son, 1878.

Wilford, John Noble. "Chilean Field Yields New Clues to People of Americas." *New York Times*, August 25, 1998. http://www.nytimes.com/library/national/science/082598sci-monteverde .html.

Williams, Stephen L., Suzanne B. McLaren, and Marion A. Burgwin. "Paleo-Archaeological and Historical Records of Selected Pennsylvania Mammals." *Annals of the Carnegie Museum* 54 (April 23, 1985): 77–188.

Wintroub, Michael. *A Savage Mirror*. Stanford, CA: Stanford University Press, 2006.

Wiseman, Frederick Matthew. *The Voice of Dawn*. Lebanon, NH: University Press of New England, 2001.

Worth, John E. "Guale." In *Handbook of North American Indians*, edited by Raymond D. Fogelson, 14:238–44. Washington, DC: Smithsonian Institution, 2004.

———. "Razing Florida: The Indian Slave Trade and the Devastation of Spanish Florida, 1659–1715." In *The Mississippi Shatter Zone*, edited by Robbie Ethridge and Sheri M. Shuck-Hall, 295–311. Lincoln: University of Nebraska Press, 2009.

———. *The Timucuan Chiefdoms of Spanish Florida*. Gainesville: University Press of Florida, 1998.

———. "Yamasee." In *Handbook of North American Indians*, edited by Raymond D. Fogelson, 14:245–53. Washington, DC: Smithsonian Institution, 2004.

Yesner, David R. "Human Dispersal into Interior Alaska: Antecedent Conditions, Modes of Colonization, and Adaptation." *Quaternary Science Reviews* 20 (January 2001): 315–27.

Young, Alexander, ed. *Chronicles of the First Planters of the Colony of Massachusetts Bay*. Boston: C. C. Little & J. Brown, 1846.

Zuk, Marlene. "A Great Pox's Greatest Feat: Staying Alive." *New York Times*, April 29, 2007. http:// www.nytimes.com/2008/04/29/health/29essa.html.

Index

Mather, Cotton
  alliance with French, 154
  on intermarriage among, 304
  on punishment of Philip's son, 174
  recounting of Hannah Duston's story,
    216–18
  on Satan's appearance as Indian, 211
  sermon at Elizabeth Emerson's execu-
    tion, 210
Mather, Increase, 175
Matoaka (Pocahontas), 90, 93, 199
Mattamuskeet, 242
Mattawamkeag (Madahamcouit), ME,
    197, 237
Mawooshen, 15, 27, 392n
Maxwell, Hu, 68–69
McNallen, Stephen A., 44
Meadowcroft Rockshelter, PA, 40, 45
Medfield, MA, 167
Meductic, NB, 182–83, 196f, 198
Memeskia ("La Demoiselle"; "Old
    Briton"), 310
*Memoirs of Major Robert Stobo of the Vir-
    ginia Regiment*, 338, 372, 374
Memoirs of Odd Adventures . . .(Gyles),
    208
Menamesit, MA, 168, 171
Merchant Adventurers of London, 120
Merrell, James, 306
Merrimack River settlements, 209
Merrymeeting Bay, ME, 190
Mesandowit, 188
Messamouet, 102
Metacom (Philip, "King Philip," Meta-
    comet). *See also* King Philip's War
  complaints about Praying Indians,
    154
  insurrection by, 150
  as military leader, 161–62
  murder of, 174
  names given, 404–5n
  Native supporters and opposition to,
    160
  Paul Revere's portrait of, 157f
  relationship to Massasoit, 30, 158
  view of English, 158

Mexican Inquisition, 74–75
Mexico, 53, 98
Miami (tribe)
  alliance with French, 261, 344
  French and/or Indian attacks on, 310,
    412–13n
  homeland, 60
  at Lancaster treaty council, 291–92
  reception of Céloron's expedition by,
    301
  relocations of, 291–92
Miantonomi
  alliance with the Bay Colony, 135, 137,
    146
  capture and murder, 148
  efforts to influence English, 140,
    147–48
  warnings about Pequot, 139
Michif language, 305
Michilimackinac, Roger's Rangers at,
    384
migrations, Paleolithic, 36, 44–45
Mi'kmaq (*mi'k'makik*)
  early contact with Europeans, 8, 21
  epidemic among, 102
  homeland, 63
  relationship with Wapánahki, 5
  scalping of, 83
  trade with Europeans, xvi, 18, 105
  treatment of captives by, 202
Mingo
  attacks on Pennsylvania settlers, 351
  at Lancaster treaty council, 291–92
  relationship with Croghan, 260
  relocations and removals, 261, 327
Minion (ship), 73–74, 75n
Minuit, Peter, 116
Miranda, Angel de, 222
missionaries, Spanish, 105
mission Indians, 227
Mission Santa Catalina de Guale, 84
Mississippi Valley
  development of complex social struc-
    tures, 53–54
  Mississippian cultures, 227
  population collapse, 98–99